THE GEOPOLITICS READER

The twentieth ce ... ng of the term at the
beginning of the ... conflict and change,
the problematic ... great struggles over
power and spac ... in the Great Power
imperialism of th ... Germany, geopolitics
provides an insi ... of global political
space. During th ... used to describe a
permanent globa ... d the Soviet Eastern
bloc that finally e

In today's new w ... nt than ever as new
global struggles ... the post-Cold War
world. Everyday ... by a geopolitical
imagination that ... previously ignored
during the Cold ... movement struggles,
global finance a ... this post-Cold War
world, the doma ... most hotly debated
issues of our tim ... no-nationalism and
post-modernism.

The Geopolitics ... election of readings
on geopolitics, fr ... es fresh perspectives
on the key issues ... g discourse, power,
gender and political economy.

Gearóid Ó Tuathail is Associate Professor of Geography at Virginia Polytechnic Institute and State University, **Simon Dalby** is Associate Professor in Geography at Carleton University and **Paul Routledge** is a Lecturer in Geography at Glasgow University.

THE GEOPOLITICS READER

edited by

Gearóid Ó Tuathail, Simon Dalby
and
Paul Routledge

LONDON AND NEW YORK

First published 1998
by Routledge
11 New Fetter Lane, London EC4P 4EE

Simultaneously published in the USA and Canada
by Routledge
29 West 35th Street, New York, NY 10001

Reprinted 1999, 2001, 2003 (twice)

Routledge is an imprint of the Taylor & Francis Group

Typeset in Sabon by Solidus (Bristol) Limited
Printed and bound in Great Britain by
TJ International Ltd, Padstow, Cornwall

British Library Cataloguing in Publication Data
A catalogue record for this book is available from the British Library

Library of Congress Cataloging in Publication Data
The Geopolitics Reader / [edited by] Gearóid Ó Tuathail, Simon Dalby, and Paul Routledge.
p. cm.
Includes bibliographical references and index.
(pbk.: alk. paper)
1. Geopolitics. I. Ó Tuathail, Gearóid. II. Dalby, Simon.
III. Routledge, Paul.
JC319.G49 1997
302.1′2—dc21 97-13460

ISBN 0–415–16270–X (hbk)
0–415–16271–8 (pbk)

To the alternative geopoliticians,
E.P. Thompson, Petra Kelly and Ken Sara Wiwa

CONTENTS

FIGURES AND TABLES

CARTOONS AND MAPS

ACKNOWLEDGEMENTS

This book was willed into existence by Tristan Palmer and we all wish to acknowledge his initial inspiration in getting this project started. As we started to gather the readings, we quickly became dependent on a number of hard-working graduate students, administrative assistants and others. Gearóid Ó Tuathail wishes to thank Derek McCormack for his diligent work in scanning in the readings that make up the first three sections (and a few others as well). The Department of Geography at Virginia Tech provided the important computer facilities and support that made this project possible. Timothy Luke and Douglas Borer in the Department of Political Science at Virginia Tech helped out along the way with comments, reactions and references. Finally, Michela Verardo tolerated the wired moments this project generated and made mutual laughter a production value.

Paul Routledge thanks Teresa Flavin for reality control; Mike Shand of the Department of Geography, University of Glasgow for the scanning of documents at short notice; Helen Boyd, Eilidh Douglas and Sheena Glassford of the Department of Geography, University of Glasgow for secretarial support.

Simon Dalby thanks Steve Gwynne-Vaughan for help with the scanner and some comments on the introduction to Part 4; Laurie-Ann Hossein for a useful review of Part 4; Susan Tudin for background research, and Cara Stewart who once again lived through the domestic disruptions of academic writing.

All three of us would like to thank the institutions, publishing houses, agents and authors who kindly provided us with the copyright permissions needed to compose this Reader. Thanks also to Sarah Lloyd at Routledge who supported the project once Tristan moved on. Since geopolitics is not a science but a field of political contestation, we have decided to dedicate this book to three inspiring geopolitical activists who struggled throughout their lives to resist the dictatorship of a geopolitics wielded by the powerful over the weak.

INTRODUCTION

Thinking critically about geopolitics

Gearóid Ó Tuathail

All concepts have histories and geographies and the term "geopolitics" is no exception. Coined originally in 1899 by a Swedish political scientist named Rudolf Kjellen, the word "geopolitics" has had a long and varied history in the twentieth century, moving well beyond its original meaning in Kjellen's work to signify a general concern with geography and politics (geo-politics). Coming up with a specific definition of geopolitics is notoriously difficult, for the meaning of concepts like geopolitics tends to change as historical periods and structures of world order change. Geopolitics is best understood in its historical and discursive context of use. Back in the early years of the twentieth century, Kjellen and other imperialist thinkers understood geopolitics as that part of Western imperial knowledge that dealt with the relationship between the physical earth and politics. Associated later with the notorious Nazi foreign policy goal of *Lebensraum* (the pursuit of more "living space" for the German nation), the term fell out of favor with many writers and commentators after World War II (O'Loughlin, 1994). During the later years of the Cold War, geopolitics was used to describe the global contest between the Soviet Union and the United States for influence and control over the states and strategic resources of the world. Former US Secretary of State Henry Kissinger almost single-handedly helped to revive the term in the 1970s by using it as a synonym for the superpower game of balance-of-power politics played out across the global political map (Hepple, 1986).

Since then geopolitics has enjoyed a revival of interest across the world as foreign policy makers, strategic analysts, transnational managers and academics have struggled to make sense of the dynamics of the world political map. One reason why geopolitics has become popular once again is that it deals with comprehensive visions of the world political map. Geopolitics addresses the "big picture" and offers a way of relating local and regional dynamics to the global system as a whole. It enframes a great variety of dramas, conflicts and dynamics within a grand strategic perspective, offering an Olympian viewpoint that many find attractive and desirable. Furthermore, while unavoidably textual, it nevertheless promotes a spatial way of thinking that arranges different actors, elements and locations simultaneously on a global chessboard. It has a multidimensional global cachet – global both in a geographical (worldwide) and a conceptual (comprehensive and total) sense – and *appears* more visual than verbal, more objective and detatched than subjective and ideological. In addition, geopolitics is of interest to certain people because it seems to promise unusual insight into the future direction of international affairs and the coming shape of the world political map. Many decision makers and analysts come to geopolitics in search of crystal ball visions of the future, visions that get beyond the beclouded confusion of the immediate to offer glimpses of a future where faultlines of conflict and cooperation are clear. In a shrinking and speeding world of intense time–space compression wrought by telecommunication revolutions and globalizing economic networks and webs, the desire for perspectives offering "timeless insight" is stronger than ever. In complex post-modern times, in sum, geopolitical visions and visionaries seem to thrive.

In today's *new* world order, specifications of the post-Cold War relationship between geography, power and world order vary considerably as geopolitical visionaries vie with each other to delimit a "new geopolitics." For some, the end of the Cold War has allowed the emergence of a new geopolitical order dominated by geo-economic questions and issues, a world where the globalization of economic activity and global flows of trade, investment, commodities and images are re-making states, sovereignty and the geographical structure of the planet. For others, the "new geopolitics" describes a world dominated no longer by territorial struggles between competing blocs but by emerging transnational problems like terrorism, nuclear proliferation and clashing civilizations. For yet others, the relationship of politics to the earth is more important than ever as states and peoples struggle to deal with environmental degradation, resource depletion, transnational pollution and global warming. For the environmentally minded intellectual and policy maker, the "new geopolitics" is not geo-economics but ecological politics or ecopolitics. Clearly, there are many competing visions of the "new geopolitics."

In compiling a Geopolitics Reader for the very first time, we have tried to collect the most illuminating examples of the old and the new geopolitics, the historical geopolitics of the early twentieth century as well as the multi-dimensional new geopolitics of the late twentieth century. A simple contrast between an old and a new or a classic and contemporary geopolitics, however, is inadequate as a means of grasping the heterogeneity of geopolitical discourses in both the past and the present. Respecting the significance of historical dimensions of geopolitics, yet wishing also adequately to convey historical and contemporary contestations around geopolitics, we have composed a Reader of five parts, two of which address geopolitics historically and two of which deal with the geopolitics of today, while the final section addresses resistance to geopolitics both historically and contemporaneously. Inevitably, because of the limits of space, we have had to leave out certain readings, perspectives and regional geopolitical rivalries, a decision that does not mean we consider their significance marginal.

Part 1 of the Reader is the shortest in terms of readings. It addresses the imperialist origins of geopolitical thought, documenting the entwining of geopolitical visions with imperialist strategy and racist white supremist thinking in the period leading up to World War II. While all the imperial powers of this time had geopolitical philosophies marked by racist attitudes and beliefs, we have chosen to concentrate on the key rivalry between the British Empire and the German state in the early twentieth century, a rivalry at the heart of World Wars I and II.

Part 2 addresses Cold War geopolitics, documenting the origins, consequences and eventual passing of the Cold War as a structure of world order and a complex of geopolitical discourses and practices. Again, we have chosen to focus on the key rivalry, this time between the United States and the Soviet Union.

In Part 3 we provide an introduction to the geopolitical debates over the nature and meaning of the "new world order" that was officially proclaimed as such by President Bush during the Gulf crisis and subsequent war against Iraq in 1990–1991. Because the end of the Cold War effectively left the United States as the sole remaining superpower, we have concentrated on US-centered attempts to give this "new world order" meaning.

Part 4 is devoted exclusively to environmental geopolitics. With rainforest depletion continuing unabated, pollution levels in many cities reaching dangerous new highs, and atmospheric ozone depletion taking place at alarming rates, the politics of how the earth is (ab)used and managed are now more important than ever. The readings we have chosen provide an introduction to the many political struggles over the nature, meaning and cause of contemporary environmental change.

Part 5 is an innovative section that is devoted to the theme of resistance and geopolitics. Although we stress the essentially contested nature of geopolitical discourses throughout the other parts of the book and include many critical readings within them, we felt it was important to document the often overlooked or ignored underside to geopolitics. Since, as we shall see, so much of geopolitics in the past was concerned with imperialist expansion and ideological struggles between competing territorial states, we felt it was important to acknowledge and document the attempt by many critical intellectuals and social movements throughout history to resist the international "geopolitics from above" of hegemonic states and to assert, in opposition, their own localized "geopolitics from below." Since geopolitics has for so long been a militaristic practice monopolized by statist elites, conservative politicians and geopolitical "experts," it is important that we broaden the debate and consider the many different voices – minority civil rights, postcolonial, indigenous, feminist, trade unionist, etc. – opposing the dominant understanding and practice of geopolitics by foreign policy "statesmen" and so-called "wise men" (Enloe, 1990; Isaacson and Thomas, 1986). Finally, the Reader concludes with a reflection on the many different dimensions to geopolitics as knowledge and power at the end of the twentieth century.

Each section of the Reader has a comprehensive introduction to the readings that follow. These introductions place the readings within their historical and geographical context, and discuss their significance within the history of international politics and world order. Whenever possible, we have tried to include readings that directly comment and/or critique each other. In this way, you will be able to appreciate the essentially contested nature of geopolitical readings and texts. To further this goal, we have also chosen to illustrate the Reader with images and political cartoons that are themselves "geopolitical texts" of a graphically visual nature. Some of these images are disturbing while others are the type of humorous images that disclose the unacknowledged psychic anxieties and investments that often motivate geopolitical theory and practice. The best of these cartoons, such as the ones by Tony Auth, Steve Bell and Matt Wuerker (1992), are acts of transgression that call into question dominant relations of power, truth and knowledge. In contrast to the Olympian eye of the geopolitician, they deploy an anti-geopolitical eye (Dodds, 1996; Ó Tuathail, 1996a).

Informing and organizing the Reader as a whole is a critical vision of geopolitics, a perspective that has come to be known as "critical geopolitics" (Ó Tuathail, 1996b). Concisely defined, critical geopolitics seeks to reveal the hidden politics of geopolitical knowledge. Rather than defining geopolitics as an unproblematic description of the world political map, it treats geopolitics as a discourse, as a culturally and politically varied way of describing, representing and writing about geography and international politics. Critical geopolitics does not assume that "geopolitical discourse" is the language of truth; rather, it understands it as a discourse seeking to establish and assert its own truths. Critical geopolitics, in other words, politicizes the creation of geopolitical knowledge by intellectuals, institutions and practicing statesmen. It treats the production of geopolitical discourse as part of politics itself and not as a neutral and detached description of a transparent, objective reality (Dodds and Sidaway, 1994).

In order to help you think critically about the multifaceted and fascinating dimensions of geopolitics, we wish to outline two "methods of study" that critical geopoliticians bring to bear upon the study of geopolitics. Each of these "methods" will help you develop a deeper understanding of the readings that comprise this volume. They provide a conceptual framework for evaluating the arguments and claims made in the readings. They will also reinforce the central argument this Reader seeks to make, namely that *the production of geopolitical knowledge is an essentially contested political activity*. Geopolitics, in short, is about politics!

GEOPOLITICS, DISCOURSE AND "EXPERTS"

The French philosopher Michel Foucault once stated that "the exercise of power perpetually creates knowledge and, conversely, knowledge constantly induces effects of power" (Foucault, 1980: 52). Throughout his many challenging historical and philosophical works, Foucault sought to document how structures of power in society (the military, police, doctors and judicial systems, for example) create structures of knowledge that justify their own power and authority over subject populations. The military, for example, explains and justifies its power in society by promoting a discourse concerning "national security," a discourse in which it claims to be authoritative and expert. This important and constantly changing discourse in turn, as Foucault suggests, induces its own effects of power. If most "military experts" proclaim their agreement that "we" need to control this region or buy that weapon system to "safeguard our national security," then there is a good chance that the military institutions of the state will receive increased resources from political leaders for new missions and new weapons systems. This does not always happen, of course, because other "experts" might disagree with the military or other institutions and interests might protest at the large amounts of money being spent on the military at the expense of pressing social needs The military's discourse of "national security" often clashes with the "social security" discourse of other intellectuals and interest groups. Controlling the meaning of the concept of "security" – defining it again and again in military and not social terms, for example – by controlling the dominant discourse about it, therefore, becomes an extremely important means of exercising power within a state. Monopolizing the right to speak authoritatively about "security" in name of everyone – the ability to evoke the "national interest" or a universal "we" – is at the crux of the practice of power. The exercise of power, Foucault astutely observed, is always deeply entwined with the production of knowledge and discourse.

The idea of geopolitics has been implicated in many different structures of power/knowledge throughout the twentieth century (see Table 1). Even before the term geopolitics was even coined, there were a number of important intellectuals who wrote about the influence of geography on the conduct of global strategy in the late nineteenth century. The American naval historian Alfred Mahan (1840–1914), for example, wrote about the importance of the physical geography – territorial mass and physical features in relation to the sea – in the development of seapower by expanding states in his classic study *The Influence of Seapower Upon History*, which was first published in 1890 (Mahan, 1957). The road to national greatness, not surprisingly for the professional naval officer Mahan, was through naval expansionism. The German geographer Friedrich Ratzel (1844–1904) also wrote about the importance of the relationship between territory or soil and the nation in the development of imperial strength and national power. In his book *Political Geography* (1897), Ratzel, who was deeply influenced by social Darwinism, considered the state to be a living organism engaged in a struggle for survival with other states. Like a living organism, the state needs constantly to expand or face decay and death. Ratzel's social Darwinism celebrated the German nation and German soil as superior to all others. Germany, he argued, should expand at the expense of "inferior" states (organisms) to secure more *Lebensraum* or living space for itself.

The writings of Mahan and Ratzel were not unusual. As we shall see in Part 1, the theme of imperial expansionism was also central to the writings of Halford Mackinder, Karl Haushofer, Adolf Hitler and others. *It is within imperialist discourse that geopolitics first emerges as a concept and practice.* In the early part of the twentieth century, geopolitics is a form of power/knowledge concerned with promoting state expansionism and securing empires. All the leading geopoliticians were conservative white male imperialists who sought, in their own way, to explain and justify imperial expansionism by their own particular national state or, as they and others often termed it, their "race." As one can well imagine, the writings of this elite caste of men were full of the hubris of empire and national exceptionalism: their country represented the zenith of civilization; their way of life was superior to that of others; their ideals were the ideals of all of "mankind" or humanity. Geopoliticians considered themselves to be masters of the globe. They thought in terms of continents and strategized in worldwide terms, labeling huge swaths of the globe with names like "heartland" and "rimlands" (Spykman, 1942). Present also were multiple supremacist arguments, sometimes overtly expressed but more often tacitly assumed: the

Table 1 Discourses of geopolitics

Discourse	*Key intellectuals*	*Dominant lexicon*
Imperialist geopolitics	Alfred Mahan	Seapower
	Friedrich Ratzel	*Lebensraum*
	Halford Mackinder	Landpower/Heartland
	Karl Haushofer	Landpower/Heartland
	Nicholas Spykman	Rimlands
Cold War geopolitics	George Kennan	Containment
	Soviet and Western political and military leaders	First/Second/Third World countries as satellites and dominos
		Western vs. Eastern bloc
New world order geopolitics	Mikhail Gorbachev	New political thinking
	Francis Fukuyama	The end of history
	Edward Luttwak	Statist geo-economics
	George Bush	US led new world order
	Leaders of G7, IMF, WTO	Transnational liberalism/neoliberalism
	Strategic planners in the Pentagon and NATO	Rogue states, nuclear outlaws and terrorists
	Samuel Huntington	Clash of civilizations
Environmental geopolitics	World Commission on Environment and Development	Sustainable development
	Al Gore	Strategic environmental initiative
	Robert Kaplan	Coming anarchy
	Thomas Homer-Dixon	Environmental scarcity
	Michael Renner	Environmental security

supposed "natural" supremacy of men over women; the white race over other races; European civilization over non-European civilizations. One particularly virulent form of this entwined sexism, racism and national chauvinism was the ideology of the Nazi Party in Germany which celebrated idealized visions of "Aryan manhood" while persecuting and vilifying what it constructed as "Jewish Bolshevism." In this case, the power of discourse was to become murderous as those who were corralled into the category "Jewish Bolshevism" were at first persecuted and later sent to their death in concentration camps and death factories like Auschwitz (Mayer, 1988).

The outbreak of a Cold War between the United States and the Soviet Union provided a new context for the production of geopolitical power/knowledge in the post-war period. *It is within Cold War discourse that geopolitics matures as both theory and practice.* Whereas the imperialist geopolitics of the early part of the twentieth century tended to emphasize the conditioning or determining influence of physical geography on foreign policy and global strategy, the Cold War geopolitics that came to be produced around the US–Soviet antagonism entwined geography so closely with ideology that it was difficult to separate the two. Halford Mackinder described part of the Russian landmass as the "heartland," a geographical and territorial region, but to George Kennan, the architect of the US post-war policy of "containment" of the Soviet Union, Russia was never simply a territory but a constantly expanding threat (Reading 6). The very geographical terminology used to describe the world map was also a description of ideological identity and difference. The West was more than a geographical region; it was an imaginary community of democratic states that supposedly represented the very highest standards of civilization and development. Even historically "Eastern" powers like Japan and South Korea were part of this imaginary and symbolic "West." The Soviet Union was represented as an "Eastern power," the mirror image of the West. It was, in the crude cinematically influenced vision of President Ronald Reagan, "the evil empire." The regions and peoples of Eastern Europe were known as "the Eastern bloc." All states with Communist governments were said to belong to the "Second World" which contrasted with the "First World" which was, of course, the West.

In distinction to both the First and Second World, geopolitical and social science experts from both capitalist and communist countries defined a so-called Third World of poor and developing countries out of the heterogeneous rest that fitted into neither camp. Distinguished not only by its traditionalism and underdevelopment, the Third World was conceptualized as a zone of competition between the West and the East and a distinct object of study within post-war social science (Pletsch, 1981). Across the diverse states of the Third World, certain geographical regions became zones of fierce competition and geopolitical strategizing. Geopolitical "experts" from both sides constantly evaluated and surveyed the strategic value of such regions as the Middle East, the Horn of Africa, southern Africa, Indochina, the Caribbean and Central America. Geopolitics became a game of superpower politics played out across the world map. A new Cold War hubris developed in Washington and Moscow as their competing geopolitical experts designated spaces of the world as belonging either to "us" or to "them," to the "free world" as opposed to the "totalitarian world" in the discourse of Western Cold War geopolitics, to the "people's democracies" as opposed to the "capitalist and imperialist West" in the discourse of Soviet Cold War geopolitics. Both the American and the Soviets were preoccupied with the "fall" of certain states to the enemy. This fear was particularly acute in the United States after the so-called "fall" of China to the Soviet camp in 1949 and it soon spawned the anti-communist hysteria of McCarthyism within the United States. What this in turn helped produce was the "domino theory," a form of geopolitical reasoning that conceptualized states as no more than potentially falling dominoes in a great superpower game between the communist East and the capitalist West. The domino theory marked the apotheosis of Cold War geopolitics as a type of power/knowledge that completely ignored the specific geographical characteristics of places, peoples and regions. Complex countries like Vietnam were no more than abstract "stakes" in a global geopolitical power game (Reading 8). The tragedy of the triumph of this type of discourse in US political culture – a triumph made possible by McCarthyism destroying the careers of many of the US's best "regional experts" on Vietnam and China – was that it ended with the US and other Western soldiers fighting for one side in a bloody civil war in a country they knew very little about and whose real strategic significance was marginal (Halberstam, 1972). The Korean, Vietnam and subsequent Reagan sponsored Central American wars in the 1980s are vivid instances of the "power effects" and murderous consequences of the discourse of Cold War geopolitics. The same can be said for the Soviet interventions in Hungary in 1956, Czechoslovakia in 1968 and Afghanistan in 1979.

As a consequence of the end of Cold War in the early 1990s, international politics has experienced a crisis of meaning. The old defining struggle between a capitalist West and a communist East has passed. No overarching defining struggle of international politics has taken its place. Many experts, nevertheless, have tried to define what they claim to be the essential contours of the new world order. *It is within discourse on the new world order that geopolitics is being renewed and re-specified as an approach and practice.* Before even the breakup of the Soviet Union, intellectual experts like Francis Fukuyama and Edward Luttwak offered different visions of the post-Cold War new world order. Offering a late twentieth-century version of the long-standing Western hubris towards the rest of the world, Fukuyama claimed that humanity was reaching "the end of history," for Western liberalism was triumphing across most of the planet. Current Western states were at the pinnacle of history; most of the rest of the world were, at last, realizing this (Reading 13). In contrast to Fukuyama's idealist West-and-the-rest vision, Edward Luttwak foresaw a world where states as territorial entities would continue to compete with each other, though now in geo-economic and not geopolitical conflicts (Reading 14). He stressed trade conflicts between the United States and Japan in a way that suggested a new West (United States) versus the East (Japan) faultline developing in world affairs.

Luttwak's vision of geo-economics is strongly statist but other geo-economic visions stress the relative decline of states and the importance of transnational flows and institutions. "Transnational liberalism" or "neoliberalism" is a doctrine that holds that the globalization of trade, production and markets is both a necessary and desirable development in world affairs (Agnew and Corbridge, 1995). It is most notably articulated (with varying degrees of enthusiasm) by the leaders of the Group of Seven (G7) industrialized states (Canada, France, Germany, Italy, Japan, the United Kingdom and the United States) and by neoliberal economic "experts" in the International Monetary Fund (IMF), the World Bank and the World Trade Organization (WTO). In contrast to the optimism about globalization articulated by neoliberals, the American neoconservative political scientist Samuel Huntington stresses the power of transnational geocultural blocs over transnational geo-economic flows in his vision of the future of world order. Huntington argues that ancient civilizational blocs underpin world affairs. Obscured by the Cold War, they are emerging once again as the faultlines of a West-versus-the-rest clash of civilizations (Reading 19).

Co-existing with these discourses on the new world order are related discourses and ensembles of experts who address the politics of environmental change. Initially an issue of little concern, the "environment" has over the last few decades emerged as an object of considerable focus and concern, an objectified externality in need of study and management and a dynamic system that is the source of many of our newest discourses of threat and danger. Entwined with many other issues like development, population growth and the structures of inequality within the world, the question of "nature" has become the "problem of the environment" while the scale of this problem – initially local and national – has become conceptualized as global. A new object of discourses that did not exist a few decades ago, the "global environment" is now the subject of considerable scientific research efforts in the advanced industrialized world, of transnational conferences and legal statutes and of the newest discourses on the global by a caste of intellectuals we can describe as "environmental geopoliticians." *It is within discourses on global environmental change that the relationship between the earth and the human within the geopolitical tradition is being re-negotiated and a new "environmental geopolitics" is being created.* Like other geopolitical discourses, this relatively new domain of knowledge has its own particular systems of expertise, institutions of governance, caste of "green" intellectuals, perspectivalist visions of the globe and relations of power. As the readings in Part 4 demonstrate, the definition, delimitation and geographical dimensions of "the global environmental problem" are essentially contested. Knowledge of "the global environment" is never neutral and value-free. Many of the measures proposed to address global environmental degradation and pollution reflect vested interests and protect certain structures of power that are deeply implicated in the creation and perpetuation of environmental problems.

INTELLECTUALS, INSTITUTIONS AND IDEOLOGY

In specifying geopolitics as we have within larger discursive formations – imperialist discourse, Cold War discourse, and discourses on the new world order and global environmental change – we have noted the importance of so-called "experts" in specifying and proclaiming certain "truths" about international politics. The processes by which certain intellectual figures become "expert" and get promoted or certified as such by institutions like the media, academia and the state, whereas other intellectual voices and perspectives get marginalized, vary considerably over time and across space. In most instances, these processes are quite complicated, involving as they do factors like schooling and socialization, gender and social networks, place, personality and political beliefs. As a critical tool for thinking about these issues, the triangle of intellectuals, institutions and ideology is one you should bear in mind when thinking about geopolitics as power/knowledge (see Figure 1). Let us consider each point of this triangle in detail.

The practice of statecraft has long produced its own intellectuals, those theorists and former practitioners who wrote and continue to write "how to" books about international politics. One of the most famous "how to" books is Machiavelli's *The Prince* in which he outlines a series of practices (many quite criminal) that the prince should follow if he wishes to remain in power. This "advice to the prince" literature is the specialization of *intellectuals of statecraft*, those intellectuals who offer normative and imperative rules for the conduct of strategy and statecraft by the rulers of the state. Intellectuals of statecraft take a "problem-solving" approach to theory, taking the existent institutions and organization of state power as they find them and theorizing from the perspective of these institutions and relations of power. Their goal is not to change the organization of power within a state but to augment and facilitate its smooth operation (Cox, 1986). In dominant states like Great Britain during the nineteenth century and the United States after World War II, intellectuals of statecraft are the functionaries who think strategically with the interests of the state in mind and address its problems of "hegemonic management."

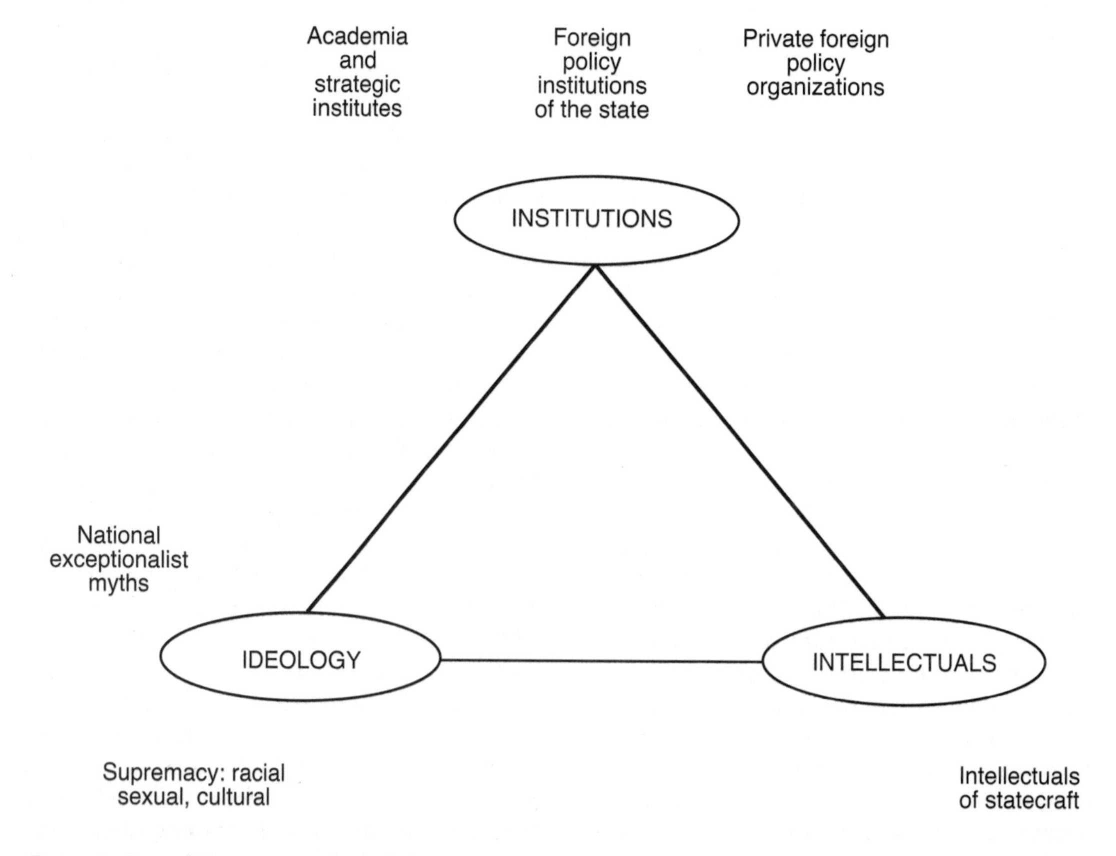

Figure 1 Geopolitics as power/knowledge

Over the years, the literature produced by aspiring, established and retired intellectuals of statecraft has become a publishing industry in its own right, with numerous books and journals each year devoted to debate on the conduct of statecraft. Within this community of "problem solvers" for the state, a few figures are usually promoted, represented and treated as "master strategists" or "gray eminences" by the publishing industry and mass media. One such figure is Henry Kissinger, another Kissinger's boss, former President Richard Nixon. A politician who first came to prominence in the United States as an anti-communist crusader alongside Joseph McCarthy, Nixon craved acceptance as a "senior statesman" from US political society and the media, especially given the disgrace of his resignation of the presidency. To achieve this end, Nixon regularly produced geopolitical books and articles in which he pontificated on international politics and what the current president needed to do. Despite his crimes (both domestic and international), Nixon achieved a considerable measure of success, becoming an occasional advisor for President Bush and even President Clinton. His books also were widely read by influential diplomats and journalists.

Geopolitics and geopoliticians need to be understood within the context of the long tradition of "advice to the prince" literature. Historically, geopoliticians were intellectuals of statecraft who emphasized the role of geographical constraints and opportunities on the conduct of foreign policy. While many early geopoliticians liked to think of themselves as "scientific" and "objective," they were far from being detached and apolitical. In fact, the opposite was most often the case. Geopoliticians craved power. Some academics, like Halford Mackinder, sought it out by entering the political system while others, like Karl Haushofer, contented themselves with being professors and occasional advisors to political leaders. Other geopoliticians as practicing diplomats and foreign policy decision makers were already within positions of power. Even when not in direct positions of power, key intellectuals of statecraft can influence foreign policy debates and agenda from their position within civil society as prominent professors, journalists and media commentators.

Intellectuals, of course, are not free-floating thinkers in society but thinkers embedded within certain institutional structures and social networks of power, privilege and access. In thinking critically about geopolitics, we must consider not simply intellectuals alone but the institutions and social networks that enabled them to become intellectuals and "experts" on geopolitics. In many cases, there are layers of interlocking institutions involved: universities, private foreign policy research institutes, think-tanks, the media establishment and government agencies. For the early imperialist geopoliticians, the key institutional structures were usually universities and learned societies. Halford Mackinder, for example, earned his living as one of the first professors in the discipline of geography in the United Kingdom. His career within geography was made possible by his association with the Royal Geographical Society (RGS), an all-male explorers and travelers club established in London in 1830 that served as a social gathering place and network for the ruling establishment of the British Empire.

During the Cold War, prominent geopoliticians were usually associated with and circulated between a variety of different institutions. George Kennan, for example, came to prominence as a career foreign service officer, went on to direct the US state's new post-war Policy Planning agency and subsequently became an academic historian and professional writer. Like many members of the American foreign policy establishment, he became a member of the Council on Foreign Relations, a private meeting club of New York bankers established at the beginning of the twentieth century that subsequently developed into the quintessential foreign policy establishment institution in US civil society (Schulzinger, 1984). In more recent years, many other private foreign policy think-tanks and strategic studies institutes have been established like the Rand Corporation in Santa Monica, the Foreign Policy Research Institute in Philadelphia or the Heritage Foundation in Stanford, each with their own journals and publishing operations (Crampton and Ó Tuathail, 1996). In nearly all cases, these private institutions are the creation of powerful conservative individuals, interest groups and foundations, some with quite extremist views and ambitions. As one might expect, these institutions represent only the interests of the powerful and privileged and not that of the poor and weak. The foreign policy perspective of bankers and defense contractors will be articulated by the intellectuals of statecraft they hire; the foreign policy perspective of peasants or social workers will not.

This process of selecting certain intellectuals as "expert" is, as one might suspect, highly political and politicized. As a general rule, the most powerful institutions in any state or society will tend to sponsor those intellectuals who hold the same ideological viewpoint as they do. Ideologies are important, for states are governed and held together by certain widely shared systems of belief. At their most elemental, these ideological systems of belief include adherence to the common "national exceptionalist" myths of a state and support for the existing structures of power within a state and society (Agnew, 1983). In general, geopoliticians are usually strong national chauvinists and also entrenched conservatives. Historically at least, they have operated within and given voice to multiple Western ethnocentric discourses of power, articulating national and personal variations of racial, sexual and cultural supremacy in the name of "common sense," "reason" and an "objective perspective" (Haraway, 1991). The specific nature of their ideological worldview, of course, can be quite nuanced. Halford Mackinder, Karl Haushofer and George Kennan were very different types of "nationalist" intellectuals yet they also shared a general revulsion towards industrial modernity, though their attitudes and arguments on this were quite distinctive. Nevertheless, general support for the prevailing economic and cultural establishments of one's own state and society is common to most geopoliticians. Challenging the ethnocentrism, racism and sexism of geopoliticians both historically and today, is hazardous, however, for many groups seek to freeze intellectual inquiry by labeling it "politically correct." Ironically, it is those who use this label who are working in a politically correct way in the interests of the powerful for they seek to safeguard discourses of power – from national exceptionalist myths to implicit racial hierarchies, civilizational ethnocentrism, unreflective universalism, and patriarchy – from any kind of challenge and scrutiny.

Not all of the readings collected in this volume are those of conservative and nationalist geopoliticians. An alternative figure to the intellectual of statecraft is the *dissident intellectual*, the critically minded intellectual who is less interested in obtaining and exercising power than in challenging the prevailing "truths" of geopolitics and the structures of power, political economy and militarism they justify. Whereas the intellectual of statecraft or geopolitician is an insider who wants to be even more inside, the dissident intellectual is an outsider, one who usually challenges the ruling nationalist orthodoxy, in particular states and societies. In some cases, these intellectuals gain a certain media celebrity or, perhaps more accurately, a notoriety because of their questioning ways. The English historian E.P. Thompson, a leader in the British Campaign for Nuclear Disarmament (CND) in the early 1980s, was vilified by both his own government and by the Soviet Union for his attempt to think beyond the Cold War logic of mutual antagonism and militarism that dominated politics in the post-World War II period and divided the continent of Europe in two. The Cold War, Thompson argued in 1982:

> has become a habit, an addiction. But it is a habit supported by very powerful material interests in each bloc: the military-industrial and research establishments of both sides, the security services and intelligence operations, and the political servants of these interests. These interests command a large (and growing) allocation of the skills and resources of each society; they influence the direction of each society's economic and social development; and it is in the interest *of* these interests to increase that allocation and to influence this direction even more (1982: 169).

Amongst the intellectual servants of the military-industrial complex at this time in the West were Cold War geopoliticians like Henry Kissinger, Zbigniew Brzezinski, General Alexander Haig and others. Thompson's arguments echo those of other dissident figures like the physicist Andrei Sakharov in Russia – a scientist in the Soviet military-industrial complex who later became its leading critic – and the linguistic scholar Noam Chomsky in the United States. During the Vietnam War, Chomsky was a fierce critic of those he dubbed the "new mandarins" in the US national security state, the military-bureaucratic intellectuals who justified and prosecuted the US war against radical nationalism in Vietnam and elsewhere in the Third World (Chomsky, 1969).

Throughout this volume, we have deliberately tried to illustrate the essentially contested nature of geopolitical knowledge by presenting readings that directly comment and critique each other. Our overall aim is to provoke debate and reflection on the politics of geopolitical knowledge in the twentieth century. Geopolitics, as this volume makes clear, is not an objective, scientific form of knowledge. It is about the operation of discourse and power/knowledge, and it is also about how intellectuals, institutions and ideology create structures of power within states. Too often in the past, geopolitics has been treated not as discourse but as detached and objective description of how the world "really is." In challenging this approach in this book, we are seeking to render the relations of power embedded in geopolitical discourses visible and manifest. For Foucault, wherever there is power, there is also resistance. *It is within discourses of resistance that the power effects of geopolitical discourses are problematized.* Since this intellectual and political aim is central to this volume, it is imperative that we consider not only the discourses forged by the powerful, the hegemonic and the privileged but also the counter-hegemonic discourses of those who are marginalized, ignored and silenced by dominant discourses. We all live within ensembles of power, knowledge and expertise. Gaining an appreciation of this is the first step on the path towards a critical understanding of the discourses of geopolitics that currently enframe our own locations, identities and worlds.

REFERENCES AND FURTHER READING

Agnew, J. (1983) "An Excess of 'National Exceptionalism'," *Political Geography Quarterly*, 2: 151–166.

Agnew, J. and Corbridge, S. (1995) *Mastering Space: Hegemony, Territory and International Political Economy*, London: Routledge.

Chomsky, N. (1969) *American Power and the New Mandarins*, New York: Pantheon.

Cox, R. (1986) "Social Forces, States and World Orders," in Keohane, R. (ed.) *Neorealism and Its Critics*, New York: Columbia University Press.

Crampton, A. and Ó Tuathail, G. (1996) "Intellectuals, Institutions and Ideology: The Case of Robert Strausz-Hupé and 'American Geopolitics,'" *Political Geography*, 15: 533–555.

Dodds, K. (1996) "The 1982 Falklands War and a Critical Geopolitical Eye: Steve Bell and the If . . . Cartoons," *Political Geography*, 15: 571–591.

Dodds, K. and Sidaway, J. (1994) "Locating Critical Geopolitics," *Society and Space*, 12: 515–524.

Enloe, C. (1990) *Bananas, Beaches, and Bases: Making Feminist Sense of International Politics*, Berkeley, Calif.: University of California, 1990.

Foucault, M. (1980) *Power/Knowledge*, New York: Pantheon.

Halberstam, D. (1972) *The Best and the Brightest*, New York: Penguin.

Haraway, D. (1991) *Simians, Cyborgs, and Women: The Reinvention of Nature*, New York: Routledge.

Hepple, L. (1986) "The Revival of Geopolitics," *Political Geography Quarterly*, 5 (supplement): S21–S36.

Isaacson, W. and Thomas, E. (1986) *The Wise Men: Six Friends and the World They Made: Acheson, Bohlen, Harriman, Kennan, Lovett, McCloy*, New York: Simon and Schuster.

Mahan, A. (1957) *The Influence of Seapower Upon History*, New York: Hill and Wang.

Mayer, A. (1988) *Why Did the Heavens Not Darken?*, New York: Pantheon.

O'Loughlin, J. (ed.) (1994) *Dictionary of Geopolitics*, Westport, Conn.: Greenwood Press.

Ó Tuathail, G. (1996a) "An Anti-Geopolitical Eye: Maggie O'Kane in Bosnia, 1992–93," *Gender, Place and Culture*, 3 (2): 171–185.

——(1996b) *Critical Geopolitics*, London: Routledge.

Pletsch, C.E. (1981) "The Three Worlds, or the Division of Social Scientific Labor, circa 1950–1975," *Comparative Studies in Society and History*, 23: 565–590.

Thompson, E.P. (1982) *Beyond the Cold War*, New York: Pantheon.

Schulzinger, R. (1984) *The Wise Men of Foreign Affairs: The History of the Council on Foreign Relations*, New York: Columbia University Press.

Spykman, N. (1942) *America's Strategy in World Politics*, New York: Harcourt Brace.

Wuerker, M. (1992) *Standing Tall in Deep Doo-Doo: A Cartoon Chronicle of the Campaign '92 and the Bush/Quayle Years*, New York: Thunder Mouth Press.

PART 1
Imperialist Geopolitics

PART 1

INTRODUCTION

Gearóid Ó Tuathail

Geopolitics, as a form of power/knowledge, was born in the era of imperialist rivalry between the decades from 1870s to 1945 when competing empires clashed and fought numerous wars, two of which were worldwide wars, all the time producing, arranging and then altering and revising the lines of power that were the borders of the world political map. An era characterized by colonial expansionism abroad and industrial modernization at home, it was a time of tremendous technological achievement, social upheaval and cultural transformation. The dominant imperialist structure of the age was the British Empire which, despite its increasing territorial size over the decades, was poorly adjusting to the transforming conditions of world power, particularly those in the early twentieth century. The other "great" imperial powers of the time – Russia, France, Italy, the United States, Germany and later Japan – were its general rivals and sought to profit from its difficulties and relative decline. Each of these imperialist states produced their own leading intellectuals of statecraft and came to develop their own distinctive cultural variants of geopolitics, congealments of geographical knowledge and imperialist power strategizing.

The most historically and geographically fated imperialist rivalry of the period was that between the British Empire and the rising imperial aspirations of the German state in central Europe, a rivalry that was at the crux of the two murderous worldwide wars that destroyed the lives of millions of people in the twentieth century. It is this rivalry that we examine here through an investigation of the geopolitical writings of the British geographer Halford Mackinder, the German general turned geopolitician Karl Haushofer and the political agitator who became the German *Führer*, Adolf Hitler. To remind us that geopolitics was not a European monopoly, we shall briefly examine US president Theodore Roosevelt's 1905 corollary to the Monroe Doctrine as an example of Western hemispheral geopolitics in the early twentieth century.

Halford Mackinder began his career teaching geography in 1887 at Oxford University thanks to the influence and sponsorship of the Royal Geographical Society. Mackinder had impressed a number of fellows of the RGS earlier that year when, at the young age of 25, he addressed the society and made the case for a "new geography" of academic synthesis to supersede the "old geography" of exploration and discovery that largely defined geography in the nineteenth century. Not everyone was impressed, however. One crusty old Admiral sat in the front row muttering "damn cheek, damn cheek" as he spoke (Blouet, 1987: 40). To those traditionalists who saw geography as a "manly science" of military adventuring and "lion hunting," Mackinder must have appeared as a young bookish upstart. Many on the leadership council of the society, however, were sympathetic to his arguments and subsequently championed him for a position at Oxford, agreeing to pay half his salary for five years.

Because geography was seen by many within the RGS as a "manly" outdoorish science, Mackinder felt over the subsequent years the need to prove his scientific manhood to the traditionalists in the all-male RGS. Consequently, Mackinder and some social acquaintances undertook in 1899 an expedition to climb Mount Kenya in what was then "British East

Africa" and is now the independent state of Kenya. From Mackinder's point of view, the expedition was a ripping success – he successfully climbed the mountain, gave a triumphant address at the RGS upon his return, and became a manly "geographical expert" – though it was hardly that for those colored natives who were shot by Mackinder's party during the course of the African expedition (Kearns, 1997).

As a geographical expert who had proved his worth in the colonies, Halford Mackinder felt strongly about the role geographical knowledge could play in addressing the relative decline of the British Empire in the early twentieth century, a relative decline dramatically illustrated by the difficulties the British army had in winning the Boer War (1899–1902). Mackinder supported the imperial reform movement of Joseph Chamberlain, the former Colonial Secretary who sought to modernize the British Empire by imposing a common external tariff against the products of other "Great Powers" at this time (Ó Tuathail, 1992). Like many of his compatriots, Mackinder worried about the rising power of the German Empire on the European continent. Geographical education, for him, was an important weapon in the struggle for "relative efficiency" between the Great Powers, particularly between Great Britain and the German state of Kaiser Wilhelm II. Geography, he argued, was a necessary subject in educating "the children of an Imperial race" (Mackinder, 1907: 36). Most of the British masses were of "limited intelligence" so it was the duty of an elite of experts to educate them to think like the rulers of a vast overseas Empire. It is essential, he argued, "that the ruling citizens of the worldwide Empire should be able to visualize distant geographical conditions .... Our aim must be to make our whole people think Imperially – think that is to say in spaces that are world wide – and to this end our geographical teaching should be directed" (1907: 37–38). Geography was a discipline that *disciplined*; it taught the uneducated masses to think in the political way that experts like Mackinder wanted them to think.

Geography could also educate the political leaders of the Empire about the geographical factors that Mackinder claimed conditioned human history and the conduct of strategy. On a cold January evening in 1904, Mackinder gave an address to the RGS on precisely this theme. Mackinder's talk on "The Geographical Pivot of History" (Reading 1) created little stir at the time – few political leaders heard him speak – but it was destined to make him famous decades later when, during World War II, the American and British public discovered German geopolitics and the reverence it accorded the ideas of Mackinder. Mackinder's address is important in the history of geopolitics for three reasons: for its god's eye global view; for its division of the globe into vast swaths of territory, and for its sweeping story of geography's conditioning influence on the course of history and politics. These three "innovations" in Mackinder's text account for its subsequent influence and appeal.

First, though he does not use the word geopolitics, Mackinder's essay "invents" geopolitics as a new detached perspective that surveys the globe as "closed" political space. *Geopolitics is a new way of seeing international politics as a unified worldwide scene.* Mackinder, and subsequent geopoliticians, adopt a god's eye view which looks down on what he calls "the stage of the whole world":

> For the first time we can perceive something of the real proportion of features and events on the stage of the whole world and may seek a formula which shall express certain aspects, at any rate, of geographical causation in history (1904: 421).

This sentence is extremely important. The "we" it invokes is the community of geographical or geopolitical experts, educated and privileged white men like Mackinder who, by virtue of these social privileges, can adopt an Olympian perspective on the world, perceive the real proportion of features and events, and seek formulae or laws to explain history. All the elements of imperialist geopolitics are in this one sentence: the divine eye gaze upon the world; the implicit claim that only "objective" experts can perceive the real, and the desire to find underlying laws to explain all of history. Over and over again, these elements are found in imperialist and more modern forms of geopolitics. What they mark is the arrogance of an elite of geopolitical experts who play god by claiming to see objectively, perceive the real and explain all.

Critical thinkers challenge all of these claims and argue that geopolitical knowledge is *essentially contested*. Geopolitical experts are never detached but embedded in economic, political, racial and sexual relations of power (as Mackinder certainly was). They do not see objectively but within the structures of meaning provided by their socialization into certain (usually privileged) backgrounds, intellectual contexts, political beliefs and culture. They do not see "the real" but see that which their culture *interprets and constructs* as "the real." Their so-called "laws" of strategy are often no more than self-justifications for their own political ideology and that of those in power within their state. Their production of knowledge about international politics, in other words, is a form of power which they wield to serve their own political ends.

Second, Mackinder's text is remembered for its map of "The Natural Seats of Power" and its invention of the game of labeling huge swaths of the world's territory with a singular identity. Like all maps, this Mercator projection map is an *interpretation of the earth* and not a true representation of it. Mercator projections radically distort the size of the Northern latitudes, enlarging Greenland and Russia, for example, and shrinking the Australian and African continents. The centering of the map on Eurasia inevitably renders that region pivotal and North and South America marginal. To illustrate his thesis graphically, Mackinder labels enormous tracts of territory with simple identities like "pivot area," "inner or marginal crescent" and "lands of the outer or insular crescent." The great irony of this god-like labeling of the earth is that, in so doing, Mackinder eliminates the tremendous geographical diversity and particularity of places on the surface of the earth. Difference becomes sameness. Geographical heterogeneity becomes geopolitical homogeneity. This "loss of geography" is, as we shall see, a recurrent feature of those geopolitical discourses that play the game of earth labeling.

Third, Mackinder's address is remembered because of the sweeping story Mackinder tells about "the geographical causation of history." At the center of this story is the relationship between physical geography and transportation technology. Mackinder claims that there are three epochs of history (represented below in Table 2) which he names after the explorer Christopher Columbus. Each epoch is defined by dominant dramas – remember his stage metaphor! – and "mobilities of power." With the era of geographical exploration and discovery at an end, Mackinder suggests that history is now entering the post-Columbian epoch, an epoch of closed space where events in one part of the globe will have ripple effects across the globe. More significantly, from a British imperial point of view, "trans-continental railways are now transmuting the conditions of land-power, and nowhere can they have such effect as in the closed heart-land of Euro-Asia." This is alarming to Mackinder's reading of the interests of the British Empire because it threatens to change the balance of power between landpower (continental Europe, particularly Germany) and seapower (the British Empire) in Eurasia.

> The oversetting of the balance of power in favour of the pivot state ... would permit of the use of vast continental resources for fleet-building, and the empire of the world would then be in sight. *This might happen if Germany were to ally herself with Russia* (1904: 436; emphasis added).

It was this latter scenario that was Mackinder's greatest fear. The political leaders of the British Empire must do everything in their power to prevent an alliance between Germany and the "heart-land" of world power he identifies within the then Czarist Empire.

In a subsequent book called *Democratic Ideals and Reality* (1919), written immediately after World War I with a view to influencing the Versailles Peace negotiations, Mackinder made his strategic recommendation to the victorious political leaders even more explicit. Renaming what he called Euro-Asia the "World-Island" and the "pivot area" the "Heartland," he declared:

> Who rules East Europe commands the Heartland;
> Who rules the Heartland commands the World-Island;

Table 2 Halford Mackinder's geopolitical story

Epoch	*Dominant drama*	*Dominant mobility of power*	*Ascendant region and power type*
Pre-Columbian	Asiatic invasions of Europe	Horse and camel	The landpower of the Asian steppes
Columbian	European overseas expansionism	Sailing vessels and sea transportation	The seapower of the European colonial empires
Post-Columbian	Closed space and the struggle for relative efficiency	Railways	The landpower of those who control the heartland

> Who rules the World-Island commands the World (1919: 150).

Again, behind this sloganistic strategy is a simple recommendation. What must be prevented is German expansionism in Eastern Europe and a German alliance with what was the old Czarist Empire before the war but became the Soviet Union in the early 1920s.

Mackinder's ideas had little impact on British foreign policy during his lifetime though they did, as we shall see, earn the admiration of a school of German militarist geographers led by Karl Haushofer. One reason for their lack of impact at home is that his arguments had many flaws. His thesis was too sweeping, his interpretation of human history too simplistic and geographically deterministic, and his claims about the importance of mobility in the development of power onesided. Mackinder neglected the importance of organization in the development of power, he missed the revolutionary implications of airpower for the twentieth century, and, most significantly, he underestimated the emergent power of the United States (which he strangely describes as an eastern power!) while overestimating the strategic significance of the vast spaces of the Russian "heartland."

By 1904 the United States had emerged as a significant player on the world's stage. Consequent to its humiliating defeat of the Spanish Empire in 1898, the United States acquired the Philippines as a colony and became the imperial overlord of Cuba, imposing upon it the Platt Amendment which granted the United States the legal right to interfere in Cuban political life and territorial control over the strategic naval base of Guantanemo in perpetuity (a base it still occupies to this day). Motivated, in part, by the seapower doctrine of Alfred Mahan, which stressed the significance of acquiring overseas naval bases, the United States also acquired the Hawaiian islands and Guam. Mahan and other prominent imperialists in the United States like Brook Adams, Henry Cabot Lodge and Theodore Roosevelt justified such imperialist expansionism in a variety of ways. Throughout his voluminous writings, Admiral Mahan argued in an institutionally self-serving way that the path to national greatness lay in commercial and naval expansionism. All truly great powers were naval powers. It was not necessary to acquire whole territories and formally occupy them (this, after all, was colonialism and the United States liked to think of itself as an anti-colonial nation); what the United States needed was an informal empire based on "open door" trade and a string of overseas naval bases that would give its navy the ability to project power in a troublesome region whenever it needed to do so.

Implicit within Mahan's naval expansionist creed, and even more explicit in the dovetailing visions of Lodge and Roosevelt, was a social Darwinian ideology that held that all states, peoples and so-called "races" were in a struggle for survival with each other and only the fittest and most aggressive survived. The supremacy of particular peoples and races, Theodore Roosevelt believed, was best expressed in war, an activity he romanticized intensely as manly, vigorous, exciting and fundamental to greatness. "There is no place in the world," he wrote, "for nations who have become enervated by soft and easy life, or who have lost their fiber of vigorous hardiness and manliness" (quoted in Beale, 1956: 52). Roosevelt's obsession with demonstrating "manliness" made him a crusading militarist, a forward-charging "rough rider" who gloried in his heroic exploits in Cuba during the brief Spanish–American war. Like many other imperialists of his time, Roosevelt was a white supremacist, one who believed that there was a natural hierarchy of "races" with white Anglo-Saxons at the top and a whole series of "inferior races" like the Chinese, Latin Americans and Negroes well below them (Hunt, 1987). The category of "race" was a rather flexible one which referred to nationality, language, culture and manners as much as it did to skin color and biological inheritance. Roosevelt's racism, in contrast to that later exalted by the Nazis, was more civilizational and ethnographic than it was biological and genetically determinist. Certain races could, with help and effort, be "raised up" to a higher level of civilization.

All of these different elements – seapower imperialism, bellicose masculinity, anti-colonial commercial expansionism and civilizational racism – came together when Theodore Roosevelt became president in 1901 after the assassination of William McKinley. Full of national pride and imperial hubris, Roosevelt argued that America was a "masterful race" which should "speak softly" but carry a "big stick" in the Pacific, Caribbean and Latin America. In Central America, Roosevelt practiced an aggressive form of geopolitical interventionism which gave birth to the state of Panama. Formerly a province of Columbia, Roosevelt's administration fermented an independence movement in the region in order to secure the territory necessary to construct a canal linking the Atlantic and Pacific oceans. Just after his resounding election to the presidency in 1905, Roosevelt sought to formalize his geopolitical thinking into a so-called "corollary" to the Monroe Doctrine (Reading 2), the grandiloquent declaration by President James Monroe in 1823 that European powers should not "extend their system to any portion of this hemisphere." Roosevelt's corollary sought to give notice that the American hemisphere was the special preserve of the United States. As, according to Roosevelt, the most civilized and superior state in the hemisphere, the United States had a right, indeed an obligation, to "exercise an international police power" in the region to keep troublesome and uncivilized states in line. Intervention in the affairs of unruly and immature states in order to enforce the rule of law and restore discipline was part of what Rudyard Kipling called "the white man's burden," the so-called "burden" that comes from being superior and more civilized than everyone else, an arrogant philosophy Roosevelt's corollary perfectly articulates.

The white supremacist sentiment that was common to the practice of British and American imperialist geopolitics in the early twentieth century also found distinct expression in Germany, where a school of German geopolitical thought was first codified after World War I by Karl Haushofer (1869–1946). Haushofer was a former military commander who became a political geographer at 50 years of age after retiring from the German army with the rank of Major General. Born in Munich, Haushofer's military career took him to Japan from 1908 to 1910 where he admired the national unity of a Japanese state that was strongly anti-democratic and increasingly militarist in orientation (Dorpalen, 1942). Haushofer, the military officer, in particular admired the discipline of Japanese life and the blind obedience and devotion with which the Japanese people followed their leaders. His stay in Japan provoked him to write a book and doctoral dissertation on the German influence on the development of the Japanese state (which was indeed significant).

During World War I, Haushofer served as a field commander for the German army on the Eastern front, with Rudolf Hess, later deputy leader of the Nazi Party, as his aide-de-camp. Devastated by Germany's defeat, Haushofer turned to academia and with the help of friends obtained a lecturing post in political geography at the University of Munich. Hess soon enrolled as one of his students. Munich, at this time, was a city of revolutionary and counter-revolutionary ferment. In 1919, a group of revolutionary socialists, rebeling against the wartime slaughter and material hardship brought down upon them by the Kaiser and his generals, established a socialist republic in Bavaria, a political experiment that was soon violently crushed by the military. In 1923, a new violently nationalist party called the National Socialist Workers Party (the 'Nazis' for short), headquartered in Munich and made up largely of disaffected ex-soldiers, attempted to seize power in a Beer Hall Putsch. Hess, a senior member of the new party, fled in the wake of the failure and was hidden by Haushofer in his summer home in the Bavarian mountains. When Hess eventually gave himself up and was imprisoned, Haushofer visited him in Landsberg prison where Hess introduced Haushofer to the leader of the Nazi Party, Adolf Hitler.

Like many of the veterans of World War I for whom military service was the formative experience of their manhood, Haushofer, Hess and Hitler had a deep hatred of the peace treaty that took away Germany's colonies and part of its national territory after the war: the Treaty of Versailles. All felt that this treaty had emasculated Germany, a natural world power with a large advanced population that was reduced to living on a "narrow" territorial area. After Versailles, they believed that Germany's need for *Lebensraum* or living space was greater than ever. Consequently, they all worked, in their different ways, to overthrow the Treaty of Versailles and "make Germany a world power again" which they, as male militarists, understood to mean making Germany a dominant *military* power capable of expanding territorially at the expense of its neighbors. Greatness, for male militarists, is invariably tied up with fantasies of martial glory and territorial triumph.

Karl Haushofer's crusade to overthrow the Treaty of Versailles led him to found the journal *Zeitschrift fur Geopolitik* (*Journal of Geopolitics*) in 1924. This journal was to serve as the flag ship for the new school of geography Haushofer helped create: German *Geopolitik* (geopolitics). Like Mackinder in Great Britain, Haushofer believed that the leaders of the state should be educated in the geographical relationships he claimed governed international politics. Mixing the social Darwinist ideas of his intellectual hero, Friedrich Ratzel, and the ideas of Mackinder (Haushofer greatly admired Mackinder's writings describing "The Geographical Pivot of History," as "a geopolitical masterwork"; Weigert, 1942: 116), Haushofer reduced the complexity of international relations to a few basic laws and principles which he tirelessly promoted in the *Zeitschrift* and numerous books. In a book on frontiers, Haushofer outlines the Ratzelian organic theory of the state and uses this to polemicize against the Treaty of Versailles. International politics was a struggle for survival between competing states. In order to survive, the German state must achieve *Lebensraum*. The best means of achieving this, following Mackinder (no doubt to his own horror), is for Germany to develop an alliance with the heartland power, the Soviet Union. Furthermore, Haushofer argued, Germany should align itself with Japan and strive to create a continental-maritime block stretching from Germany through Russia to Japan against the global maritime empires of France and Great Britain, empires Haushofer believed were weak and in decay.

In "Why Geopolitik?" (Reading 3) which was published in 1925, Haushofer claims that the reason Germany lost World War I was because its leaders did not study geopolitics. Geopolitics, for Haushofer, is the study of the "earth-boundedness" of political processes and institutions. Like Mackinder, he attributes special power to the god-like geopolitician, treating geopolitics as a faith that offers divine revelations. Geopolitics can make certain predictions. It can provide "realistic insight into the world picture as it presents itself from day to day." It will help "our statesmen ... see political situations as they really are." Only the geopolitician can "see what is"! Haushofer's persistent emphasis on the need for geopolitical "training" is nothing more than a legitimation for the right-wing militarist foreign policy he and others promoted at the expense of the fragile democracy of the Weimar Republic. Haushofer justifies this "training" by declaring that our enemies study geopolitics so "we" had better start too!

Haushofer discussed his ideas with Adolf Hitler at Landsburg prison, where Hitler enjoyed a rather comfortable imprisonment. He presided over a midday meal, had as many visitors as he wanted and spent much of his time outside in the garden. From July 1924 onwards, he began dictating *Mein Kampf* to Rudolf Hess and also to another secretary, Emil Maurice. Volume 1, dictated in prison, was published in 1925. Volume 2, which Hitler dictated in his villa on Obersalzberg after his early release from prison, was published at the end of 1926. The book did not sell widely until after the Nazis were handed power by conservative elements in the German state in 1933, fearful as they were of communism and social revolution (Bullock, 1992: 140). Whereas Haushofer sought to advise leaders, Hitler sought to become the one that would restore Germany's greatness. Hitler had used the Beer Hall Putsch trial to project himself as the man of action and destiny who would lead "the revolution against the revolution" (i.e. a nationalist and militarist counter-revolution against communism and the "bourgeois" Weimar Republic).

Hitler's *Mein Kampf* is a despicably racist book full of hatred in which Hitler outlines his crude social Darwinist vision of the world. He describes a racial struggle for survival between the pure and the impure (hybrid), the healthy and the parasitic, the national and the international, the noble and the treacherous. This basic set of distinctions is mapped onto the fundamental distinction Hitler makes between identity and difference, "Us" and "Them," the Self and the Other (see Table 3). Using two pseudoscientific racial categories with no basis in fact, "the Aryan" and "the Jew," Hitler defines a positive insider identity which he champions in opposition to a negative outsider identity. He invents "the German" in opposition to "the Jew." That people of Jewish faith and heritage could also be good German citizens was a contradiction in terms to the Nazis (even though many had fought and died during the war for Germany). As the historian Alan Bullock has noted, the identity category

> "the Jew" as one encounters it in the pages of *Mein Kampf* and Hitler's ravings bears no resemblance to flesh-and-blood human beings of Jewish descent: [It] is an invention of Hitler's obsessional fantasy, a Satanic creation, expressing his need to create an object on which he could concentrate his feelings of aggression and hatred (Bullock, 1992: 145).

Like similar racist categories, "the Jew" is an eminently flexible archetype. "The Jew" could represent both an ultra-capitalist (bankers, financiers, industrial and department store owners) and an ultra-communist (a "Bolshevik," Marxist or German leftist), two totally opposite identities. Logical contradictions such as this are common in racist reasoning. The Other is whatever the racist decides it is. Both identities are present in the composite category, "Jewish Bolshevism" which represents "rootlessness," "internationalism" and "decay" in Hitler's worldview. The opposite of "Jewish Bolshevism" is "folkish nationalism," Hitler's racist version of German nationalism that imagined Germany as an idealized community of healthy and racially pure Aryan peasants rooted in the soil and ruled over by "natural leaders" like Hitler.

There are three fundamentally racist discourses of danger championed by Hitler in *Mein Kampf*. The first is the external threat posed to his idealized German nation by a supposed "international Jewish conspiracy" against all nation-states, particularly Germany. The headquarters of this conspiracy are, again contradictorily, the radically different states of the United States (international finance capitalism) and the Soviet Union (communism). The second is the internal threat posed by German leftist organizations and political parties, and the general supposed "decay" of modern urban life (miscegenation, degenerate art, prostitution and mental illness). The third threat, which combines elements of all of Hitler's obsessions, is one that finds expression in Chapter XIV of Volume II of *Mein Kampf* (Reading 4). In this chapter, Hitler borrows from both Ratzel and Haushofer (without citation) to claim that

Table 3 Hitler's racist map of identity and difference

Identity	*Difference*
Self: "Us"	Other: "Them"
Insider, friend, citizen	Outsider, stranger, foreigner
Aryan German nation	Jewish Bolshevism
Rooted in traditional organic soil and society Idealized image of peasant society with "natural leaders"	Rootlessness; do not have any soil of their own and so are "parasitic" on the territory of others. Associated with the vices of modernity and urban life
A folkish community of the beautiful, healthy and racially pure	A collection of the impure, the hybrid and unhealthy, the dirty and the degenerate

Germany is currently an "impotent" nation without adequate territorial resources to feed its people. The Treaty of Versailles has "constricted" the German nation and left it without adequate space, especially in comparison to the other world powers. Since nations are competitive organic entities that gain nourishment from the soil, the German nation must begin pursuing *Lebensraum* or else face decay and further decline. In Hitler's view of the world, great power is only possible if the state controls great territorial spaces.

Hitler argues that it is up to the National Socialist movement (the Nazi Party) to "endeavor to eliminate the discrepancy between our population and our area," "to bring the land into consonance with the population." This "discrepancy," this need for harmony between land and people, is a "truth" generated by Hitler's appropriation of Ratzelian discourse, an example of the power of discourse to invent "the real" and, on the basis of this suppossed "truth" or "reality" (which, in actuality, is only an ideological construction), to legitimate and justify certain political visions, in this case Hitler's militaristic ambitions. Because of his racism, Hitler has utter contempt for the Soviet Union, viewing it as a state in decay. Yet, paradoxically, the Jewish Bolshevism of the Soviet Union is a mortal threat. Again, we encounter a contradiction. The Other is weak and degenerate yet the Other is also an implacable and dangerous enemy. As he makes clear, Hitler does not want to return to the 1914 borders of Germany, to the territory of Germany before the Treaty of Versailles. Hitler's plans are much more radical. He scorns those, like Karl Haushofer, calling for an "Eastern orientation" or an alliance between Russia and Germany. Hitler's program is an "eastern policy in the sense of aquiring the necessary soil for our German people." From this one line, first published in 1926, we can see Hitler's megalomaniac desire to play god with the map of continental Europe and re-arrange it completely. His imperialist vision was for the colonization of the East by a renewed German Empire. The German "Aryan" master race would enslave the sub-human "Slavs" of the East. Hitler's "eastern policy" ultimately led to the genocidal war against the Soviet Union and "Jewish Bolshevism" that begain in 1941 (Mayer, 1988). Ironically, one of the reasons for Hitler's downfall was his racist assumption that the Soviet Union would collapse in a few months after its invasion by the German military war machine.

It is worth noting that there were important differences between the German geopolitics of Karl Haushofer and the Nazi geopolitics of Adolf Hitler. Haushofer nationalism was more conservative-aristocratic than counterrevolutionary fascist. Haushofer considered the British Empire the ultimate enemy of Germany and urged an alliance with the Soviet Union, whereas Hitler admired the British Empire and ultimately wanted to conduct a crusade against the Soviet Union and Jewish Bolshevism. In Haushofer's Ratzelian schema, space not race is the ultimate determinate of national destiny, whereas for Hitler race is more important than space. Racists believe that destiny is internal and biological not external and environmental (Bassin, 1987).

Nevertheless, these differences should not detract from the fundamental support Haushofer gave to Hitler and the Nazi regime both before and after it was handed power by the conservative establishment in 1933. Although he never became a Nazi Party member, Haushofer promoted Nazi ideology, writing a book called *National Socialist Thought in World Politics* to mark the Nazi ascent to power, even denouncing Jews despite the fact that his own wife was Jewish. Together with his son, Albrecht, Haushofer helped facilitate the German–Japanese cooperation that eventually resulted in the Anticomintern Pact of 1936. Both were advisors to Hitler during the Munich conference of 1938. The Nazi–Soviet pact of 1939 seemed to represent Haushofer's thinking. However, the strange flight of Rudolf Hess to England in May 1941 ended the influence of the Haushofer family (Heske, 1987). Karl was even imprisoned for eight weeks in Dachau after Hess's flight. Albrecht, too, was imprisoned. After his release, Albrecht maintained links with those aristocratic elements in the German establishment belatedly planning the overthrow of Hitler. Their attempt to assassinate Hitler on July 20, 1944 failed, however, and Albrecht Haushofer was imprisoned once again. Upon his release from Moabit prison in April 1945, he was murdered by a roaming SS squad.

Despite the limited role of the Haushofer's in the last few years of the Nazi state, many sensational press stories in the allied countries, particularly the United States, presented him as the scientific brain behind the Nazi blueprint for "world conquest." Haushofer was said to run an enormous Institute of Geopolitics at the University of Munich which supposedly gathered information from all over the world. This was then used by Haushofer and his colleagues to make predictions about the course of world politics and give advice to Nazi leaders about the most opportune times to invade countries and the like. All of this sensationalism, which Haushofer complained about after the war, was exaggerated and largely untrue (Ó Tuathail, 1996). Nevertheless, the admiration the German geopoliticians had for Mackinder was noticed by the allies and some of his works were re-read and re-published.

After the fall of Berlin and the end of the war, the American Jesuit priest Father Edmund Walsh, the founder of the School of Foreign Service at Georgetown University in Washington DC, was flown to Germany to interrogate Karl Haushofer on his teachings and possible influence on Nazi foreign policy. The question as to whether Haushofer should be tried for war crimes in Nuremberg had to be decided. Walsh and the American army found a frail and disillusioned old man. In a statement before Father Walsh and the American army on November 2, 1945, Haushofer tried to explain his teachings and writings within the context of post-war Germany. As might be expected, Haushofer's "Defense of German Geopolitics" (Reading 5) is a self-serving document in which he seeks to disassociate himself from the horrors of Nazism. He claims that he was interested in educating and training German youth about the world, that he occasionally overstepped the boundary separating pure and practical science but that a scholar "should have the right to stand at the side of his people with all his mental power." Pointing to his own family's suffering, he claims that he opposed "imperialistic plans of conquest," a highly questionable interpretation.

What is interesting and disturbing about Haushofer's "Defense," however, is his claim that much of what he did was "legitimate" geopolitics. Haushofer points out that his captives have acknowledged this, that many of his lectures correspond to what Walsh taught at Georgetown, that many British and American thinkers where "the basic inspirers of his teaching," and that the original goals of German geopolitics were quite similar to "legitimate American geopolitics"! Haushofer's basic defense is that the "legitimacy" of German geopolitics was corrupted by the Nazis. This line of argumentation is much too convenient for it seeks to save geopolitics and Haushofer's reputation by blaming everything on the Nazis. What Haushofer does not acknowledge and recognize is *the essentially contested and political nature of all geopolitical discourse*, whether it be German, American or British. In trying to take refuge in the concept of "legitimate" geopolitics, the life of Karl Haushofer brings into question the very legitimacy of all geopolitics. Geopolitics is never objective up to a certain point, scientific to a certain borderline, or legimate up to a balance of a certain percentage as he suggests (and as his interrogators believed too). *Geopolitical discourse is political from the very outset.* At this time, it was invariably entwined with the dominant ideologies and culture of nationalist chauvinism in the states where it was produced. Haushofer is thus right to claim that he was merely doing what other geopoliticians (like Edmund Walsh and Isaiah Bowman in the United States) were doing in their particular countries. But this does not mean that Haushofer's geopolitics is morally equivalent to that of these other geopoliticans. Haushofer was guilty of propagandizing a militarist and imperialistic version of German nationalism. He was complicitious with many of the aims of Adolf Hitler and the Nazi Party. As such, he was guilty of lending support to one of the most murderous and brutal state regimes in the twentieth century.

Haushofer's "Defense" raises some fundamental questions about the practice of geopolitics as a whole. As a form of power/knowledge, geopolitics was clearly complicitous with many chauvinist, racist and imperialist ideologies in the first half of the twentieth century. It justified oppressive European colonial empires that were premised on white supremist assumptions, imperialist interventionism, and, in Hitler's geopolitics, brought imperialist thinking and racist brutality to the European continent. It encouraged statesmen to play god with the world political map and justified appalling state violence, the culmination of which was World War II.

Geopolitics did not go away after World War II and the fall of Nazi Germany. It was about to change form. Late on Sunday night, March 10, 1946, Karl Haushofer and his wife Martha walked to a secluded hollow on their country estate in the Bavarian mountains. Both took an arsenic drink and then Karl helped his wife hang herself to make sure of death. Karl himself fell dead soon afterwards, his hands, as Edmund Walsh describes it, "clutching the Bavarian soil which he so passionately loved and so often described in his writings on *Lebensraum*" (Walsh, 1948: 34). A year later Halford Mackinder died in England. Six days after his death, President Harry Truman of the United States addressed a joint session of Congress and requested economic and military aid to help the governments of Greece and Turkey fight against the worldwide communist threat. The imperialist geopolitics of old was giving way to a newly emergent Cold War geopolitics.

REFERENCES AND FURTHER READING

On Mackinder

Blouet, B. (1987) *Halford Mackinder: A Biography*, College Station: Texas A & M.

Kearns, G. (1997) "The Imperial Subject: Geography and Travel in the Work of Mary Kingsley and Halford Mackinder," *Transactions, Institute of British Geographers*, N.S., forthcoming.

Mackinder, H.J. (1907) "On Thinking Imperially," *Lectures on Empire*, ed. M. E. Sadler, London: Privately Printed.

—— (1919) *Democratic Ideals and Reality*, New York: Henry Holt.

Ó Tuathail, G. (1992) "Putting Mackinder in his Place: Material Transformations and Myth," *Political Geography Quarterly*, 11: 100–118.

Parker, W.H. (1982) *Mackinder: Geography as an Aid to Statecraft*, Oxford: Clarendon Press.

On Theodore Roosevelt and American imperialism

Beale, H. (1956) *Theodore Roosevelt and the Rise of America to World Power*, New York: Collier.

Hunt, M. (1987) *Ideology and U.S. Foreign Policy*, New Haven: Yale University Press.

LaFeber, W. (1963) *The New Empire: An Interpretation of American Expansion, 1860–1898*, Ithaca, NY: Cornell University Press.

Miller, N. (1992) *Theodore Roosevelt: A Life*, New York: Morrow.
Roosevelt, T. (1913) *Theodore Roosevelt: An Autobiography*, New York: Macmillan.
Williams, W.A. (1980) *Empire as a Way of Life*, New York: Oxford University Press.
—— (1962) *The Tragedy of American Diplomacy*, New York: Dell.

On Haushofer

Bassin, M. (1987) "Race Contra Space: The Conflict Between German *Geopolitik* and National Socialism," *Political Geography Quarterly*, 6: 115–134.
Dorpalen, A. (1942) *The World of General Haushofer: Geopolitics in Action*, New York: Farrar and Rinehart.
Heske, H. (1987) "Karl Haushofer: His Role in German Geopolitics and in Nazi Politics," *Political Geography Quarterly*, 6: 135–144.
Ó Tuathail, G. (1996) *Critical Geopolitics: The Politics of Writing Global Space*, London: Routledge.
Parker, G. (1985) *Western Geopolitical Thought in the Twentieth Century*, New York: St Martin's Press.
Walsh, E. (1948) *Total Power*, New York: Doubleday.
Weigert, H. (1942) *Generals and Geographers: The Twilight of Geopolitics*, London: Oxford University Press.

On Hitler and Nazi ideology

Bauman, Z. (1989) *Modernity and the Holocaust*, Cambridge: Polity.
Bullock, A. (1992) *Hitler and Stalin: Parallel Lives*, New York: Knopf.
Mayer, A. (1988) *Why Did The Heavens Not Darken?*, New York: Pantheon.

Cartoon 1 Hands off!
This British postcard dates from the beginning of the century and features the British lion rebuking the threatening advance of the German eagle towards the globe.
Source: Courtesy of Dr Peter Taylor's private collection

1 HALFORD J. MACKINDER

"The Geographical Pivot of History"

from *Geographical Journal* (1904)

When historians in the remote future come to look back on the group of centuries through which we are now passing, and see them foreshortened, as we today see the Egyptian dynasties, it may well be that they will describe the last 400 years as the Columbian epoch, and will say that it ended soon after the year 1900. Of late it has been a commonplace to speak of geographical exploration as nearly over, and it is recognized that geography must be diverted to the purpose of intensive survey and philosophic synthesis. In 400 years the outline of the map of the world has been completed with approximate accuracy, and even in the polar regions the voyages of Nansen and Scott have very narrowly reduced the last possibility of dramatic discoveries. But the opening of the twentieth century is appropriate as the end of a great historic epoch, not merely on account of this achievement, great though it be. The missionary, the conqueror, the farmer, the miner, and, of late, the engineer, have followed so closely in the traveller's footsteps that the world, in its remoter borders, has hardly been revealed before we must chronicle its virtually complete political appropriation. In Europe, North America, South America, Africa, and Australasia there is scarcely a region left for the pegging out of a claim of ownership, unless as the result of a war between civilized or half-civilized powers. Even in Asia we are probably witnessing the last moves of the game first played by the horsemen of Yermak the Cossack and the shipmen of Vasco da Gama. Broadly speaking, we may contrast the Columbian epoch with the age which preceded it, by describing its essential characteristic as the expansion of Europe against almost negligible resistances, whereas mediaeval Christendom was pent into a narrow region and threatened by external barbarism. From the present time forth, in the post-Columbian age, we shall again have to deal with a closed political system, and none the less that it will be one of worldwide scope. Every explosion of social forces, instead of being dissipated in a surrounding circuit of unknown space and barbaric chaos, will be sharply re-echoed from the far side of the globe, and weak elements in the political and economic organism of the world will be shattered in consequence. There is a vast difference of effect in the fall of a shell into an earthwork and its fall amid the closed spaces and rigid structures of a great building or ship. Probably some half-consciousness of this fact is at last diverting much of the attention of statesmen in all parts of the world from territorial expansion to the struggle for relative efficiency.

It appears to me, therefore, that in the present decade we are for the first time in a position to attempt, with some degree of completeness, a correlation between the larger geographical and the larger historical generalizations. For the first time we can perceive something of the real proportion of features and events on the stage of the whole world, and may seek a formula which shall express certain aspects, at any rate, of geographical causation in universal history. If we are fortunate, that formula should have a practical value as setting into perspective some of the competing forces in current international politics. The familiar phrase about the westward march of empire is an empirical and fragmentary attempt of the kind. I propose this evening describing those physical features of the world which I believe to have been most coercive of human action, and presenting some of the chief phases of history as organically connected with them, even in the ages when they were unknown to geography. My aim will not be to discuss the influence of this or that kind of feature, or yet to make a study in regional geography, but rather to exhibit human history as part of the life of the world organism. I recognize that I can only arrive at one aspect of the truth, and I have no wish to stray into excessive materialism. Man and not nature initiates, but nature in large measure controls. My concern is with the general physical control, rather than the causes of universal history. It is obvious that only a first approximation to truth can be hoped for. I shall be humble to my critics.

The late Professor Freeman held that the only history which counts is that of the Mediterranean and European races. In a sense, of course, this is true, for it is among these races that have originated the ideas which have rendered the inheritors of Greece and Rome dominant throughout the world. In another and very important sense, however, such a limitation has a cramping effect upon thought. The ideas which go to form a nation, as opposed to a mere crowd of human animals, have usually been accepted under the pressure of a common tribulation, and under a common necessity of resistance to external force. The idea of England was beaten into the Heptarchy by Danish and Norman conquerors; the idea of France was forced upon competing Franks, Goths, and Romans by the Huns at Chalons, and in the Hundred Years' War with England; the idea of Christendom was born of the Roman persecutions, and matured by the Crusades; the idea of the United States was accepted, and local colonial patriotism sunk, only in the long War of Independence; the idea of the German Empire was reluctantly adopted in South Germany only after a struggle against France in comradeship with North Germany. What I may describe as the literary conception of history, by concentrating attention upon ideas and upon the civilization which is their outcome, is apt to lose sight of the more elemental movements whose pressure is commonly the exciting cause of the efforts in which great ideas are nourished. A repellent personality performs a valuable social function in uniting his enemies, and it was under the pressure of external barbarism that Europe achieved her civilization. I ask you, therefore, for a moment to look upon Europe and European history as subordinate to Asia and Asiatic history, for European civilization is, in a very real sense, the outcome of the secular struggle against Asiatic invasion . . .

[. . .]

For a thousand years a series of horse riding peoples emerged from Asia through the broad interval between the Ural mountains and the Caspian sea, rode through the open spaces of southern Russia, and struck home into Hungary in the very heart of the European peninsula, shaping by the necessity of opposing them the history of each of the great peoples around – the Russians, the Germans, the French, the Italians, and the Byzantine Greeks. That they stimulated healthy and powerful reaction, instead of crushing opposition under a widespread despotism, was due to the fact that the mobility of their power was conditioned by the steppes, and necessarily ceased in the surrounding forests and mountains.

A rival mobility of power was that of the Vikings in their boats. Descending from Scandinavia both upon the northern and the southern shores of Europe, they penetrated inland by the river ways. But the scope of their action was limited, for, broadly speaking, their power was effective only in the neighbourhood of the water. Thus the settled peoples of Europe lay gripped between two pressures – that of the Asiatic nomads from the east, and on the other three sides that of the pirates from the sea. From its very nature neither pressure was overwhelming, and both therefore were stimulative. It is noteworthy that the formative influence of the Scandinavians was second only in significance to that of the nomads, for under their attack both England and France made long moves towards unity, while the unity of Italy was broken by them. In earlier times, Rome had mobilized the power of her settled peoples by means of her roads, but the Roman roads had fallen into decay, and were not replaced until the eighteenth century...

[...]

Mobility upon the ocean is the natural rival of horse and camel mobility in the heart of the continent. It was upon navigation of oceanic rivers that was based the Potamic stage of civilization, that of China on the Yangtze, that of India on the Ganges, that of Babylonia on the Euphrates, that of Egypt on the Nile. It was essentially upon the navigation of the Mediterranean that was based what has been described as the Thalassic stage of civilization, that of the Greeks and Romans. The Saracens and the Vikings held sway by navigation of the oceanic coasts.

The all important result of the discovery of the Cape road to the Indies was to connect the western and eastern coastal navigations of Euro-Asia, even though by a circuitous route, and thus in some measure to neutralize the strategical advantage of the central position of the steppe-nomads by pressing upon them in rear. The revolution commenced by the great mariners of the Columbian generation endowed Christendom with the widest possible mobility of power, short of a winged mobility. The one and continuous ocean enveloping the divided and insular lands is, of course, the geographical condition of ultimate unity in the command of the sea, and of the whole theory of modern naval strategy and policy as expounded by such writers as Captain Mahan and Mr Spenser Wilkinson. The broad political effect was to reverse the relations of Europe and Asia, for whereas in the Middle Ages Europe was caged between an impassable desert to south, an unknown ocean to west, and icy or forested wastes to north and north-east, and in the east and south-east was constantly threatened by the superior mobility of the horsemen and camelmen, she now emerged upon the world, multiplying more than thirty fold the sea surface and coastal lands to which she had access, and wrapping her influence round the Euro-Asiatic land-power which had hitherto threatened her very existence. New Europes were created in the vacant lands discovered in the midst of the waters, and what Britain and Scandinavia were to Europe in the earlier time, that have America and Australia, and in some measure even Trans-Saharan Africa, now become to Euro-Asia. Britain, Canada, the United States, South Africa, Australia, and Japan are now a ring of outer and insular bases for sea-power and commerce, inaccessible to the land-power of Euro-Asia.

But the land-power still remains, and recent events have again increased its significance. While the maritime peoples of Western Europe have covered the ocean with their fleets, settled the outer continents, and in varying degree made tributary the oceanic margins of Asia, Russia has organized the Cossacks, and, emerging from her northern forests, has policed the steppe by setting her own nomads to meet the Tartar nomads. The Tudor century, which saw the expansion of Western Europe over the sea, also saw Russian power carried from Moscow through Siberia. The eastward swoop of the horsemen across Asia was an event almost as pregnant with political consequences as was the rounding of the Cape, although the two movements long remained apart.

It is probably one of the most striking coincidences of history that the seaward and the landward expansion of Europe should, in a sense, continue the ancient opposition between Roman and Greek. Few great failures have had more far-reaching consequences than the failure of Rome to Latinize the Greek. The Teuton was civilized and Christianized by the Roman, the Slav in the main by the Greek. It is the Romano-Teuton who in later times embarked upon the ocean; it was the Graeco-Slav who rode over the steppes, conquering the Turanian. Thus the modern land-power differs from the sea-power no less in the source of its ideals than in the material conditions of its mobility.

In the wake of the Cossack, Russia has safely emerged from her former seclusion in the northern forests. Perhaps the change of greatest intrinsic importance which took place in Europe in the last century was the southward migration of the Russian peasants, so that, whereas agricultural settlements formerly ended at the forest boundary, the centre of the population of all European Russia now lies to south of that boundary, in the midst of the wheat-fields which have replaced the more western steppes. Odessa has here risen to importance with the rapidity of an American city.

A generation ago steam and the Suez canal appeared to have increased the mobility of sea-power relatively to land-power. Railways acted chiefly as feeders to ocean-going commerce. But trans-continental railways are now transmuting the conditions of land-power, and nowhere can they have such effect as in the closed heartland of Euro-Asia, in vast areas of which neither timber nor accessible stone was available for road-making. Railways work the greater wonders in the steppe, because they directly replace horse and camel mobility, the road stage of development having here been omitted . . .

[. . .]

The Russian railways have a clear run of 6000 miles from Wirballen in the west to Vladivostok in the east. The Russian army in Manchuria is as significant evidence of mobile land-power as the British army in South Africa was of sea-power. True, that the Trans-Siberian railway is still a single and precarious line of communication, but the century will not be old before all Asia is covered with railways. The spaces within the Russian Empire and Mongolia are so vast, and their potentialities in population, wheat, cotton, fuel, and metals so incalculably great, that it is inevitable that a vast economic world, more or less apart, will there develop inaccessible to oceanic commerce.

As we consider this rapid review of the broader currents of history, does not a certain persistence of geographical relationship become evident? Is not the pivot region of the world's politics that vast area of Euro-Asia which is inaccessible to ships, but in antiquity lay open to the horse-riding nomads, and is today about to be covered with a network of railways? There have been and are here the conditions of a mobility of military and economic power of a far-reaching and yet limited character. Russia replaces the Mongol Empire. Her pressure on Finland, on Scandinavia, on Poland, on Turkey, on Persia, on India, and on China replaces the centrifugal raids of the steppe men. In the world at large she occupies the central strategical position held by Germany in Europe. She can strike on all sides and be struck from all sides, save the north. The full development of her modern railway mobility is merely a matter of time. Nor is it likely that any possible social revolution will alter her essential relations to the great geographical limits of her existence. Wisely recognizing the fundamental limits of her power, her rulers have parted with Alaska; for it is as much a law of policy for Russia to own nothing over seas as for Britain to be supreme on the ocean.

Outside the pivot area, in a great inner crescent, are Germany, Austria, Turkey, India, and China, and in an outer crescent, Britain, South Africa, Australia, the United States, Canada, and Japan. In the present condition of the balance of power, the pivot state, Russia, is not equivalent to the peripheral states, and there is room for an equipoise in France. The United States has recently become an eastern power, affecting the European balance not directly, but through Russia, and she will construct the Panama canal to make her Mississippi and Atlantic resources available in the Pacific. From this point of view the real divide between east and west is to be found in the Atlantic ocean.

The oversetting of the balance of power in favour of the pivot state, resulting in its expansion over the marginal lands of Euro-Asia, would permit of the use of vast continental resources for fleet-building, and the empire of the world would then be in sight. This might happen if Germany were to ally herself with Russia. The threat of such an event should, therefore, throw France into alliance with the over-sea powers, and France, Italy, Egypt, India, and Korea would become so many bridge heads where the outside navies would support armies to compel the pivot allies to deploy land forces and prevent them from concentrating their whole strength on fleets. On a smaller scale that was what Wellington accomplished from his sea-base at Torres Vedras in the Peninsular War. May not this in the end prove to be the strategical function of India in the British Imperial system? Is not this the idea underlying Mr. Amery's conception that the British military front stretches from the Cape through India to Japan ?

The development of the vast potentialities of South America might have a decisive influence upon the system. They might strengthen the United States, or, on the other hand, if Germany were to challenge the Monroe doctrine successfully, they might detach Berlin from what I may perhaps describe as a pivot policy. The particular combinations of power brought into balance are not material; my contention is that from a geographical point of view they are likely to rotate round the pivot state, which is always likely to be great, but with limited mobility as compared with the surrounding marginal and insular powers.

I have spoken as a geographer. The actual balance of political power at any given time is, of course, the product, on the one hand, of geographical conditions, both economic and strategic, and, on the other hand, of the relative number, virility, equipment, and organization of the competing peoples. In proportion as these quantities are accurately estimated are we likely to adjust differences without the crude resort to arms. And the geographical quantities in the calculation are more measurable and more nearly constant than the human. Hence we should expect to find our formula apply equally to past history and to present politics. The social movements of all times have played around essentially the same physical features, for I doubt whether the progressive desiccation of Asia and Africa, even if proved, has in historical times vitally altered the human environment. The westward march of empire appears to me to have been a short rotation of marginal power round the south-western and western edge of the pivotal area. The Nearer, Middle, and Far Eastern questions relate to the unstable equilibrium of inner and outer powers in those parts of the marginal crescent where local power is, at present, more or less negligible.

In conclusion, it may be well expressly to point out that the substitution of some new control of the inland area for that of Russia would not tend to reduce the geographical significance of the pivot position. Were the Chinese, for instance, organized by the Japanese, to overthrow the Russian Empire and conquer its territory, they might constitute the yellow peril to the world's freedom just because they would add an oceanic frontage to the resources of the great continent, an advantage as yet denied to the Russian tenant of the pivot region.

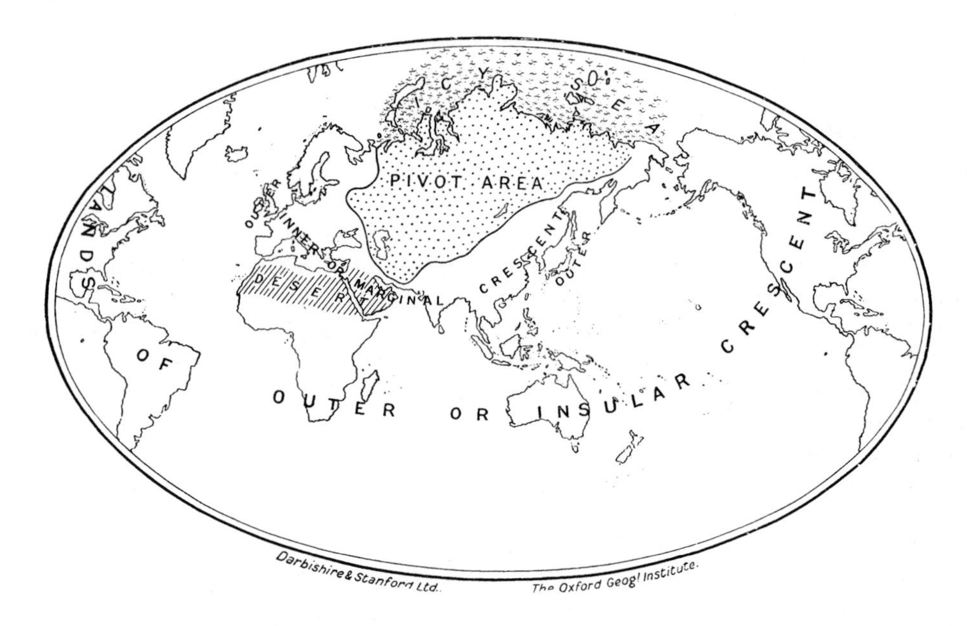

Map 2 The natural seats of power
Source: Mackinder 1904

2 THEODORE ROOSEVELT

"The Roosevelt Corollary"

from *A Compilation of Messages and Papers of the Presidents* (1905)

... It is not true that the United States feels any land hunger or entertains any projects as regards the other nations of the Western Hemisphere save such as are for their welfare. All that this country desires is to see the neighboring countries stable, orderly, and prosperous. Any country whose people conduct themselves well can count upon our hearty friendship. If a nation shows that it knows how to act with reasonable efficiency and decency in social and political matters, if it keeps order and pays its obligations, it need fear no interference from the United States. Chronic wrongdoing or an impotence which results in a general loosening of the ties of civilized society, may in America, as elsewhere, ultimately require intervention by some civilized nation, and in the Western Hemisphere the adherence of the United States to the Monroe Doctrine may force the United States, however reluctantly, in flagrant cases of such wrongdoing or impotence, to the exercise of an international police power. If every country washed by the Caribbean Sea would show the progress in stable and just civilization which with the aid of the Platt amendment Cuba has shown since our troops left the island and which so many of the republics in both Americas are constantly and brilliantly showing, all question of interference by this Nation with their affairs would be at an end. Our interests and those of our southern neighbors are in reality identical. They have great natural riches, and if within their borders the reign of law and justice obtains, prosperity is sure to come to them. While they thus obey the primary laws of civilized society they may rest assured that they will be treated by us in a spirit of cordial and helpful sympathy. We would interfere with them only in the last resort, and then only if it became evident that their inability or unwillingness to do justice at home and abroad had violated the rights of the United States or had invited foreign aggression to the detriment of the entire body of American nations. It is a mere truism to say that every nation, whether in America or anywhere else, which desires to maintain its freedom, its independence, must ultimately realize that the right of such independence cannot be separated from the responsibility of making good use of it.

In asserting the Monroe Doctrine, in taking such steps as we have taken in regard to Cuba, Venezuela, and Panama, and in endeavoring to circumscribe the theater of war in the Far East, and to secure the open door in China, we have acted in our own interest as well as in the interest of humanity at large. There are, however, cases in which while our own interests are not greatly involved, strong appeal is made to our sympathies.... But in extreme cases action may be justifiable and proper. What form the action shall take must depend upon the circumstances of the case; that is, upon the degree of the atrocity and upon our power to remedy it. The cases in which we could interfere by force of arms as we interfered to put a stop to intolerable conditions in Cuba are necessarily very few.

3 KARL HAUSHOFER

"Why Geopolitik?"

from *The World of General Haushofer* (1942)

While the theoretical foundations of Geopolitik were laid only in recent times, its practical application – the instinctive sense for geopolitical possibilities, the realization of its deep influence on political development – is as old as history itself. Geopolitical vision inspired daring leaders who guided their people along novel never-before-travelled roads. Powerful new states emerged because their creators, with the sensitivity of the true statesman, understood the geopolitical demands of the hour. Without such insight, violence and arbitrariness would have charted the course of history. Nothing with lasting value could have been created. All structures of state which might have been erected would sooner or later have crumbled into dust and oblivion before the eternal forces of soil and climate.

To be sure, the powerful will of a great and strong man may tear masses and nations away from soil-bound existence into roads other than nature had provided for them. But such actions are short-lived. In the end every people will sink back into its accustomed ways; its lasting earth-bound traits will eventually win out.

GEOPOLITIK AS EDUCATION IN STATECRAFT

Although our eyes can not penetrate the darkness of the future, scientific geopolitical analysis enables us to make certain predictions. Should we not therefore, attempt to explore the field of Geopolitik more fully than we – and especially our diplomats – have thought necessary? To pose the question is to answer it. Our statesmen in particular ought to familiarize themselves with all those aspects of politics that can be determined scientifically before piloting the destiny of state and nation into the mists of the unknown future. Jurisprudence and political science, which have been considered the sole prerequisites of education in statecraft, do not provide adequate training. A sound knowledge of geography and history is just as important. Above all, our future leaders must be schooled in geopolitical analysis.

Only this can give them the needed realistic insight into the world picture as it presents itself from day to day. Not by accident is the word "Politik" preceded by that little prefix "geo." This prefix means much and demands much. It relates politics to the soil. It rids politics of arid theories and senseless phrases which might trap our political leaders into hopeless utopias. It puts them back on solid ground. Geopolitik demonstrates the dependence of all political developments on the permanent reality of the soil.

A whole body of literature has grown around this thesis. For the Alpine countries, Ratzel has traced the interdependence between politics and geographical environment in his *Alps as the Center of Historical Movements*. Krebs has given us an equally valuable work in his *Contributions to the Political Effects of Climate* in which he reveals the connection between lack of rain, aridity, and social and political unrest in East Asia. Kjellen, in his *Problem of the Three Rivers* (Rhine, Danube, Vistula), has shown us how the unhappy fate of Central Europe is inseparably tied up with the course of these rivers. And H.J. Mackinder, in his "Geographical Pivot of History," has attempted to review the entire world geopolitically and to forecast in

1904 what would happen between 1914 and 1924.

Why did our leading statesmen fail to see what this student of geopolitics realized as early as 1904? Most likely because they lacked geopolitical training. In spite of excellent legal education and great administrative experience, they were unable to realize the effects of political-geographical trends. "Geographical ignorance may cost us dearly," warned Sir Thomas Holdich, one of England's most experienced students and drawers of boundaries.

GEOPOLITIK AND PRACTICAL POLITICS

Geopolitik has come to stay. We arrive at this conclusion from the fact that its application is gaining a growing following all over the world, while disregard of its teachings becomes increasingly dangerous. Some political successes can doubtless be attributed to geopolitical groundwork, among them the skillful selection of such English bases as Hong Kong, Singapore, and Penang. The reorganization of the Australian Commonwealth and the foundation of its new capital, Canberra, are likewise the result of geographical considerations. Geopolitically, even the choice of Tsingtao was a good one, provided one considers the establishment of a German base in China as geopolitically justifiable.

THE MISSION OF GEOPOLITIK

Geopolitik will serve our statesmen in setting and attaining their political objectives. It will present them with the scientific equipment of concrete facts and proven laws to help them see political situations as they really are. As an exact science, Geopolitik deserves serious consideration. Our leaders must learn to use all available tools to carry on the fight for Germany's existence – a struggle which is becoming increasingly difficult due to the incongruity between her food production and population density.

For our future foreign policy we therefore need Geopolitik. We need the same thorough training in this discipline as developed by England – though not under that name – with one-sided purposefulness, as adopted by France [in the *Institut de France* and the *Ecole de Politique*], and as it is beginning to be used by Japan. Geopolitik is a child of geography; whoever takes up its study should therefore be trained geographically. To teach it requires first-hand knowledge; teachers of Geopolitik must know from practical experience not only the country they are teaching about but also the one in which they are teaching. We must, moreover, study Geopolitik with a view to the present and future rather than to the past. As a nation governed by lawyers, we Germans have been too much under the influence of the *lex lata*. We considered politics more in terms of dead history than of living science: we looked back rather than ahead. In this manner we lost contact with the future. Making retrospective instead of precautionary future politics, we were left out of the realignment of the world when it occurred at the turn of the century.

This policy was doomed to failure. *Ducunt volentem, nolentem trahunt fata!* [Only those who are willing are guided by fate; the unwilling ones are dragged !] Nowhere does this maxim of Roman wisdom apply more truly than in the realm of politics. We learned our lesson . . .

[. . .]

Germany must emerge out of the narrowness of her present living space into the freedom of the world. We must approach this task well equipped in knowledge and training. We must familiarize ourselves with the important spaces of settlement and migration on earth. We must study the problem of boundaries as one of the most important problems of Geopolitik. We ought to devote particular attention to national self-determination, population pressure, living space, and changes in rural and urban settlement, and we must closely follow all shifts and transfers of power throughout the world.

The smaller the living space of a nation, the greater the need for a far-sighted policy to keep the little it can still call its own. A people must know what it possesses. At the same time, it should constantly study and compare the living spaces of other nations. Only thus will it be able to recognize and seize any possibility to recover lost ground.

"We must see foreign nations as they really are, not as we would like them to be." This occasional remark of Erich von Drygalski [Haushofer's academic mentor and thesis supervisor at the University of Munich] has served me as a beacon in my geopolitical work. Let us not stake our future foolishly on one card, let us not choose allies which others – better trained geopolitically – have considered doomed a half-century earlier. By prudent, courageous analysis of our world-political situation we shall always be able to preserve our sacred soil from shameful defeat. The admonitions "see what is," and "keep away from whatever our national honor cannot tolerate," are the pilot lights of our voyage. They are modest enough and even hardly sufficient to help our ship of state gain the open sea.

And yet – "I have neither men, arms, munitions, nor instructions . . .," the future commander of France's Army of the North wrote desperately to the Chief of National Defense on October 21, 1870. A victorious enemy was pressing him in front, and he was standing with his back against the wall – neutral Belgium that was already within gun range. Yet half a century later his grandsons stood east of the Rhine in a defenseless Germany, masters of the world's third largest colonial empire. During those fifty years France had taken up the study of geopolitics!

4

ADOLF HITLER

"Eastern Orientation or Eastern Policy?"

from *Mein Kampf* (1942)

There are two reasons which induce me to submit to a special examination the relation of Germany to Russia: 1. Here perhaps we are dealing with the most decisive concern of all German foreign affairs; and 2. This question is also the touchstone for the political capacity of the young National Socialist movements to think clearly, and to act correctly. . . .

[. . .]

If under foreign policy we must understand the regulation of a nation's relations with the rest of the world, the manner of this regulation will be determined by certain definite facts. As National Socialists we can, furthermore, establish the following principle concerning the nature of the foreign policy of a folkish state:

The foreign policy of the folkish state must safeguard the existence on this planet of the race embodied in the state, by creating a healthy, viable natural relation between the nation's population and growth on the one hand and the quantity and quality of its soil on the other hand.

As a healthy relation we may regard only that condition which assures the sustenance of a people on its own soil. Every other condition, even if it endures for hundreds, nay, thousands of years, is nevertheless unhealthy and will sooner or later lead to the injury if not annihilation of the people in question.

Only an adequately large space on this earth assures a nation of freedom of existence. . . .

Germany today is no world power. Even if our momentary military impotence were overcome, we should no longer have any claim to this title. What can a formation, as miserable in its relation of population to area as the German Reich today, mean on this planet? In an era when the earth is gradually being divided up among states, some of which embrace almost entire continents, we cannot speak of a world power in connection with a formation whose political mother country is limited to the absurd area of five hundred thousand square kilometers.

From the purely territorial point of view, the area of the German Reich vanishes completely as compared with that of the so called world powers. Let no one cite England as a proof to the contrary, for England in reality is merely the great capital of the British world empire which calls nearly a quarter of the earth's surface its own. In addition, we must regard as giant states, first of all the American Union, then Russia and China. All are spatial formations having in part an area more than ten times greater than the present German Reich. And even France must be counted among these states. Not only that she complements her army to an ever-increasing degree from her enormous empire's reservoir of colored humanity, but racially as well, she is making such great progress in negrification that we can actually speak of an African state arising on European soil . . .

Thus, in the world today we see a number of power states, some of which not only far surpass the strength of our German nation in population, but whose area above all is the chief support of their political power. Never has the relation of the German Reich to other existing world states been as unfavorable as at the beginning of our history two thousand years ago and again today. Then we were a young people, rushing headlong into a world of great crumbling state formations, whose last giant, Rome, we ourselves helped to fell. Today we find ourselves in a world of great power states in process of formation, with our own Reich sinking more and more into insignificance.

We must bear this bitter truth coolly and soberly in mind. We must follow and compare the German Reich through the centuries in its relation to other states with regard to population and area. I know that everyone will then come to the dismayed conclusion which I have stated at the beginning of this discussion: Germany is no longer a world power, regardless of whether she is strong or weak from the military point of view.

We have lost all proportion to the other great states of the earth, and this thanks only to the positively catastrophic leadership of our nation in the field of foreign affairs, thanks to our total failure to be guided by what I should almost call a testamentary aim in foreign policy, and thanks to the loss of any healthy instinct and impulse of self-preservation.

If the National Socialist movement really wants to be consecrated by history with a great mission for our nation, it must be permeated by knowledge and filled with pain at our true situation in this world; boldly and conscious of its goal, it must take up the struggle against the aimlessness and incompetence which have hitherto guided our German nation in the line of foreign affairs. Then, without consideration of "traditions" and prejudices, it must find the courage to gather our people and their strength for an advance along the road that will lead this people from its present restricted living space to new land and soil, and hence also free it from the danger of vanishing from the earth or of serving others as a slave nation.

The National Socialist movement must strive to eliminate the disproportion between our population and our area – viewing this latter as a source of food as well as a basis for power politics – between our historical past and the hopelessness of our present impotence. And in this it must remain aware that we, as guardians of the highest humanity on this earth, are bound by the highest obligation, and the more it strives to bring the German people to racial awareness so that, in addition to breeding dogs, horses, and cats, they will have mercy on their own blood, the more it will be able to meet this obligation . . .

[. . .]

We National Socialists must never under any circumstances join in the foul hurrah patriotism of our present bourgeois world. In particular it is mortally dangerous to regard the last pre-War developments as binding even in the slightest degree for our own course. From the whole historical development of the nineteenth century, not a single obligation can be derived which was grounded in this period itself. In contrast to the conduct of the representatives of this period, we must again profess the highest aim of all foreign policy, to wit: to bring the soil into harmony with the population. Yes, from the past we can only learn that, in setting an objective for our political activity, we must proceed in two directions: Land and soil as the goal of our foreign policy, and a new philosophically established, uniform foundation as the aim of political activity at home.

I still wish briefly to take a position on the question as to what extent the demand for soil and territory seems ethically and morally justified. This is necessary, since unfortunately, even in so called folkish circles, all sorts of unctuous big-mouths step forward, endeavoring to set the rectification of the injustice of 1918 as the aim of the German nation's endeavors in the field of foreign affairs, but at the same time find it necessary to assure the whole world of folkish brotherhood and sympathy.

I should like to make the following preliminary remarks: The demand for restoration of the frontiers of 1914 is a political absurdity of such proportions and consequences as to make it seem a crime. Quite aside from the fact that the Reich's frontiers in 1914 were anything but logical. For in reality they were neither complete in the sense of embracing the people of German nationality, nor sensible with regard to geo-military expediency. They were not the result of a considered political action, but momentary frontiers in a political struggle that was by no means concluded; partly, in fact, they were the results of chance.

As opposed to this, we National Socialists must hold unflinchingly to our aim in foreign policy, namely, to secure for the German people the land and soil to which they are entitled on this earth. And this action is the only one which, before God and our German posterity, would make any sacrifice of blood seem justified: before God, since we have been put on this earth with the mission of eternal struggle for our daily bread, beings who receive nothing as a gift, and who owe their position as lords of the earth only to the genius and the courage with which they can conquer and defend it; and before our German posterity in so far as we have shed no citizen's blood out of which a thousand others are not bequeathed to posterity. The soil on which someday German generations of peasants can beget powerful sons will sanction the investment of the sons of today, and will some day acquit the responsible statesmen of blood-guilt and sacrifice of the people, even if they are persecuted by their contemporaries . . .

But we National Socialists must go further. The right to possess soil can become a duty if without extension of its soil a great nation seems doomed to destruction. And most especially when not some little nigger nation or other is involved, but the Germanic mother of life, which has given the present-day world its cultural picture. Germany will either be a world power or there will be no Germany. And for world power she needs that magnitude which will give her the position she needs in the present period, and life to her citizens . . .

[. . .]

Never forget that the rulers of present-day Russia are common blood-stained criminals; that they are the scum of humanity which, favored by circumstances, overran a great state in a tragic hour, slaughtered and wiped out thousands of her leading intelligentsia in wild blood lust, and now for almost ten years have been carrying on the most cruel and tyrannical regime of all time. Furthermore, do not forget that these rulers belong to a race which combines, in a rare mixture, bestial cruelty and an inconceivable gift for lying, and which today more than ever is conscious of a mission to impose its bloody oppression on the whole world. Do not forget that the international Jew who completely dominates Russia today regards Germany, not as an ally, but as a state destined to the same fate. And you do not make pacts with anyone whose sole interest is the destruction of his partner. Above all, you do not make them with elements to whom no pact would be sacred, since they do not live in this world as representatives of honor and sincerity, but as champions of deceit, lies, theft, plunder, and rapine . . .

In Russian Bolshevism we must see the attempt undertaken by the Jews in the twentieth century to achieve world domination. Just as in other epochs they strove to reach the same goal by other, though inwardly related processes. Their endeavor lies profoundly rooted in their essential nature. No more than another nation renounces of its own accord the pursuit of its impulse for the expansion of its power and way of life, but is compelled by outward circumstances or else succumbs to impotence due to the symptoms of old age, does the Jew break off his road to world dictatorship out of voluntary renunciation, or because he represses his eternal urge. He, too, will either be thrown back in his course by forces lying outside himself, or all his striving for world domination will be ended by his own dying out. But the impotence of nations, their own death from old age, arises from the abandonment of their blood purity. And this is a thing that the Jew preserves better than any other people on earth. And so he advances on his fatal road until another force comes forth to oppose him, and in a mighty struggle hurls the heaven-stormer back to Lucifer.

The fight against Jewish world Bolshevization requires a clear attitude toward Soviet Russia. You cannot drive out the Devil with Beelzebub. If today even folkish circles rave about an alliance with Russia, they should just look around them in Germany and see whose support they find in their efforts. Or have folkish men lately begun to view an activity as beneficial to the German people which is recommended and promoted by the international Marxist press? Since when do folkish men fight with armor held out to them by a Jewish squire? . . .

If the National Socialist movement frees itself from all illusions with regard to this great and all-important task, and accepts reason as its sole guide, the catastrophe of 1918 can some day become an infinite blessing for the future of our nation. Out of this collapse our nation will arrive at a complete reorientation of its activity in foreign relations, and, furthermore, reinforced within by its new philosophy of life, will also achieve outwardly a final stabilization of its foreign policy. Then at last it will acquire what England possesses and even Russia possessed, and what again and again induced France to make the same decisions, essentially correct from the viewpoint of her own interests, to wit: A political testament.

The political testament of the German nation to govern its outward activity for all time should and must be:

Never suffer the rise of two continental powers in Europe. Regard any attempt to organize a second military power on the German frontiers, even if only in the form of creating a state capable of military strength, as an attack on Germany, and in it see not only the right, but also the duty, to employ all means up to armed force to prevent the rise of such a state, or, if one has already arisen, to smash it again. See to it that the strength of our nation is founded, not on colonies, but on the soil of our European homeland. Never regard the Reich as secure unless for centuries to come it can give every scion of our people his own parcel of soil. Never forget that the most sacred right on this earth is a man's right to have earth to till with his own hands, and the most sacred sacrifice the blood that a man sheds for this earth . . .

Neither western nor eastern orientation must be the future goal of our foreign policy, but an eastern policy in the sense of acquiring the necessary soil for our German people. Since for this we require strength, and since France, the mortal enemy of our nation, inexorably strangles us and robs us of our strength, we must take upon ourselves every sacrifice whose consequences are calculated to contribute to the annihilation of French efforts toward hegemony in Europe. Today every power is our natural ally, which like us feels French domination on the continent to be intolerable.

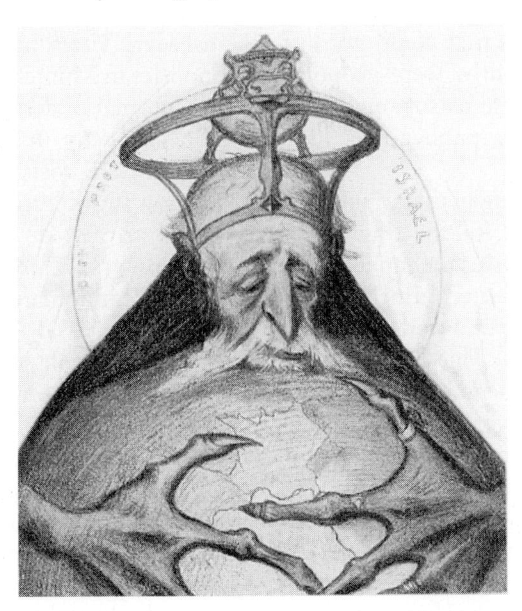

Cartoon 3 Rothschild
Anti-Semitism was never a specifically German phenomenon. This 1898 French cartoon illustrates the pervasive myth that the world was in the hands of Jewish bankers, personified by James (Jakob) Rothschild.
Source: C. Leandre 1898

5 KARL HAUSHOFER

"Defense of German Geopolitics"

from *Total Power: A Footnote to History* (1948)

Although not the originator of the technical term "geopolitics," nevertheless I have rightly been considered as the leading exponent of its manifestation in Germany.... The manner in which German geopolitics came into being is, by the same token, the justification for its appearance as a subject of higher teaching from 1919 onward; it was born of necessity.... It would be an inhuman and impossible demand to expect that a German scientist could disregard the inadequacy of the distribution of living space in central Europe, which had occurred in those times as a result of its overdeveloped industrialization and urbanization. To this must be added the dismemberment of central Europe by frontiers that could not last long and which, consequently, were geopolitically unjustifiable. For these reasons my book *Frontiers*, as well as other publications, was written.

What seemed most lacking in the resumption of the educational process for the training of German youth after the war was the capability to think in terms of wide space (in continents!) and the knowledge of the living conditions of others, namely of oceanic peoples. This broadness of thought, limited by a continental narrowness as well as by smallness in its world vision, became narrow-minded and lost in a welter of trivial controversies. It was cut off from the energizing breath of the sea and robbed of its overseas connections.... The knowledge, therefore, of the great ways of life that were essentially sea-minded – the British Empire, the United States of America, Japan, the Dutch East Indian Empire – was then even more inadequate than was the knowledge of the Near and Middle East, Eurasia, and the Soviet Union.

Therefore it seemed necessary for German geopolitics to provide knowledge about the empires that are spread over all the seas and about the Indo-Pacific space. By that means a counterweight was created against the pressure from within during the period 1919 to 1933. Later, this sense of pressure, under the tension of internal party conflicts, unfortunately served more and more to overshadow and obscure this necessary knowledge of other lands.

In meeting this obligation the faculty of foreign sciences of the University of Berlin also served, together with the only Institute for Political Geography that existed in all Germany. This was directed by my son, Professor Albrecht Haushofer. There never was any institute for geopolitics in Munich . . .

No normal understanding man of any other nation can deny that a German scholar also, after such a laborious career and with every aspiration for objectivity, should have the right to stand at the side of his people with all his mental power. This he does because of the findings in his domain of knowledge, because of conclusions arrived at honestly and legitimately in such a struggle for existence as prevailed during the years from 1919 to 1932.

Although I never claimed as my own the principle: "My country, right or wrong," in its complete consequences, nevertheless it has to be admitted that the borderline is easily crossed between pure science and practical science in such times of extreme tension. Therefore it happened (slipped in) that I occasionally overstepped those borders. This I also admitted and regretted openly to the interrogators; it was recognized on their part also that from 1933 onward I could work only under pressure, since my oral and written expressions were subject to four types of censorship.

Since the interrogators acknowledge that, in comparison with the United States conception of 'legitimate" geopolitics, German geopolitics worked its way up to a balance of knowledge 60 to 70 per cent of which could be generally accepted as valid science, an exact differentiation will likewise have to be made between all that was printed about geopolitics before 1933 and after 1933.

If my whole scientific working material had not been broken up and in part carried off at the beginning of May by [the US army] I could point to numerous lectures, dating from the years 1919 to 1933, which correspond in their development, for example, with Scheme II "Methodology" of a course on geopolitics of the School of Foreign Service at Georgetown University in use there on 1 July 1944. Among my requisitioned papers was the collected and fully developed groundwork of my lectures.

All that was written and printed after 1933 was "under pressure" and must be judged accordingly. How the effects of this pressure (in which Rudolf Hess, who tried rather to protect, did not participate) eventually worked out can be proved by nearly three years either of imprisonment or of limitation on freedom imposed on my family, also by my own confinement in Dachau concentration camp, the murder of my eldest son by the Gestapo on 23 April 1945, the severe control over and later the suspension of the *Journal of Geopolitics*.

In the Third Reich the party in power lacked any official organ receptive to or understanding the doctrines of geopolitics. Therefore they only used and wrongly understood catchwords which they did not even comprehend. Only Rudolf Hess, from the time when he was my pupil, before even the NSDAP [Nazi Party] ever existed, and the Minister for Foreign Affairs, Von Neurath, had a certain understanding for geopolitics without being able to apply it successfully...

Those theories, originally deriving from Friedrich Ratzel (*The Earth and Life; Political Geography; Anthropogeography*) and from those who continued his theories in the United States (Semple) and in Sweden (Rudolf Kjellen), were formed to a larger extent from sources among English-speaking peoples than from continental peoples. They were presented to German circles in the form of the principle: "Let us educate our masters."

Mahan, Brook Adams, Joe Chamberlain ...; Sir Thomas Holdrich (*The Creator of Frontiers*); Sir Halford Mackinder (*The Geographical Pivot of History*); Lord Kitchener (1909); later I[saiah] Bowman (*The New World*, and other writings) were the basic inspirers of my teachings and were quoted again and again...

Imperialistic plans of conquest were never favored, neither by me in my writings nor in my lectures. As in my book on Frontiers I also protested against the crippling of Germany through the border decisions of the Versailles Treaty, so in my public lecture activities I stood up for the Germans in South Tyrol. I welcomed the incorporation of Sudeten German territories, but I never approved of annexation of territories alien to our people and which had no German settlements.

I always regarded dreams of such annexations as dangerous dreams and therefore disapproved them.

The fact that thousands of German settlers were repatriated to Germany at much expense and suffering through VDA [Association of Germans Living Abroad] under my leadership, proves in the best manner that at that time, in any case, an occupation of those territories was not planned or, at least that the desirability of such an occupation was not known. If National Socialism had revealed, by the way it published its ideals in the early years of its development, that they included the conquest of alien-blooded peoples and their territories, it would have brought about its own retirement from power. This I stressed on every occasion, among others on 8 November 1938, and I opposed such plans of conquest. I believed in the promise of saturation made in 1938.

A truly equitable determination of frontiers which would satisfy everybody and which does not impose hardships on parts of any people is practically impossible because of the immense complicated overlapping of border languages and economic centers that have developed in the course of time, especially in eastern Europe. I, therefore, as well as my son Albrecht, and others of my pupils and co-workers tried in long discussions, without success, to work out completely just and lasting principles for such a delimitation of borders. In that, my efforts always were focused on the task of not creating irredentas in any form.

Therefore it is self-evident that the charge of planning conquest, including carefully worked out maps to infiltrate into continents, such as South America, was manufactured from thin air. In such matters the sensation-loving press was raving without let or hindrance, even using detailed forgeries of maps . . .

The book *Mein Kampf* I saw for the first time when the first edition was already in print. I refused to review this book because it had nothing to do with geopolitics. For me, at that time, it seemed to be one of the many ephemeral publications for purposes of agitation. It is self-evident that I had no part in its origin and I believe I am protected against the suspicion of participation, mentioned in the yellow press, if one makes a scientific comparison of my style of writing and the style of that book. I never saw Hitler alone. The last time I saw him was in the presence of witnesses on 8 November 1938, and I then had a sharp disagreement with him. From then on I was in disgrace. Since Rudolf Hess's flight in May 1941 I was exposed to the persecution of the Gestapo which ended only at the end of April 1945 with the murder of my eldest son because he shared the secret of 20 July 1944 [the plot against Hitler's life]. He also was in contact with English-speaking peoples. My friendship with Rudolf Hess had its origin in 1918 and is, in common with his attendance at my lectures at the University, four years older than the foundation of the National Socialist party. I saw Hitler for the first time in 1922, when he was one of the many popular platform orators who were then mushrooming from the overheated soil of the German people and from the multiplicity of societies and political movements . . .

From autumn 1938 onward was the Way of Sorrow for German geopolitics. The individual fate of father and son is illustrated by my imprisonment and his death. This happened within the framework of the suffering of "political science" in all central Europe under the pressure of the autocracy of one party down to the misuse and misinterpretation [of geopolitics] by state officials. Despite all that, German geopolitics had originally – from 1919 to 1932 – goals quite similar to American geopolitics.

In the program of geopolitics, on its first appearance, one finds a statement saying that it aspired to be "the geographical conscience of the state." It should then, for instance, have demanded in 1938 that Germany be satisfied and grateful for the solution reached at Munich. When I actually tried to put this into effect – after my return from Italy and when I finally reached the head of the state on 8 November 1938 – I fell into his disfavor for it and never saw him again. Until that date, therefore, this representative of German geopolitics may well regard himself as a legitimate pre-defender, even in the sense of American geopolitics.

The goal of German geopolitics originally had been, in common with legitimate American geopolitics – to achieve the possibility of excluding disorders in the future, like those of 1914 to 1918, through mutual understanding of peoples and their potentialities to develop on the basis of their cultural foundations and living space; also to obtain for minorities the highest measure of justice and politico-cultural autonomy – as was the case in Estonia, for instance, and for a time seemed to be accomplished in Transylvania.

This presupposed a geographically correct picture of the world; it required mutuality, moreover, and respect of one nationality and race by others as well as recognition of the human right to "personality." It demanded the highest degree of indulgence and tolerance, of which my lectures and activities were replete, for instance, from 1919 to 1932 . . .

In the memorandum which was written as answers to the questioning of General Eisenhower's staff and which lay before the interrogators, I specified in detail that an international geopolitics could become one of the best means to prevent future world catastrophes. It would have to be built on a lively exchange of ideas and persons, of professors, teachers, assistants, and students.

In the spirit of its name and by the political art of its leadership it could restore to due honor the "sacrament of the earth," the holiness of the soil which supports humanity.

German geopolitics, between the earthquakes of 1914 to 1919 and from 1938 to 1945, endeavored to build a road toward this exalted goal.

Granting that errors and mistakes accompanied the course of geopolitics, they can be turned to profit by the wisdom of that saying in the English language: "All human progress resolves itself into the building of new roads."

PART 2

Cold War Geopolitics

PART 2

INTRODUCTION

Gearóid Ó Tuathail

Questions of geography were always deeply implicated in the Cold War that developed between the Soviet Union and the United States after World War II. By the end of the war, the states of Eastern Europe had become part of a Soviet sphere of influence. The Stalinist regime that governed the Soviet Union was a bureaucratic dictatorship that was determined to create a security zone for itself to prevent yet another invasion of its territory by Western powers. This had happened immediately after the Bolshevik revolution and again when Hitler invaded in June of 1941. Approximately twenty million Russians had died defending their homeland against Hitler's racist crusade and rolling back the German war machine until it was destroyed and Berlin captured. Peace with security was thus foremost on Stalin's mind, which in practical terms meant a peace with the security of a substantial Soviet sphere of influence in Eastern and Central Europe.

The United States' experience in World War II was considerably different. Its national territory and civilian population escaped the horrific destruction and indiscriminate mass murder of the total war waged on the European continent, in North Africa and in Asia. With all its leading competitors in ruins, the United States was the single most powerful state in the world, a state with supreme confidence in its nationalist myths and ideals (see Reading 11). During the war, the American stance had combined a maximalist statement of its political and economic ideals with a minimalist program of war aims. Once the war was over, the American state – led by an inexperienced president in Harry Truman who suddenly had an awesome weapon, the atomic bomb, at his command – found it difficult to resist envisioning all of the world according to its ideals of political democracy and capitalist economics. Like other world powers before it, America's leaders claimed that America's ideals were the universal ideals of all. A clash between the Soviet Union and the United States over the future of Eastern Europe was probably inevitable; a Cold War between both powers, however, was not.

Why the antagonism between these states developed in such a way as to eventually divide the European continent in two was a consequence of the geopolitical reasoning that became dominant in the USA in 1946 and 1947 and the reaction it provoked from the Stalinist regime. While certain groups within the Truman administration favored diplomacy and a certain amount of *realpolitik* dealing with Stalin, others championed an implacable view of the Soviet Union as an inherently expansionist power that quickly became orthodoxy. An early defining statement of this essentialist conception of the Soviet Union was provided by the United States' *chargé d'affairs* in Moscow, George Kennan. In February 1946, Kennan, ill and bedridden, dictated an 8,000 word communique to Washington that became known as the Long Telegram. In it Kennan expounded his conception of the Soviet Union as an historically and geographically determined power with an unfolding necessity to constantly expand. This, Kennan argued, was the essence of the Soviet Union and nothing really could be done about it. Most significantly, no deals can or should be struck with the Soviet Union.

Kennan's views were pounced upon by more hardline anti-communist elements in the Truman administration and widely circulated.

Kennan himself was recalled to Washington to head up a new Policy Planning agency within the "national security state" being created by the Truman administration at this time (Yergin, 1978). The Truman administration's attitude to the Soviet Union became more belligerent as the Soviet Union sought to manipulate internal politics in various Eastern European states to its own advantage. In March 1946 former British prime minister Winston Churchill strengthened the hardline forces in the Truman administration by charging that an "iron curtain" has descended across the continent of Europe. Churchill had his own agenda of preempting an anti-imperialist alliance of the USA and the USSR against the British Empire in favor of an Anglo-American anti-communist alliance that would commit the United States to aid the British Empire in a joint struggle against the Soviet Union (Taylor, 1990). To the Soviets, this alliance appeared to be already in existence when over the course of 1946 the Soviet army was eased out of Iran, a country on its borders, while Anglo-American oil companies gained control over that country's valuable and strategic oil reserves.

Mutual suspicion and antagonism deepened in 1947 when the British government informed the Americans that they could no longer afford to aid the reactionary Greek monarchy trying to re-establish itself in power after the war. The need for the Truman administration to convince a reluctant US Congress to provide aid to the corrupt Greek monarchy in its fight against leftist guerrillas and to Turkey in a lingering squabble with the Soviets over control of the Dardanelles provided the occasion for a speech in which President Truman outlined what became known as the Truman Doctrine (Reading 6).

The Truman Doctrine is the first significant public statement of American Cold War geopolitics. In it Truman uses the local situation of the civil war in Greece and the long-standing dispute over the Dardanelles to enunciate a more universal struggle between freedom and totalitarianism across the globe. Dwelling not on the geographical specificity of the conflicts in question, Truman's speech strives to articulate abstract and absolutist truths. In a dramatic crescendo, Truman declares: "At the present moment in world history nearly every nation must choose between alternative ways of life." This "choice," however, is not a free choice but a worldwide struggle between two ways of life which are simplistically represented by Truman as freedom versus totalitarianism.

Truman's rhetorical leap from the local to the universal, from the particular to the absolute, was to become characteristic of American Cold War geopolitics. Within such a discourse, the geographical complexities of particular places and specific conflicts soon became displaced by the Manichean categories and formulaic terms of crude Cold War geopolitics. Like the imperialist geopoliticians before the war, Truman adopts a god's eye view of the globe and implicitly uses the abstract categories of "the free world" and "the enslaved world" to mentally construct a black and white map of international politics. The geographical kaleidoscope of the map becomes the geopolitical monochrome of good versus evil, capitalism versus communism, the West versus the East, America versus the Soviet Union. All places and conflicts are to be interpreted within the binary terms of this Manichean map.

The deployment of such simplistic black and white reasoning to read the international political scene begins with the Truman Doctrine, as does another element of American Cold War geopolitical discourse. In reasoning that foreshadowed what would become known as the domino theory, Truman's Secretary of State, Dean Acheson, explained before Congress that like

> apples in a barrel infected by one rotten one, the corruption of Greece would infect Iran and all to the east. It would also carry infection to Africa through Asia Minor and Egypt, and to Europe through Italy and France, already threatened by the strongest domestic Communist parties in Western Europe (Acheson, 1969: 219).

That such a representation of complex geographically embedded states as "apples in a barrel" was possible is a mark of the hubris present in the reasoning of the new "masters of the globe," the American intellectuals of statecraft within the Truman administration. It is also indicative of the triumph of an anti-geographical form of reasoning in Cold War geopolitical discourse. The geographical specificity and complexity of particular conflicts, such as that in Greece or Turkey, were not important. What was all important was the "higher truth" of the struggle between freedom and totalitarianism across the world map. Thus when Truman declares in his speech that it is "necessary only to glance at a map," the map he has in mind is one where states (or "nations" as he calls them) are equivalent to dominoes about to "fall." Only physical proximity is seen as geography and nothing else.

George Kennan was privately critical of the crude and alarmist tone of the Truman Doctrine which successfully scared Congress, as Truman hoped it would, into providing aid to the embattled Greek and Turkish governments. Many professional foreign policy experts considered Truman's declaration dangerous because it contained no rational calculation of means and ends for US foreign policy. His statement that "it must be the policy of the United States to support free peoples who are resisting attempted subjugation by armed minorities or by outside pressures" placed no geographic limits on US foreign policy. Implicitly, the Truman Doctrine envisioned a world-wide anti-communist crusade: an unlimited totalitarian threat required an unlimited global commitment by the United States.

While George Kennan was cognizant of this danger (and became even more so as he got older), an essay written by him called "The Sources of Soviet Conduct" and published a few months after the Truman Doctrine in the main journal of the Council on Foreign Relations, *Foreign Affairs*, reinforced rather than questioned the crude geopolitical vision articulated in Truman's speech (Reading 7). Published initially under the pseudonym "Mr X" before Kennan's identity was disclosed, this essay is the intellectual foundation of the post-war American foreign policy of "containment" of the Soviet Union. Expanding ideas he had developed in the Long Telegram and elsewhere, Kennan argued that Soviet communism was the ideology of a maladjusted group of fanatics who had seized power in 1917 and were driven by a perpetual insecurity to destroy "all competing power" both inside and outside the country. Communist ideology is ultimately a "fig leaf" for Kennan, the primordial sources of Soviet conduct being internal to and determined by Russian history and geography: "From the Russian-Asiatic world out of which they had emerged they [Soviet communists] carried with them a skepticism as to the possibilities of permanent and peaceful coexistence of rival forces" (1947: 570). Soviet communist caution and flexibility are precepts

> fortified by the lessons of Russian history: of centuries of obscure battles between nomadic forces over the stretches of a vast unfortified plain. Here caution, circumspection, flexibility and deception are the valuable qualities; and their value finds natural appreciation in the Russian or the oriental mind (1947: 576).

Again, like in Truman's speech, we are dealing with essences and absolute truths. Soviet communists are insecure fanatics. Their ideology in tandem with the primordial patterns of Russian history and geography have produced a Soviet state that is inherently expansionist. The problem with such absolutist truths, however, is that they are overly deterministic and functionally anti-historical and anti-geographical. Essentialist assertions dominate historical contingencies and geographical particularities. As the historian Anders Stephanson (1989: 76) has argued, Soviet foreign policy in fact

> varied substantially over time in both magnitude and target (as Kennan should have known), depending precisely on which powers seemed to pose the greatest danger.... Who is actually out there doing, doing what to whom, were important questions to Moscow.

The position Kennan articulates, however, absolves Western leaders and intellectuals of statecraft from actually engaging with the practical specifics of Soviet foreign policy at particular times and places. It promotes retreat to absolute truths and geopolitical slogans about communists being fanatics and the Soviet Union being implacably expansionist. Kennan's argument, in other words, objectifies the Soviet

Union as a predetermined expansionist entity that needs containment "by the adroit and vigilant application of counter-force at a series of constantly shifting geographical and political points. . . ." Kennan's argument ironically precludes his own profession: diplomacy. Since the Soviets are supposedly fanatics, there is no real possibility of dialogue and negotiation with them. They are Other.

Kennan's call for "a policy of firm containment, designed to confront the Russians with unalterable counter-force at every point where they show signs of encroaching upon the interests of a peaceful and stable world" echoes the unlimited rhetoric found in the Truman Doctrine. Kennan's conclusion about the Soviet challenge being "a test of the over-all worth of the United States as a nation among nations" evokes long-standing American myths of manifest destiny and national exceptionalism and reverberates with the global anti-communist crusade envisioned by Truman (see Reading 11). The unlimited and universalist nature of this crusade, however, was profoundly unsettling to some. In a series of newspaper articles that subsequently became the book *The Cold War* that named the era, the political journalist Walter Lippmann described the "X" article's recommendations as a "strategic monstrosity" (Lippmann, 1947: 18). It makes no distinctions between places and commits the United States to confront the Russians with counterforce "at every point" across the globe, "instead of at those points which we have selected because, there at those points, our kind of sea and air power can best be exerted" (1947: 19). Furthermore, it gives a "blank check" from the American people to its military institutions and to those regimes the US government decides are allies in its global crusade against communism. Lippmann concludes by emphasizing diplomacy, noting that for "a diplomat to think that rival and unfriendly powers cannot be brought to a settlement is to forget what diplomacy is all about" (1947: 60).

Lippmann's worst fears, however, were largely realized as diplomacy became sidelined and containment militarism became the guiding principle of US foreign policy. The Soviet response to the hardening Western attitude and the Marshall Plan aid program for select states in Europe was to fall back on its own Manichean vision of the world. As articulated by the Soviet intellectual of statecraft Andrei Zhdanov in September 1947, the world was divided into "two camps," an "imperialist and anti-democratic camp" led by the United States with the British Empire as its leading ally versus an "anti-imperialist and democratic camp" led by the Soviet Union and the "new democracies" in Eastern Europe (Reading 8). These "new democracies," however, were in reality Soviet-inspired regimes that were prohibited by the Soviet Union from participating in the Marshall Plan. Though not all beholden to Moscow, their domestic political structure became increasingly Stalinist over the years as the Cold War between the US and the USSR deepened and polarized the political map of Europe.

Frozen on the map over the next four decades, the Cold War came to describe a geopolitical system with two constituent geopolitical orders, each of which was characterized by a particular organization of domestic, allied and "Third World" space. The very term "Third World," as we noted at the outset, is a product of the Cold War's division of global space into a First World of capitalist states, a Second World of communist states and a Third World of developing states where capitalism and communism, the US and the USSR, were in competition with each other. The geopolitical order established by the Americans after World War II was geographically more extensive than the Soviet order. First, domestic politics within the United States was organized by containment militarism which was legitimated by exaggerated visions of the "Soviet threat" (Wolfe, 1984). This facilitated the creation and expansion of a national security state and a confinement of the discourse of US political culture within limits established by the right. Those on the left critical of either US militarism or corporate capitalism were constantly red-baited by politicians on the right like Joseph McCarthy and Richard Nixon. Patriotism became defined as anti-communism, and politicians from Truman to Kennedy and Nixon to Reagan rode anti-communist crusades all the way to the White House.

American Cold War geopolitical discourse also had an important economic dimension. The Cold War, according to Wolfe (1982), was central to the creation of a consensual "politics of growth" in post-war America. Through exaggeration of the Soviet threat, American intellectuals of statecraft were able to transform the US state from a reluctant isolationist power into a crusading interventionist power dedicated to promoting an open world economy and safeguarding the free enterprise system. "Containment" became an unquestioned imperative within American foreign policy. Cold War visions of "containment" were also extended into American domestic life and popular culture. Figures like Ronald Reagan, president of the Screen Actors Guild from 1947 to 1952, for example, sought to enforce the cultural authority of a conservative and patriarchal white establishment by "blacklisting" those whose ideas challenged this hegemonic cultural order (Campbell, 1990; May, 1989).

Second, the establishment and modernization of a global system of extended deterrence, by means of NATO in Western Europe and the Mutual Security Treaty with Japan, helped incorporate and subordinate the US's major capitalist allies into an American led military system. The economic reconstruction and recovery of Western Europe and Japan were facilitated by generous aid from the US state and its promotion of an open capitalist world economy. A convergence of interests among the ruling classes in all three regions facilitated the establishment of an American "empire by invitation" (Lundestad, 1990).

Third, the US state sought to establish for itself the freedom, in the space demarcated as the "Third World," to intervene and attack peoples and states that US intellectuals of statecraft considered a threat to their version of "American" values, institutions and economic interests (Kolko, 1988; Chomsky, 1991). The general proclivity of the US state for unilateralist interventionism to oppose radical social revolution was already manifest in Central America and the Caribbean before the Cold War. This proclivity became a global one after World War II and led the US national security state, through the work of the Central Intelligence Agency and other groups, to intervene in the domestic politics of many states, in some instances like Iran in 1953, Guatemala in 1956 and Chile in 1973, aiding the overthrow of democratically elected governments. The US also got massively involved militarily in a number of regions and fought bloody wars in Korea and Vietnam among other places against what it perceived as a worldwide communist threat.

The geopolitical order established by the Soviet communist elite in the wake of World War II was largely confined to Eastern Europe and the Soviet Union. Its order was defined by, first, the domination of domestic politics and political culture by the Communist Party. Patriotism was so entwined with communism, which was akin to an official state religion, that anyone who questioned it was automatically branded an agent of the "imperialist West." Dissident intellectuals were persecuted and sent to gulags, internal exile, and mental homes. Just as the United States built a huge military-industrial complex to support its national security state, so also did the Soviet Union, its state structure and institutions becoming even more militarized than those of the United States. Second, the Soviet geopolitical order was characterized by the maintenance of a system of extended deterrence in Eastern Europe by means of pro-Moscow ruling communist elites and the military structures of the Warsaw Pact Organization. Because it did not have nearly the resources and wealth of the capitalist West, the Soviet state intervened erratically in the Third World, selectively sponsoring a few radical states like Egypt (for a period), North Korea, Vietnam and Cuba.

Europe was the principal theater where both competing geopolitical orders faced each other and the site of its greatest militarization. Ironically both superpowers came to share a mutual interest in the Cold War as a system because it guaranteed their mutual positions on the European continent. Cox (1990: 31) notes:

> Historically . . . the Cold War served the interests of both the USSR and the United States. For this reason neither sought to alter the nature of the relationship once it had been established. Their goal, therefore, was not so much victory over the other as the maintenance of balance. In this sense

> the Cold War was more of a carefully controlled game with commonly agreed rules than a contest where there could be clear winners and losers.

Yet there were real winners and losers *but within not between* the respective geopolitical orders. Evaluated in terms of war, death and destruction, the Cold War saw the Soviet state wage war in its geopolitical zone against popular uprisings in Poland, Hungary, Czechoslovakia and Afghanistan. The US, with the help of certain allies, sought to police radical movements in its zone and waged war against radical social change in the Third World. From Vietnam to Afghanistan, the Cold War was far from being an "imaginary war" or a "long peace" (Kaldor, 1990; Gaddis, 1987).

That the United States became involved in civil wars in Korea and Vietnam, locations thousands of miles from the United States and of questionable strategic value in themselves, was a consequence of the dominance of an unlimited understanding of containment and a limited understanding of geography in US geopolitical discourse. Truman's universal crusade against communism and Kennan's call for firm containment "at every point" made the task of delimiting US interests difficult to sustain. Secretary of State Dean Acheson tried to define a "defensible perimeter" of the United States which excluded the Asiatic mainland, including Korea, yet within a year the United States was at war in Korea against what was represented as the latest front in a worldwide communist challenge. The same happened with Vietnam where Presidents Eisenhower, Kennedy and Johnson declared on various occasions that the United States should not become militarily involved.

Yet, the universalist nature of America's anticommunist crusade and the simplemindedness of its domino theory reasoning confounded these expressions of intent not to get involved. The hysterical vision of Southeast Asia falling like a row of dominoes to communism was absurd geographically yet nevertheless dominant geopolitically. The "best and brightest" US intellectuals of statecraft pushed the United States to become militarily involved in the Vietnamese civil war. As a consequence, that war became even more bloody and brutal than it already was, dragging out until the Americans eventually withdrew their forces and the South Vietnamese forces collapsed in 1975. O'Sullivan (Reading 9) traces the origins and evolution of the domino theory in US geopolitical discourse, arguing that it represents a particularly impoverished form of reasoning that fails to capture the importance of geographic uniqueness and place bound identities in international politics.

Reasoning equivalent to the domino theory can also be found in Soviet geopolitical discourse, though here the dominoes or satellite states were geographically much closer to the Soviet Union. The attempt by the communist leaders of Czechoslovakia to institute a series of reforms designed to address their deteriorating economic situation became a matter of concern for Soviet and other Eastern European leaders in 1968. In order to stimulate the economy, Czechoslovak reformers led by Alexander Dubček instituted a series of measures that loosened the firm dictatorship of the Communist Party over the economy and state. The result was increasing cultural liberalization, what reformers celebrated as "the Prague Spring" but what nervous communist bureaucratic dictators in East Germany, Poland and the Soviet Union described as the "Czechoslovakian disease." Fearing that this "disease" of political reform and cultural liberalization would spread, the Red Army invaded Czechoslovakia on August 20 with the support of smaller units from Poland, East Germany, Hungary and Bulgaria.

The justification for the invasion became known as the "Brezhnev Doctrine," a geopolitical statement originally published as an article in the official Soviet Communist newspaper *Pravda* by Politbureau leader Leonid Brezhnev under the pseudonym "Kovalev" (Reading 10). In this article, Brezhnev articulates the limits within which the communist satellite states of Eastern Europe must operate, effectively spelling out the subordination of the geographically diverse Eastern European communist dictatorships to the Soviet geopolitical order. Any decision these states make "must damage neither socialism in their country nor the fundamental interests of the other socialist countries nor the worldwide communist movement."

State communist leaders who exercise their nominal state sovereignty and national independence in a way that deviates from these principles are guilty of "one-sidedness" and "revisionism," code words for unacceptable independent thinking. Throughout the article, the Soviet invasion is justified by resort to Manichean geopolitical discourse. Its "Us" against "Them" and any group that tries to promote greater democracy and loosen the dictatorship of the Communist Party in Eastern Europe is ultimately aiding the enemy, which in Soviet geopolitical discourse is "world imperialism" and, echoing the line that West Germany is still innately fascist and expansionist, "West German revanchists." In the binary logic of Soviet geopolitical discourse, any questioning of Cold War categories or promotion of "neutrality" is objectively "antisocialist" and "counterrevolutionary." Remarkably, Brezhnev claims that the Red Army and its support units in Czechoslovakia are "not interfering in the country's internal affairs" but helping the Czechoslovak people exercise their "inalienable right to decide their destiny themselves ..." However, there is an all important qualification: "... after profound and careful consideration, without intimidation by counterrevolutionaries, without revisionist and nationalist demagoguery." In other words, the Red Army is merely helping the Czechoslovak people exercise their self-determination in a way that the Soviet Union's leadership judges to be ideologically and geopolitically correct. Such is the thin apologism for military interventionism and geopolitical domination.

O'Sullivan's argument against the "loss of geography" evident in such geopolitical reasoning is developed further by Ó Tuathail and Agnew in their study of geopolitics and discourse (Reading 11). A foundational essay in the establishment of critical geopolitics, this reading outlines four theses on geopolitical reasoning and international politics. It then turns to analyze the practical geopolitical reasoning found in American foreign policy historically, concluding with a deconstruction of ways in which Kennan represents the USSR in his Long Telegram and "X" article.

Because of its Vietnam experience, the domestic political consensus around the policy of containment militarism that the US state had pursued since the late 1940s was subject to increasing challenge and critique. The old "absolute truths" of the Soviet Union as an implacably expansionist power with which one could not negotiate and of a worldwide communist conspiracy directed from Moscow had propelled the United States into Vietnam and sent thousands of its soldiers to their death. In a bid to adjust to the changed conditions of world power in the 1970s and the breakdown of the foreign policy consensus, the Nixon administration, with Henry Kissinger as the president's leading intellectual and also practitioner of statecraft, pursued a policy of *détente* or peaceful co-existence with the Soviet Union, and accommodation with communist China. Rather than continue the ultra-militarist policy of driving for military superiority over the Soviets, the Nixon–Kissinger administration recognized that both states had the ability effectively to destroy the other. Acknowledging the reality of "mutually assured destruction" (MAD), the Nixon administration promoted the doctrine of nuclear deterrence and sought to negotiate limited arms control agreements with the Soviets. Nevertheless, Nixon and Kissinger continued the US's long-standing crusade against perceived leftist ("pro-Soviet") governments in the Third World, involving the United States in assassinations, coup d'etats and illegal wars in places like Chile, Angola and Cambodia, many of which were devestated by the resultant instability.

Not everyone within the US national security community agreed with the Nixon administration's policy of *détente* towards the Soviet Union. A band of "true believers" within the Nixon administration and, after Nixon had resigned in disgrace, within the Ford administration continued to recycle the old "absolute truths" about the Soviet Union, charging without genuine evidence but with plenty of traditional Cold War anti-communist hysteria and paranoia, that the CIA had underestimated Soviet strength and that the Soviets were engaging in a massive military buildup in a new bid for world domination. Horrified when the liberal Democratic candidate for president, Jimmy Carter, was elected to the White House in 1976, this band of "true believers" included veteran Cold Warriors like Paul Nitze, who was upset at not having an important job in the incoming Carter administration. Calling themselves "The Committee on the Present Danger" (CPD), after a similar hard right militarist group in the 1950s, the group went public at a press conference on November 11, days after Carter's election, with a manifesto called "Common Sense and the Common Danger" (Reading 12). Nitze took up office as the CPD's "Chairman, Policy Studies" in a spacious office suite at the Systems Planning Corporation, a defense contracting firm in Arlington, Virginia (Talbot, 1988: 151).

The CPD's manifesto is significant for its attempt to re-assert the old absolute truths of containment militarism at precisely the moment when it seemed that the US political system was about to move beyond them. Recycling the traditional discourses of danger for the early Cold War – "Our country is in a period of danger, and the danger is increasing" – the CPD asserted that the "threats we face" are "more subtle and indirect than was once the case" but, somewhat contradictorily, they are nevertheless massive, worldwide and unparalleled. Re-asserting the black and white world that Cold War geopoliticians like best, the manifesto declares that the "principal threat to our nation, to world peace, and to the cause of human freedom is the Soviet drive for dominance based upon an unparalleled military buildup." As one might expect from intellectuals of statecraft with ties to a domestic military-industrial complex eager to continue expanding whatever the world geopolitical situation, the CPD manifesto calls for a massive military buildup on the part of the United States to check the global communist threat and build a "strong foundation" (code for military superiority) from which to supposedly negotiate "hardheaded and verifiable agreements" with the Soviets.

Throughout the Carter years, the intellectuals of statecraft associated with the CPD worked hard to criticize and undermine the foreign policy of the Carter administration. The perceived failures of this administration's policy in Iran and elsewhere, together with the Soviet invasion of Afghanistan in 1979, were represented by the CPD as a consequence of its straying from the "absolute truths" of Cold War militarism. While these arguments were spurious, they nevertheless had the effect the CPD wanted, for late in his administration Carter began massively increasing US defense spending to the obvious glee of America's military-industrial complex.

However, it was within the Reagan administration that the CPD's recycling of old Cold War truths had their greatest impact. Ronald Reagan was a member of the founding board of directors of the committee. Upon Reagan's election, many CPD members, including Paul Nitze, took up important policy positions within the new conservative Republican administration. The Reagan administration continued Carter's late military buildup. In dollar terms, the defense budget almost doubled between 1979 and 1983, from 5.1 per cent of GNP to 6.6 per cent (Sherry, 1995: 401). Reagan began his tenure in office by declaring that "the Soviet Union underlies all the unrest that is going on. If they weren't engaged in this game of dominoes, there wouldn't be any hot spots in the world" (quoted in Sherry, 1995: 399). His administration initiated a policy of aggressive hegemonic assertionism across the globe, militarizing many conflicts and disputes at the expense of diplomacy. In Central America, for example, the Reagan administration effectively went to war against the radical nationalist Sandanista revolution which in 1979 had overthrown the US supported dictator of that country. Using the CIA and other covert and illegal means, the Reagan administration established the Nicaraguan contras who began murderous raids against Nicaragua in the early 1980s. The Reagan administration also provided massive arms supplies to the military government of El Salvador so it could murder those of its citizens who were organizing and fighting for social justice and democracy in their country (Ó Tuathail, 1986). The Reagan Doctrine of actively supporting counter-revolutionary guerillas fighting so-called "pro-Soviet" governments around the world was zealously pursued by CIA Director William Casey and "true believer" militarist males like Lt Colonel Oliver North. Operating as a shadow paramilitary government unto themselves, they provided (sometimes legally, sometimes illegally) training, weapons and money to militias from Afghanistan to Angola. Within American culture in general, fantasies of militaristic masculinity, such as that celebrated in films like the *Rambo* series, were popular (Jeffords, 1989, 1994).

In Western Europe, the Reagan administration began a new round in the militarization of the continent by pursuing the deployment of so-called "limited" nuclear weapons systems like the cruise and Pershing II missiles onto European soil. This effort by NATO to introduce medium range nuclear weapons targeting East Germany and other Central European states provoked resistance across European civil society as peace movements emerged in a number of countries in mobilization against the deployment. Working together across national frontiers and the East–West divide, many thousands of leading dissident intellectuals throughout Europe signed an Appeal for European Nuclear Disarmament (END) first launched on April 28, 1980 by a group of sponsors and the Russell Peace Foundation (Reading 13). Like the CPD manifesto, this document is also preoccupied with danger but not the danger posed by the Soviet Union as irreducible enemy. Rather, the END manifesto is concerned with the danger posed by escalating Cold War militarism and the advocates and apologists of the latest nuclear round of such militarism, like the geopoliticians of the CPD. Seeking to articulate a 'third way' beyond Cold War discourse and the mutually reinforcing militarism of the East–West confrontation, it gave voice to a European-wide "we" that is neutral and de-aligned from the Cold War blocs dominating and dividing Europe. "We must commence to act as if a united, neutral and pacific Europe already exists. We must learn to be loyal, not to 'East' and 'West,' but to each other, and we must disregard the prohibitions and limitations imposed by any national state." The END manifesto became the charter of the non-aligned European peace movement that tried, in organizing annual conventions in different European cities, to foster "détente from below" among citizen groups and diverse social movements including Christians, feminists, greens, trade unionists and democratic leftist groups. In the United States, a less radical movement to freeze nuclear weapons garnered enough popular support to provoke some politicians to question the wisdom of the Reagan administration's nuclear buildup (Sherry, 1995).

While the de-alignment associated with the END appeal had little immediate effect on the military policies of NATO and the Reagan administration, for the 'limited strike' cruise and Pershing II missiles were deployed in the early 1980s, they did nevertheless percolate via dissident Eastern European intellectuals through to a new generation of Soviet bureaucrats and communist policy officials eager to save the communist system from stagnation, corruption and imperial overstretch (most evident in the USSR's disastrous military campaign in Afghanistan). The new breed of communist politician who came to champion these new ideas was Mikhail Gorbachev. Gorbachev's foreign policy was radical for it deliberately set out to deprive the Reagan administration of its convenient "enemy image" of the USSR as an "evil empire." Launching a policy of *glasnost* or "openness" in Soviet society in 1986, Gorbachev envisioned a radical re-structuring and renewal (*perestroika*) of the USSR based on modernized and humane communist principles. Declaring that "no country enjoys a monopoly of the truth," he signalled the end of the Brezhnev Doctrine as the geopolitical principles governing the Soviet Union's relationship with the Eastern European communist regimes (Walker, 1993: 290). One Soviet commentator humorously dubbed the new geopolitical philosophy the "Sinatra Doctrine" (evoking Frank Sinatra's famous song "My Way"), the principle being that each Eastern European state can and should find its own way to reform and change without Soviet interventionism. While Gorbachev's self-interested attempt to save the communist system and the Soviet Empire from the top ultimately failed, his "new political thinking" in Soviet foreign policy helped bring about the end of the Cold War (Reading 14).

From the perspective of hardline Cold Warriors, it was the Reagan administration's military buildup that ultimately ended or, as they like to declare, "won" the Cold War. Yet such a view is difficult to sustain for without Gorbachev's push to end the Cold War peacefully, it might have been quite different, indeed horrifyingly different given the reflex militarism of the Reagan administration. Certainly many intellectuals of statecraft within the Reagan administration and subsequent Bush administration were deeply suspicious of Gorbachev and did all they could to demonize what they described as the "charm offensive" of his anti-militarist thinking about security in Europe and between the superpowers more generally. Yet, Gorbachev's new thinking was not easily demonized, especially when he backed up his words with concrete anti-militarist policies. Gorbachev's concerted push for arms reductions (not just arms control) and his refusal to intervene to save the communist dictatorships in Eastern Europe in the historic autumn of 1989 resulted in the fall of the Berlin Wall and the beginning of the end of the Cold War in Europe at last. The profound geopolitical consequences of his radical new policies eventually provoked a counter-reaction by hardliners within the Soviet military-industrial complex in August 1991, an attempted *coup* whose failure spiraled into the consequent dissolution of the USSR and the fitful emergence of the "new world order" of the 1990s.

REFERENCES AND FURTHER READING

On Truman, Kennan and origins of Cold War

Acheson, D. (1969) *Present at The Creation*, New York: Norton.

Kennan, G. (1947) "The Sources of Soviet Conduct," *Foreign Affairs*, 25: 568–582.

——(1951) *American Diplomacy*, Chicago: University of Chicago Press.

Lippmann, W. (1947) *The Cold War*, New York: Harper and Row.

Stephanson, A. (1989) *Kennan and the Art of Foreign Policy*, Cambridge: Harvard University Press.

Taylor, P. (1990) *Britain and the Cold War*, London: Pinter.

Yergin, D. (1978) *Shattered Peace: The Origins of the Cold War and the National Security State*, Boston: Houghton Mifflin.

On the Cold War as a geopolitical system

Campbell, D. (1990) *Writing Security*, Minneapolis: University of Minnesota.

Chomsky, N. (1991) *Deterring Democracy*, London: Verso.

Cox, M. (1990) "From the Truman Doctrine to the Second Superpower Détente: The Rise and Fall of the Cold War," *Journal of Peace Research*, 27: 25–41.

Gaddis, J. (1987) *The Long Peace*, New York: Oxford University Press.

Halberstam, D. (1972) *The Best and the Brightest*, London: Penguin.

Halliday, F. (1983) *The Making of the Second Cold War*, London: Verso.

Kaldor, M. (1990) *The Imaginary War*, Oxford: Basil Blackwell.

Kolko, G. (1988) *Confronting the Third World*, New York: Pantheon.

Lundestad, G. (1990) *The American 'Empire'*, London: Oxford University Press.

May, L. (ed.) (1989) *Recasting America: Culture and Politics in the Age of the Cold War*, Chicago: University of Chicago Press.

Ó Tuathail, G. (1991) "The Bush Administration and the 'End' of the Cold War: A Critical Geopolitics of U.S. Foreign Policy in 1989," *Geoforum*, 23: 437–452.

Simons, T. (1991) *Eastern Europe in the Postwar World*, New York: St Martin's Press.

Walker, M. (1993) *The Cold War: A History*, New York: Holt.

Wolfe, A. (1982) *America's Impasse: The Rise and the Fall of the Politics of Growth*, Boston: South End Press.

——(1984) *The Rise and Fall of the 'Soviet Threat'*, Boston: South End Press.

On Nixon, Kissinger, Reagan and the Committee on the Present Danger

Dalby, S. (1990) *Creating the Second Cold War*, New York: Guilford.

Jeffords, S. (1989) *The Remasculinization of America*, Bloomington: Indiana University Press.

——(1994) *Hard Bodies: Hollywood Masculinity in the Reagan Era*, New Brunswick, New Jersey: Rutgers University Press.

Litwak, R. (1984) *Détente and the Nixon Doctrine*, London: Cambridge University Press.

Ó Tuathail, G. (1986) "The Language and Nature of the 'New' Geopolitics: The Case of US–El Salvador Relations," *Political Geography Quarterly*, 5: 73–85.

Saunders, J. (1983) *Peddlers of the Crisis: The Committee on the Present Danger and the Politics of Containment*, Boston: South End Press.

Scheer, R. (1983) *With Enough Shovels: Reagan, Bush and Nuclear War*, New York: Random House.

Sherry, M. (1995) *In the Shadow of War*, New Haven: Yale University Press.

Talbot, S. (1988) *The Master of the Game: Paul Nitze and the Nuclear Age*, New York: Vintage.

On the European peace movement

Kaldor, M. (ed.) (1991) *Europe From Below*, London: Verso.

Kaldor, M. and Falk, R. (eds) (1987) *Dealignment: A New Foreign Policy Perspective*, New York: Basil Blackwell.

Kaldor, M., Holden, G. and Falk, R. (eds) (1989) *The New Detente: Re-Thinking East–West Relations*, London: Verso.

Thompson, E.P. (1985) *The Heavy Dancers*, New York: Pantheon.

Thompson, E.P. and Smith, D. (eds) (1980) *Protest and Survive*, New York: Penguin.

Thompson, E.P. *et al.* (1982) *Exterminism and Cold War*, London: Verso.

On Gorbachev's foreign policy

Gorbachev, M. (1988) *Perestroika: New Thinking for Our Country and the World*, New York: Harper and Row.

McGwire, M. (1991) *Perestroika and Soviet National Security*, Washington: Brookings.

6

PRESIDENT HARRY TRUMAN

"The Truman Doctrine"

from *Public Papers of the Presidents of the United States* (1947)

The gravity of the situation which confronts the world today necessitates my appearance before a joint session of the Congress. The foreign policy and the national security of this country are involved. One aspect of the present situation, which I wish to present to you at this time for your consideration and decision, concerns Greece and Turkey.

The United States has received from the Greek Government an urgent appeal for financial and economic assistance. Preliminary reports from the American Economic Mission now in Greece and reports from the American Ambassador in Greece corroborate the statement of the Greek Government that assistance is imperative if Greece is to survive as a free nation.

I do not believe that the American people and the Congress wish to turn a deaf ear to the appeal of the Greek Government....

[...]

The very existence of the Greek state is today threatened by the terrorist activities of several thousand armed men, led by Communists, who defy the Government's authority at a number of points, particularly along the northern boundaries. A commission appointed by the United Nations Security Council is at present investigating disturbed conditions in Northern Greece and alleged border violations along the frontier between Greece on the one hand and Albania, Bulgaria, and Yugoslavia on the other.

Meanwhile, the Greek Government is unable to cope with the situation. The Greek Army is small and poorly equipped. It needs supplies and equipment if it is to restore the authority to the Government throughout Greek territory. Greece must have assistance if it is to become a self-supporting and self-respecting democracy. The United States must supply this assistance. We have already extended to Greece certain types of relief and economic aid but these are inadequate. There is no other country to which democratic Greece can turn. No other nation is willing and able to provide the necessary support for a democratic Greek Government.

The British Government, which has been helping Greece, can give no further financial or economic aid after March 31. Great Britain finds itself under the necessity of reducing or liquidating its commitments in several parts of the world, including Greece. We have considered how the United Nations might assist in this crisis. But the situation is an urgent one requiring immediate action, and the United Nations and its related organizations are not in a position to extend help of the kind that is required....

Greece's neighbor, Turkey, also deserves our attention. The future of Turkey as an independent and economically sound state is clearly no less important to the freedom-loving peoples of the world than the future of Greece. The circumstances in which Turkey finds itself today are considerably different from those of Greece. Turkey has been spared the disasters that have beset Greece. And during the war, the United States and Great Britain furnished Turkey with material aid.

Nevertheless, Turkey now needs our support. Since the war Turkey has sought financial assistance from Great Britain and the United States for the purpose of effecting that modernization necessary for the main tenance of its national integrity. That integrity is essential to the preservation of order in the Middle East.

The British Government has informed us that, owing to its own difficulties, it can no longer extend financial or economic aid to Turkey. As in the case of Greece, if Turkey is to have the assistance it needs, the United States must supply it. We are the only country able to provide that help. I am fully aware of the broad implications involved if the United States extends assistance to Greece and Turkey, and I shall discuss these implications with you at this time.

One of the primary objectives of the foreign policy of the United States is the creation of conditions in which we and other nations will be able to work out a way of life free from coercion. This was a fundamental issue in the war with Germany and Japan. Our victory was won over countries which sought to impose their will, and their way of life, upon other nations. . . .

The peoples of a number of countries of the world have recently had totalitarian regimes forced upon them against their will. The Government of the United States has made frequent protests against coercion and intimidation, in violation of the Yalta Agreement, in Poland, Rumania, and Bulgaria. I must also state that in a number of other countries there have been similar developments.

At the present moment in world history nearly every nation must choose between alternative ways of life. The choice is too often not a free one. One way of life is based upon the will of the majority, and is distinguished by free institutions, representative government, free elections, guarantees of individual liberty, freedom of speech and religion, and freedom from political oppression. The second way of life is based upon the will of a minority forcibly imposed upon the majority. It relies upon terror and oppression, a controlled press and radio, fixed elections, and the suppression of personal freedoms.

I believe that it must be the policy of the United States to support free peoples who are resisting attempted subjugation by armed minorities or by outside pressures. I believe that we must assist free peoples to work out their own destinies in their own way. I believe that our help should be primarily through economic and financial aid, which is essential to economic stability and orderly political processes.

The world is not static and the status quo is not sacred. But we cannot allow changes in the status quo in violation of the Charter of the United Nations by such methods as coercion, or by such subterfuges as political infiltration. In helping free and independent nations to maintain their freedom, the United States will be giving effect to the principles of the Charter of the United Nations. It is necessary only to glance at a map to realize that the survival and integrity of the Greek nation are of grave importance in a much wider situation. If Greece should fall under the control of an armed minority, the effect upon its neighbor, Turkey, would be immediate and serious. Confusion and disorder might well spread throughout the entire Middle East.

Moreover, the disappearance of Greece as an independent state would have a profound effect upon those countries in Europe whose peoples are struggling against great difficulties to maintain their freedoms and their independence while they repair the damages of war. It would be an unspeakable tragedy if these countries, which have struggled so long against overwhelming odds, should lose that victory for which they sacrificed so much. Collapse of free institutions and loss of independence would be disastrous not only for them but for the world. Discouragement and possibly failure would quickly be the lot of neighboring peoples striving to maintain their freedom and independence.

Should we fail to aid Greece and Turkey in this fateful hour, the effect will be far reaching to the West as well as to the East. We must take immediate and resolute action.

I therefore ask the Congress to provide authority for assistance to Greece and Turkey in the amount of $400,000,000 for the period ending June 30, 1948. . . .

In addition to funds, I ask the Congress to authorize the detail of American civilian and military personnel to Greece and Turkey, at the request of those countries, to assist in the tasks of reconstruction, and for the purpose of supervising the use of such financial and material assistance as may be furnished. I recommend that authority also be provided for the instruction and training of selected Greek and Turkish personnel.

Finally, I ask that the Congress provide authority which will permit the speediest and most effective use, in terms of needed commodities, supplies, and equipment, of such funds as may be authorized....

This is a serious course upon which we embark. I would not recommend it except that the alternative is much more serious. The United States contributed $341,000,000,000 toward winning World War II. This is an investment in world freedom and world peace. The assistance that I am recommending for Greece and Turkey amounts to little more than one-tenth of 1 per cent of this investment. It is only common sense that we should safeguard this investment and make sure that it was not in vain.

The seeds of totalitarian regimes are nurtured by misery and want. They spread and grow in the evil soil of poverty and strife. They reach their full growth when the hope of a people for a better life has died. We must keep that hope alive. The free peoples of the world look to us for support in maintaining their freedoms. If we falter in our leadership, we may endanger the peace of the world – and we shall surely endanger the welfare of our own Nation. Great responsibilities have been placed upon us by the swift movement of events. I am confident that the Congress will face these responsibilities squarely.

7 GEORGE F. KENNAN

"The Sources of Soviet Conduct"

from *Foreign Affairs* (1947)

The political personality of Soviet power as we know it today is the product of ideology and circumstances: ideology inherited by the present Soviet leaders from the movement in which they had their political origin, and circumstances of the power which they now have exercised for nearly three decades in Russia. There can be few tasks of psychological analysis more difficult than to try to trace the interaction of these two forces and the relative role of each in the determination of official Soviet conduct. Yet the attempt must be made if that conduct is to be understood and effectively countered . . .

[. . .]

The circumstances of the immediate post-Revolution period – the existence in Russia of civil war and foreign intervention, together with the obvious fact that the Communists represented only a tiny minority of the Russian people – made the establishment of dictatorial power a necessity. The experiment with "war Communism" and the abrupt attempt to eliminate private production and trade had unfortunate economic consequences and caused further bitterness against the new revolutionary regime. While the temporary relaxation of the effort to communize Russia, represented by the New Economic Policy, alleviated some of this economic distress and thereby served its purpose, it also made it evident that the "capitalistic sector of society" was still prepared to profit at once from any relaxation of governmental pressure, and would, if permitted to continue to exist, always constitute a powerful opposing element to the Soviet regime and a serious rival for influence in the country. Somewhat the same situation prevailed with respect to the individual peasant who, in his own small way, was also a private producer.

Lenin, had he lived, might have proved a great enough man to reconcile these conflicting forces to the ultimate benefit of Russian society, though this is questionable. But be that as it may, Stalin, and those whom he led in the struggle for succession to Lenin's position of leadership, were not the men to tolerate rival political forces in the sphere of power which they coveted. Their sense of insecurity was too great. Their particular brand of fanaticism, unmodified by any of the Anglo-Saxon traditions of compromise, was too fierce and too jealous to envisage any permanent sharing of power. From the Russian-Asiatic world out of which they had emerged they carried with them a skepticism as to the possibilities of permanent and peaceful coexistence of rival forces. Easily persuaded of their own doctrinaire "rightness," they insisted on the submission or destruction of all competing power. Outside of the Communist Party, Russian society was to have no rigidity. There were to be no forms of collective human activity or association which would not be dominated by the Party. No other force in Russian society was to be permitted to achieve vitality or integrity. Only the Party was to have structure. All else was to be an amorphous mass . . .

II

. . . The Soviet concept of power, which permits no focal points of organization outside the Party itself, requires that the Party leadership remain in theory the sole repository of truth. For if truth were to be found elsewhere, there would be justification for its expression in organized activity. But it is precisely that which the Kremlin cannot and will not permit.

The leadership of the Communist Party is therefore always right, and has been always right ever since in 1929 Stalin formalized his personal power by announcing that decisions of the Politburo were being taken unanimously.

On the principle of infallibility there rests the iron discipline of the Communist Party. In fact, the two concepts are mutually self-supporting. Perfect discipline requires recognition of infallibility. Infallibility requires the observance of discipline. And the two together go far to determine the behaviorism of the entire Soviet apparatus of power. But their effect cannot be understood unless a third factor be taken into account: namely, the fact that the leadership is at liberty to put forward for tactical purposes any particular thesis which it finds useful to the cause at any particular moment and to require the faithful and unquestioning acceptance of that thesis by the members of the movement as a whole. This means that truth is not a constant but is actually created, for all intents and purposes, by the Soviet leaders themselves. It may vary from week to week, from month to month. It is nothing absolute and immutable – nothing which flows from objective reality. It is only the most recent manifestation of the wisdom of those in whom the ultimate wisdom is supposed to reside, because they represent the logic of history. The accumulative effect of these factors is to give to the whole subordinate apparatus of Soviet power an unshakeable stubbornness and steadfastness in its orientation. This orientation can be changed at will by the Kremlin but by no other power. Once a given party line has been laid down on a given issue of current policy, the whole Soviet governmental machine, including the mechanism of diplomacy, moves inexorably along the prescribed path, like a persistent toy automobile wound up and headed in a given direction, stopping only when it meets with some unanswerable force. The individuals who are the components of this machine are unamenable to argument or reason which comes to them from outside sources. Their whole training has taught them to mistrust and discount the glib persuasiveness of the outside world. Like the white dog before the phonograph, they hear only the "master's voice." And if they are to be called off from the purposes last dictated to them, it is the master who must call them off. Thus the foreign representative cannot hope that his words will make any impression on them. The most that he can hope is that they will be transmitted to those at the top, who are capable of changing the party line. But even those are not likely to be swayed by any normal logic in the words of the bourgeois representative. Since there can be no appeal to common purposes, there can be no appeal to common mental approaches. For this reason, facts speak louder than words to the ears of the Kremlin; and words carry the greatest weight when they have the ring of reflecting, or being backed up by, facts of unchallengeable validity.

But we have seen that the Kremlin is under no ideological compulsion to accomplish its purposes in a hurry. Like the Church, it is dealing in ideological concepts which are of long-term validity, and it can afford to be patient. It has no right to risk the existing achievements of the revolution for the sake of vain baubles of the future. The very teachings of Lenin himself require great caution and flexibility in the pursuit of Communist purposes. Again, these precepts are fortified by the lessons of Russian history: of centuries of obscure battles between nomadic forces over the stretches of a vast unfortified plain. Here caution, circumspection, flexibility and deception are the valuable qualities; and their value finds natural appreciation in the Russian or the oriental mind. Thus the Kremlin has no compunction about retreating in the face of superior force. And being under the compulsion of no timetable, it does not get panicky under the necessity for such retreat. Its political action is a fluid stream which moves constantly, wherever it is permitted to move, toward a given goal. Its main concern is to make sure that it has filled every nook and cranny available to it in the basin of world power. But if it finds unassailable barriers in its path, it accepts these philosophically and accommodates itself to them. The main thing is that there should always be pressure, increasing constant pressure, toward the desired goal. There is no trace of any feeling in Soviet psychology that that goal must be reached at any given time.

These considerations make Soviet diplomacy at once easier and more difficult to deal with than the diplomacy of individual aggressive leaders like Napoleon and Hitler. On the one hand it is more sensitive to contrary force, more ready to yield on individual sectors of the diplomatic front when that force is felt to be too strong, and thus more rational in the logic and rhetoric of power. On the other hand it cannot be easily defeated or discouraged by a single victory on the part of its opponents. And the patient persistence by which it is animated means that it can be effectively countered not by sporadic acts which represent the momentary whims of democratic opinion but only by intelligent long-range policies on the part of Russia's adversaries – policies no less steady in their purpose, and no less variegated and resourceful in their application, than those of the Soviet Union itself.

In these circumstances it is clear that the main element of any United States policy toward the Soviet Union must be that of a long-term, patient but firm and vigilant containment of Russian expansive tendencies. It is important to note, however, that such a policy has nothing to do with outward histrionics: with threats or blustering or superfluous gestures of outward "toughness." While the Kremlin is basically flexible in its reaction to political realities, it is by no means unamenable to considerations of prestige. Like almost any other government, it can be placed by tactless and threatening gestures in a position where it cannot afford to yield even though this might be dictated by its sense of realism. The Russian leaders are keen judges of human psychology, and as such they are highly conscious that loss of temper and of self-control is never a source of strength in political affairs. They are quick to exploit such evidences of weakness. For these reasons, it is a sine qua non of successful dealing with Russia that the foreign government in question should remain at all times cool and collected and that its demands on Russian policy should be put forward in such a manner as to leave the way open for a compliance not too detrimental to Russian prestige.

III

In the light of the above, it will be clearly seen that the Soviet pressure against the free institutions of the Western world is something that can be contained by the adroit and vigilant application of counter-force at a series of constantly shifting geographical and political points, corresponding to the shifts and maneuvers of Soviet policy, but which cannot be charmed or talked out of existence. The Russians look forward to a duel of infinite duration, and they see that already they have scored great successes. It must be borne in mind that there was a time when the Communist Party represented far more of a minority in the sphere of Russian national life than Soviet power today represents in the world community.

But if ideology convinces the rulers of Russia that truth is on their side and that they can therefore afford to wait, those of us on whom that ideology has no claim are free to examine objectively the validity of that premise. The Soviet thesis not only implies complete lack of control by the West over its own economic destiny, it likewise assumes Russian unity, discipline and patience over an infinite period. Let us bring this apocalyptic vision down to earth, and suppose that the Western world finds the strength and resourcefulness to contain Soviet power over a period of ten to fifteen years. What does that spell for Russia itself?

The Soviet leaders, taking advantage of the contributions of modern technique to the arts of despotism, have solved the question of obedience within the confines of their power. Few challenge their authority; and even those who do are unable to make that challenge valid as against the organs of suppression of the state.

The Kremlin has also proved able to accomplish its purpose of building up in Russia, regardless of the interests of the inhabitants, an industrial foundation of heavy metallurgy, which is, to be sure, not yet complete but which is nevertheless continuing to grow and is approaching those of the other major industrial countries. All of this, however, both the maintenance of internal political security and the building of heavy industry, has been carried out at a terrible cost in human life and in human hopes and energies . . .

Here is a nation striving to become in a short period one of the great industrial nations of the world while it still has no highway network worthy of the name and only a relatively primitive network of railways. Much has been done to increase efficiency of labor and to teach primitive peasants something about the operation of machines. But maintenance is still a crying deficiency of all Soviet economy. Construction is hasty and poor and in vast sectors of economic life it has not yet been possible to instill into labor anything like that general culture of production and technical self-respect which characterizes the skilled worker of the West.

It is difficult to see how these deficiencies can be corrected at an early date by a tired and dispirited population working largely under the shadow of fear and compulsion. And as long as they are not overcome, Russia will remain economically a vulnerable, and in a certain sense an impotent, nation, capable of exporting its enthusiasms and of radiating the strange charm of its primitive political vitality but unable to back up those articles of export by the real evidences of material power and prosperity.

IV

It is clear that the United States cannot expect in the foreseeable future to enjoy political intimacy with the Soviet regime. It must continue to regard the Soviet Union as a rival, not a partner, in the political arena. It must continue to expect that Soviet policies will reflect no abstract love of peace and stability, no real faith in the possibility of a permanent happy coexistence of the Socialist and capitalist worlds, but rather a cautious, persistent pressure toward the disruption and weakening of all rival influence and rival power.

Balanced against this are the facts that Russia, as opposed to the Western world in general, is still by far the weaker party, that Soviet policy is highly flexible, and that Soviet society may well contain deficiencies which will eventually weaken its own total potential. This would of itself warrant the United States entering with reasonable confidence upon a policy of firm containment, designed to confront the Russians with unalterable counter-force at every point where they show signs of encroaching upon the interests of a peaceful and stable world.

But in actuality the possibilities for American policy are by no means limited to holding the line and hoping for the best. It is entirely possible for the United States to influence by its actions the internal developments, both within Russia and throughout the international Communist movement, by which Russian policy is largely determined. This is not only a question of the modest measure of informational activity which this government can conduct in the Soviet Union and elsewhere, although that, too, is important. It is rather a question of the degree to which the United States can create among the peoples of the world generally the impression of a country which knows what it wants, which is coping successfully with the problems of its internal life and with the responsibilities of a World Power, and which has a spiritual vitality capable of holding its own among the major ideological currents of the time. To the extent that such an impression can be created and maintained, the aims of Russian Communism must appear sterile and quixotic, the hopes and enthusiasm of Moscow's supporters must wane, and added strain must be imposed on the Kremlin's foreign policies. For the palsied decrepitude of the capitalist world is the keystone of Communist philosophy. Even the failure of the United States to experience the early economic depression which the ravens of the Red Square have been predicting with such complacent confidence since hostilities ceased would have deep and important repercussions throughout the Communist world.

By the same token, exhibitions of indecision, disunity and internal disintegration within this country have an exhilarating effect on the whole Communist movement. At each evidence of these tendencies, a thrill of hope and excitement goes through the Communist world; a new jauntiness can be noted in the Moscow tread;

new groups of foreign supporters climb on to what they can only view as the bandwagon of international politics; and Russian pressure increases all along the line in international affairs.

It would be an exaggeration to say that American behavior unassisted and alone could exercise a power of life and death over the Communist movement and bring about the early fall of Soviet power in Russia. But the United States has it in its power to increase enormously the strains under which Soviet policy must operate, to force upon the Kremlin a far greater degree of moderation and circumspection than it has had to observe in recent years, and in this way to promote tendencies which must eventually find their outlet in either the break-up or the gradual mellowing of Soviet power. For no mystical, Messianic movement – and particularly not that of the Kremlin – can face frustration indefinitely without eventually adjusting itself in one way or another to the logic of that state of affairs.

Thus the decision will really fall in large measure in this country itself. The issue of Soviet–American relations is in essence a test of the over-all worth of the United States as a nation among nations. To avoid destruction the United States need only measure up to its own best traditions and prove itself worthy of preservation as a great nation.

Surely, there was never a fairer test of national quality than this. In the light of these circumstances, the thoughtful observer of Russian–American relations will find no cause for complaint in the Kremlin's challenge to American society. He will rather experience a certain gratitude to a Providence which, by providing the American people with this implacable challenge, has made their entire security as a nation dependent on their pulling themselves together and accepting the responsibilities of moral and political leadership that history plainly intended them to bear.

8 ANDREI ZHDANOV

"Soviet Policy and World Politics"

from *The International Situation* (1947)

The end of World War II brought with it big changes in the world situation. The military defeat of the bloc of fascist states, the character of the war of liberation from fascism, and the decisive role played by the Soviet Union in the vanquishing of the fascist aggressors sharply altered the alignment of forces between the two systems – the socialist and the capitalist – in favor of socialism.

What is the essential nature of these changes?

The principal outcome of World War II was the military defeat of Germany and Japan – the two most militaristic and aggressive of the capitalist countries. . . .

[Second], the war immensely enhanced the international significance and prestige of the USSR. . . .

[Third], the capitalist world has also undergone a substantial change. Of the six so-called great imperialist powers (Germany, Japan, Great Britain, the USA, France, and Italy), three have been eliminated by military defeat. France has also been weakened and has lost its significance as a great power. As a result, only two great imperialist world powers remain – the United States and Great Britain. But the position of one of them, Great Britain, has been undermined. The war revealed that militarily and politically British imperialism was not so strong as it had been. . . .

[Fourth], World War II aggravated the crisis of the colonial system, as expressed in the rise of a powerful movement for national liberation in the colonies and dependencies. This has placed the rear of the capitalist system in jeopardy. The peoples of the colonies no longer wish to live in the old way. The ruling classes of the metropolitan countries can no longer govern the colonies on the old lines. . . .

Of all the capitalist powers, only one – the United States – emerged from the war not only unweakened, but even considerably stronger economically and militarily. The war greatly enriched the American capitalists. . . . But the end of the war confronted the United States with a number of new problems. The capitalist monopolies were anxious to maintain their profits at the former high level, and accordingly pressed hard to prevent a reduction of the wartime volume of deliveries. But this meant that the USA must retain the foreign markets which had absorbed American products during the war, and moreover, acquire new markets, inasmuch as the war had substantially lowered the purchasing power of most of the countries [to do this] . . . the United States proclaimed a new frankly predatory and expansionist course. The purpose of this new, frankly expansionist course is to establish the world supremacy of American imperialism. . . .

The fundamental changes caused by the war on the international scene and in the position of individual countries have entirely changed the political landscape of the world. A new alignment of political forces has arisen. The more the war recedes into the past, the more distinct become two major trends in postwar international policy, corresponding to the division of the political forces operating on the international arena into two major camps; the imperialist and anti-democratic camp, on the one hand, and the anti-imperialist and democratic camp, on the other. The principal driving force of the imperialist camp is the USA. Allied with it are Great Britain and France. . . . The cardinal purpose of the imperialist camp is to strengthen imperialism, to hatch a new imperialist war, to combat socialism and democracy, and to support reactionary and anti-democratic pro-fascist regimes and movements everywhere.

The anti-fascist forces comprise the second camp. This camp is based on the USSR and the new democracies. It also includes countries that have broken with imperialism and have firmly set foot on the path of democratic development, such as Rumania, Hungary, and Finland....

Soviet foreign policy proceeds from the fact of the coexistence for a long period of the two systems – capitalism and socialism. From this it follows that cooperation between the USSR and countries with other systems is possible, provided that the principle of reciprocity is observed and that obligations once assumed are honored. Everyone knows that the USSR has always honored the obligations it has assumed. Britain and America are pursing the very opposite policy in the United Nations. They are doing everything they can to renounce their commitments and to secure a free hand for the prosecution of a new policy, a policy which envisages not cooperation among the nations, but the hounding of one against the other, violation of the rights and interests of democratic nations, and the isolation of the USSR....

The strategical plans of the United States envisage the creation in peacetime of numerous bases and vantage grounds situated at great distances from the American continent and designed to be used for aggressive purposes against the USSR and the countries of the new democracy....

Economic expansion is an important supplement to the realization of America's strategical plan. American imperialism is endeavoring like a usurer to take advantage of the postwar difficulties of the European countries, in particular of the shortage of raw materials, fuel, and food in the Allied countries that suffered most from the war, to dictate to them extortionate terms for any assistance rendered. With an eye to the impending economic crisis, the United States is in a hurry to find new monopoly spheres of capital investment and markets for its goods. American economic "assistance" pursues the broad aim of bringing Europe into bondage to American capital. The more drastic the economic situation of a country is, the harsher are the terms which the American monopolies endeavor to dictate to it....

Lastly, the aspiration to world supremacy and the anti-democratic policy of the United States involve an ideological struggle. The principal purpose of the ideological part of the American strategical plan is to deceive public opinion by slanderously accusing the Soviet Union and the new democracies of aggressive intentions, and thus representing the Anglo-Saxon bloc in a defensive role, and absolving it of responsibility for preparing a new war....

The unfavorable reception which the Truman doctrine was met with accounts for the necessity of the appearance of the Marshall Plan which is a more carefully veiled attempt to carry through the same expansionist policy. The vague and deliberately guarded formulations of the Marshall Plan amount in essence to a scheme to create a bloc of states bound by obligations to the United States, and to grant American credits to European countries as recompense for their renunciation of economic, and then of political, independence.

The dissolution of the Comintern, which conformed to the demands of the development of the labor movement in the new historical situation, played a positive role. The dissolution of the Comintern once and for all disposed of the slanderous allegation of the enemies of Communism and the labor movement that Moscow was interfering in the internal affairs of other states, and that the Communist Parties in the various countries were acting not in the interests of their nations, but on orders from outside....

In the course of the four years that have elapsed since the dissolution of the Comintern (1943), the Communist Parties have grown considerably in strength and influence in nearly all the countries of Europe and Asia.... But the present position of the Communist Parties has its shortcomings. Some comrades understood the dissolution of the Comintern to imply the elimination of all ties, of all contact, between the fraternal Communist Parties. But experience has shown that such mutual isolation of the Communist Parties is wrong, harmful and, in point of fact, unnatural. The Communist movement develops within national frameworks, but there are tasks and interests common to the parties of various countries. We get a rather curious state of affairs . . . the Communists even refrained from meeting one another, let alone consulting with one another on questions of mutual interest to them, from fear of the slanderous talk of their enemies regarding the "hand of Moscow." . . . There can be no doubt that if the situation were to continue it would be fraught with most serious consequences to the development of the work of the fraternal parties. The need for mutual consultation and voluntary coordination of action between individual parties has become particularly urgent at the present junction when continued isolation may lead to a slackening of mutual understanding, and at times, even to serious blunders. . . .

9

PATRICK O'SULLIVAN

"Antidomino"

from *Political Geography Quarterly* (1982)

ORIGINS AND EVOLUTION

The beginnings of domino theory can be traced to William Bullit (1947), a former US ambassador to Moscow, who voiced the fear of monolithic communism emanating from its Russian power source and engulfing the world via China and Southeast Asia. Wiens (1954), a geographer, produced a more scholarly version of this justification for American intervention in Southeast Asia. The 3000 years of Han expansion southwards from the banks of the Yang-tze was compounded with the machinations of Soviet strategists, using this inexorable process as the basis for political and military assaults on the colonial powers and building a new communist empire. In the end Wiens maintained an academic ambivalence as to whether the pressure on South Vietnam, Cambodia, Laos, Thailand, Burma and Malaya represented the historical momentum of the Han or a new force fuelled by Soviet Russia. Asprey (1975, page 708) identifies Admiral Arthur Radford as coining the domino analogy in 1953 when he was urging a carrier based bombing strike to relieve Dien Bien Phu in a meeting of the Joint Chiefs of Staff. Eisenhower took up the catchword immediately, suggesting that 'the loss of Indochina will cause the fall of Southeast Asia like a set of dominoes' (quoted in Asprey, 1975, page 711). The next month, when direct military intervention no longer looked attractive, Eisenhower and Dulles both denied the veracity of this picture, claiming that the rest of Asia could be held even if Indochina fell (Buttinger, 1967). So, from the outset there was doubt. The seed had, however, been planted in the rhetoric of US officialdom, along with the impression of South Vietnam as a strategic necessity. It is around this time that the domino model made the transition from simile to theory.

Walter Rostow and Maxwell Taylor were the chief purveyors of domino theory to the Kennedy administration, converting McNamara to this view. Within months of his inauguration Kennedy was elaborating on domino theory in a news conference, suggesting that communist control of Laos would jeopardize the West's strategic position in Southeast Asia (Department of State, 1961).

There were voices raised in opposition to domino theory. C. P. Fitzgerald (1965), an Australian historian who had lived in China for much of his life, attempted to demolish the fallacy of the dominoes, stressing the greater reality and urgency to the local people of longstanding rivalries between Burmese and Thais, Khmers and Annamese and Thais, Malays and Javanese, Filipinos and Indonesians, rather than the passing clash of communism and anticommunism. He saw the source of unrest in the region as nationalism, which is inflamed by any foreign presence. Therefore, intervention based on domino theory was generating the very forces it hoped to contain. The significance of China in events in the region was a matter of geographical fact rather than ideological geostrategy. This view was echoed by Murphy (1966) when he questioned the assumption of China's desire to expand her bounds and the insufficiency of mere adjacency as a measure of influence.

It is ironic that Fitzgerald's argument misread the future so as to predict a communist Indonesia and a compromise regime in South Vietnam to refute the significance of adjacency and spatial order in determining events. The change in attitude of Indonesian government with the death of Sukarno in 1965 and the savage repression of the Communist Party, put paid to that line of reasoning. But at the same time it provided an example of a falling domino being stood up without direct US intervention. In May 1967 McNamara recognized the error of the domino view and attempted to defuse it by reference to the Indonesian case and the first stirrings of Chinese realignment when he broached the subject of a more politic, less militarily quantitative approach to Johnson (Sheehan *et al.*, 1971, pages 271–274). Not unnaturally, the Indonesian government objected to being likened to a domino and in 1969 their spokesman in Washington made their objection to the theory plain:

> The Southeast Asian nations do not constitute lifeless entities that automatically fall one way or another depending on which way their neighbor falls. History does not operate that way. What matters is the will, the political will, the determination of a nation to preserve its own identity.... The domino theory, therefore, is to us rather a gross oversimplification of the nature of the historical processes that go on in the area. It obscures and distorts rather than illuminates our understanding and offers no guidelines for realistic policy (Soedjatmoko, 1969).

The idea did not, however, succumb to these attacks and was inherited by Nixon. In an interview with Louis Heren of *The Times* (London) published on June 29, 1970, Nixon said:

> Now I know there are those that say, 'Well, the domino theory is obsolete.' They haven't talked to the dominoes. They should talk to the Thais, Malaysians, to Singapore, to Indonesia, to the Philippines, to the Japanese, and the rest ... and if the United States leaves Vietnam ... it will be ominously encouraging to the leaders of Communist China and the Soviet Union who are supporting the North Vietnamese. It will encourage them in their expansionist policies in other areas.

Deconde (1978, page 413) cites as a more recent effort of domino theorists a one page advertisement in *The New York Times* of June 6, 1976 taken out by the Citizens Alliance for Mediterranean Freedom, headed by John Connally. In this, Ford and Kissinger urged Italian leaders to keep communists out of government in Italy to avoid endangering 'the security of the entire Mediterranean.'

Although in a more advanced text de Blij is more cautious about the theory, indicating a need for "detailed investigation and dispassionate analysis" (Glassner and de Blij, 1980), in his world textbook (1980) he offers Thailand as the current example of instability likely to fall next after Laos and Cambodia. Indeed he is so bold as to predict the event from the theory. It is worth dwelling on the events of the last few years in these cases to evaluate the worth of this simple, but seemingly powerful, model of the train of events. It was US action which brought Laos and Cambodia directly into the fray. The US invasion of Cambodia in 1970 shook Prince Sihanouk loose in favour of Lon Nol whose weak grip was replaced by the excesses of Pol Pot and the Khmer Rouge regime in 1975. The Khmers Rouges, with their draconian policy of dispersal of city dwellers, were clients of the Chinese and at odds with Hanoi and the NLF (Vietcong), by now enjoying Russian support. In 1977 the victors of Vietnam swept into Cambodia and captured Phnom Phen. The refugees in northern Cambodia and over the border in Thailand, created by Pol Pot's actions, were now infiltrated by the remnants of the Khmers Rouges. In retaliation for the Vietnamese action China, by now having achieved a rapprochment with the US to balance its antagonism to Russia, mounted a punitive attack on Hanoi. In mid-1981 Vietnam's occupation is faced with three Cambodian resistance groups, the Khmers Rouges, a rightist movement led by Son Sann and one loyal to Prince Sihanouk, who is negotiating with the other parties and China. As fear of a Vietnamese drive into Thailand and Malaysia evaporates with their poor military and economic showing, the geopolitical value of Vietnam as a bulwark against the awesome presence of China is coming to be valued by the ASEAN nations according to *The Economist*'s (1981) intelligence. The resilience of the Thai body politic under pressure of a putsch in May 1981 bodes well for the stability of the entire region. To construe these events as the collapse of a row of dominoes is to oversimplify a little.

ANALYSIS

Huff and Lutz (1974) analyzed the state to state transmission of political unrest in Central Africa as a diffusion process, finding some evidence of a contagious process from the incidence of coups d'etat. This suggests that contiguity is relevant which runs contrary to the assertion of Gyorgy (Cohen, 1963, Appendix 308) that the significance of distance in these matters had dwindled. One of the obvious failings of domino theory is that it reduces the distance dimension to a question of contiguity. The germ of a richer image of geopolitical reality can be found in Jones (1954). He introduced the concept of a force field to the study of political geography. It is with such a continuous notion in mind that we now examine the nature and validity of domino theory.

Since there is no formal statement of the theory to be found, we are forced to induce the process implied and the assumptions involved from the mechanics of its analog. A row of dominoes depends on an artificially contrived state of unstable equilibrium for the dynamic symmetry of its response to a perturbation. Potential energy is imparted by standing the pieces on their ends aligned so that each strikes the next as it falls. Should a gap greater than the length of a domino separate two, the chain reaction ceases. The dominoes have three states: initial, unstable equilibrium; in motion; final, stable equilibrium. That "falling" and "fallen" equate with "going communist" may have been satisfying to the theory's proponents. The analogists did not seem disturbed, however, with the fallen state being a stable equilibrium while standing was unstable. This red and white characterization of politics is not only naive and insulting but also runs contrary to a geographical sense of uniqueness. It utterly fails to capture the significance of regional or national identity which we see daily dominating mankind's sense of self and place.

The model treats aggression from one end of the row as the potential energy of the first domino which is translated to kinetic energy by an initial tap. It falls, registering a change to the same affiliation as the aggressor, and, in so doing, imparts this character to the next domino as it strikes it down and so forth. What the necessities of similar size and appropriate spacing translate into in geographical terms is unclear. Obviously in order to land on the beaches of San Diego some very large dominoes would have to be stationed on the Philippines, Wake Island and Hawaii. The existence of a gap like the Pacific should quiet fears of the red menace wading ashore in the west. In his exposition de Blij leaps continents and oceans to Tanzania, Mozambique, Angola, Zimbabwe and Namibia for further evidence of the domino "effect" (1980, page 536). In this manifestation it is demoted from a theory to an effect. Obviously, oceans or intervening nations are not seen as gaps containing the contagion but can be erased conveniently. The nature of the contamination process is not made very clear by the analogy. "Knocked over" is redolent of liquor stores rather than nations and hardly provides a rich enough description of the process to prescribe preventative action. "Propping up" has been used to indicate one type of solution, but has proven difficult to translate into successful political, military and economic operations. "Knocking out," i.e., lateral displacement of one or more pieces to provide a firebreak to check the progress of the conflagration, does appeal to some military minds as a feasible action. Certainly on a local scale towns and villages were wiped out to "save" them from the NLF in Vietnam. . . .

Whether or not precise conclusions could be drawn, it is evident from the vehement affirmation of separateness of smaller groups within the confines or adjacent to the territories of world power cores that the limited notion of nationality is still the most powerful geopolitical force. Those seeking *de facto* or *de jure* independence often turn to the chief competitor of their dominator for support. Cuba is the obvious example of this behavior. An understanding of the relationship of North Vietnam and China in these terms might have been more enlightening to US foreign policy in the second half of this century than domino theory.

A more felicitous mental picture of the potential for interaction of the three major power cores, their several more diffuse subsidiaries and the multiplicity of client, satellite, independent and refractory entities of international politics is conjured up by Henry Kissinger's term "linkage." Once again this verbal abstraction has not been formalized anywhere. Landau (1972, pages 118–120) perceives the model behind this "most characteristic device in Nixon–Kissinger policy" as being of a network connecting all the world's trouble spots to the Soviet Union and the USA. The resolution of particular conflicts then depends not on the merits of the case alone but on the overall balance of power between the two sides. It is clear from recent events that China was and is considered a card in this game rather than a player.

The geographic defect lay in the peculiar configuration of the network which Kissinger seems to have employed. The graph had two terminal nodes with all others connected to them by equal valued edges. According to Landau, Kissinger viewed the links from all US/Soviet points of confrontation throughout the globe as being of equal significance. In practice the original linkage theory was "little more than unreconstructed Cold Warriorism" and "a formula for perpetuating confrontation all over the world." Landau contrasts this with a "ripple" approach which sees:

> Events in the world are only as strongly interconnected as their geographic or conceptual distances are brief (1972, page 125).

A modified linkage model incorporating the friction of distance, with nations as nodes of a more fully connected network of economic and political links weighted in terms of the ease of communication and influence between states, would suggest the efficacy of indirect methods of achieving geopolitical goals. This would encourage the search for solutions by the diplomatic manipulation of second or greater order paths of influence rather than head on military posturing of action.

CONCLUSION

Linkage theory is potentially a far more satisfactory model than domino theory, but its finite structure leaves much to be desired as a useful representation of geographical reality. Geography instills a predeliction for continuous formulations. The spaceless nature of the nodes of a network is likely to downgrade the significance of internal differentiation and politics in the workings of the state. Alistair Cooke remarked in one of his "Letters from America" that you only had to spend a while with a map of Southeast Asia to dismiss domino theory. Although linkage gets you from the one dimension of domino theory into two dimensions, it still needs to be supplemented with a more elaborate intellectual picture of our world. Neither dominoes nor dyads will do as an adequate impression of geographic, cultural entities. Until such time as we shake off the coils of distance friction, our motives and actions will be influenced by where we are and where everybody else is and statesmen would do well to keep this in mind.

REFERENCES

Asprey, R. B. (1975) *War in the Shadows*. New York: Doubleday.

Bullit, W. C. (1947) A report to the American people on China. *Life*, October 13.

Buttinger, J. (1967) *Vietnam: A Dragon Embattled*. New York: Random House.

Cohen, S. B. (1963) *Geography and Politics in a World Divided*. New York: Praeger.

De Blij, H. (1980) *Geography: Region and Concept*. 3rd Edition. New York: Wiley.

Deconde, A. (1978) *A History of American Foreign Policy, Volume II Global Power (1900 to the Present)*. New York: Charles Scribner's Sons.

Department of State (1961) *Bulletin XLIV*, April 17, 543.

Economist, The (1981) Cambodia: the story so far. May 9, 48–9.

Fitzgerald, C. P. (1965) The fallacy of the dominoes. *The Nation*, June 2, 700–712.

Glassner, M. I. and de Blij, H. (1980) *Systematic Political Geography*. 3rd Edition. New York: Wiley.

Huff, D. L. and Lutz, J. M. (1974) The contagion of political unrest in independent Black Africa. *Economic Geography*, 50(4), 353–367.

Jones, S. B. (1954) A unified field theory of political geography. *Annals of the Association of American Geographers*, 44(2), 111–123.

Landau, D. (1972) *Kissinger: The Uses of Power*. Boston: Houghton Mifflin.

Murphy, R. (1966) China and the dominoes. *Asian Survey*, 6, 510–515.

Sheehan, N., Smith, H., Kenworthy, E. W. and Butterfield, F. (1971) *The Pentagon Papers*. New York: Bantam.

Soedjatmoko (1969) Southeast Asia and Security. *Survival*, October.

Wiens, H. J. (1954) *China's March towards the Tropics*. Hamden, Conn.: The Shoe String Press.

Cartoon 4 America's Indochina quagmire (by Picha)
Originally conceptualized geopolitically within the terms of the domino theory, Vietnam and South East Asia became conceptualized as a quagmire in the US geopolitical imagination as US interventionism in the region escalated and US soldiers got bogged down in a war far from home.
Source: Picha, Cartoon & Writers' Syndicate

10 LEONID BREZHNEV

"The Brezhnev Doctrine"

from *Pravda* (1968)

SOVEREIGNTY AND THE INTERNATIONALIST OBLIGATION OF SOCIALIST COUNTRIES

In connection with the events in Czechoslovakia the question of the relationship and interconnection between the socialist countries' national interests and their internationalist obligations has assumed particular urgency and sharpness. The measures taken jointly by the Soviet Union and other socialist countries to defend the social gains of the Czechoslovak people are of enormous significance for strengthening the socialist commonwealth, which is the main achievement of the international working class.

At the same time it is impossible to ignore the allegations being heard in some places that the actions of the five socialist countries contradict the Marxist–Leninist principle of sovereignty and the right of nations to self-determination.

Such arguments are untenable primarily because they are based on an abstract, nonclass approach to the question of sovereignty and the right of nations to self-determination.

There is no doubt that the peoples of the socialist countries and the Communist Parties have and must have freedom to determine their country's path of development. However, any decision of theirs must damage neither socialism in their country nor the fundamental interests of the other socialist countries nor the worldwide workers' movement, which is waging a struggle for socialism. This means that every Communist Party is responsible not only to its own people but also to all the socialist countries and to the entire Communist movement. Whoever forgets this in placing sole emphasis on the autonomy and independence of Communist Parties lapses into one-sidedness, shirking his internationalist obligation.

The Marxist dialectic opposes one-sidedness; it requires that every phenomenon be examined in terms of both its specific nature and its overall connection with other phenomena and processes. Just as, in V. I. Lenin's words, someone living in a society cannot be free of that society, so a socialist state that is in a system of other states constituting a socialist commonwealth cannot be free of the common interests of that commonwealth.

The sovereignty of individual socialist countries cannot be counterposed to the interests of world socialism and the world revolutionary movement. V. I. Lenin demanded that all Communists "struggle *against* petty national narrowness, exclusivity and isolation, and for taking into account the whole, the overall situation, for subordinating the interests of the particular to the interests of the general" (*Complete Collected Works* [in Russian], Vol. XXX, p. 45).

Socialist states have respect for the democratic norms of international law. More than once they have proved this in practice by resolutely opposing imperialism's attempts to trample the sovereignty and independence of peoples. From this same standpoint they reject left-wing, adventurist notions of "exporting revolution" and "bringing bliss" to other peoples. However, in the Marxist conception the norms of law, including the norms governing relations among socialist countries, cannot be interpreted in a narrowly formal way, outside the general context of the class struggle in the present-day world.

Socialist countries resolutely oppose the

export and import of counterrevolution. Each Communist Party is free in applying the principles of Marxism–Leninism and socialism in its own country, but it cannot deviate from these principles (if, of course, it remains a Communist Party). In concrete terms this means primarily that every Communist Party cannot fail to take into account in its activities such a decisive fact of our time as the struggle between the two antithetical social systems – capitalism and socialism. This struggle is an objective fact that does not depend on the will of people and is conditioned by the division of the world into two antithetical social systems. "Every person," V. I. Lenin said, "must take either this, our, side or the other side. All attempts to avoid taking sides end in failure and disgrace" (Vol. XLI, p. 401).

It should be stressed that even if a socialist country seeks to take an "extrabloc" position, it in fact retains its national independence thanks precisely to the power of the socialist commonwealth – and primarily to its chief force, the Soviet Union – and the might of its armed forces. The weakening of any link in the world socialist system has a direct effect on all socialist countries, which cannot be indifferent to this. Thus, the antisocialist forces in Czechoslovakia were in essence using talk about the right to self-determination to cover demands for so-called neutrality and the CSR's withdrawal from the socialist commonwealth. But implementation of such "self-determination," i.e., Czechoslovakia's separation from the socialist commonwealth, would run counter to Czechoslovakia's fundamental interests and would harm the other socialist countries. Such "self-determination," as a result of which NATO troops might approach Soviet borders and the commonwealth of European socialist countries would be dismembered, in fact infringes on the vital interests of these countries' peoples, and fundamentally contradicts the right of these peoples to socialist self-determination. The Soviet Union and other socialist states, in fulfilling their international duty to the fraternal peoples of Czechoslovakia and defending their own socialist gains, had to act and did act in resolute opposition to the antisocialist forces in Czechoslovakia.

Comrade W. Gomulka, First Secretary of the Central Committee of the Polish United Workers' Party, used a metaphor to illustrate this point:

> To those friends and comrades of ours from other countries who believe they are defending the just cause of socialism and the sovereignty of peoples by denouncing and protesting the introduction of our troops in Czechoslovakia, we reply: If the enemy plants dynamite under our house, under the commonwealth of socialist states, our patriotic, national and internationalist duty is to prevent this using any means that are necessary.

People who "disapprove" of the actions taken by the allied socialist countries ignore the decisive fact that these countries are defending the interests of worldwide socialism and the worldwide revolutionary movement. The socialist system exists in concrete form in individual countries that have their own well-defined state boundaries and develops with regard for the specific attributes of each such country. And no one interferes with concrete measures to perfect the socialist system in various socialist countries. But matters change radically when a danger to socialism itself arises in a country. World socialism as a social system is the common achievement of the working people of all countries, it is indivisible, and its defense is the common cause of all Communists and all progressive people on earth, first and foremost the working people of the socialist countries.

The Bratislava statement of the Communist and Workers' Parties on socialist gains says that "it is the common internationalist duty of all socialist countries to support, strengthen and defend these gains, which were achieved at the cost of every people's heroic efforts and selfless labor."

What the right-wing, antisocialist forces were seeking to achieve in Czechoslovakia in recent months was not a matter of developing socialism in an original way or of applying the principles of Marxism–Leninism to specific conditions in that country, but was an encroachment on the foundations of socialism and the fundamental principles of Marxism–Leninism. This is the "nuance" that is still incomprehensible to people who trusted in the hypocritical cant of the antisocialist and revisionist elements.

Under the guise of "democratization" these elements were shattering the socialist state step by step; they sought to demoralize the Communist Party and dull the minds of the masses; they were gradually preparing for a counterrevolutionary coup and at the same time were not being properly rebuffed inside the country.

The Communists of the fraternal countries naturally could not allow the socialist states to remain idle in the name of abstract sovereignty while the country was endangered by antisocialist degeneration.

The five allied socialist countries' actions in Czechoslovakia are consonant with the fundamental interests of the Czechoslovak people themselves. Obviously it is precisely socialism that, by liberating a nation from the fetters of an exploitative system, ensures the solution of fundamental problems of national development in any country that takes the socialist path. And by encroaching on the foundations of socialism, the counterrevolutionary elements in Czechoslovakia were thereby undermining the basis of the country's independence and sovereignty.

The formal observance of freedom of self-determination in the specific situation that had taken shape in Czechoslovakia would signify freedom of "self-determination" not for the people's masses and the working people, but for their enemies. The antisocialist path, the "neutrality" to which the Czechoslovak people were being prodded, would lead the CSR straight into the jaws of the West German revanchists and would lead to the loss of its national independence. World imperialism, for its part, was trying to export counterrevolution to Czechoslovakia by supporting the antisocialist forces there.

The assistance given to the working people of the CSR by the other socialist countries, which prevented the export of counterrevolution from the outside, is in fact a struggle for the Czechoslovak Socialist Republic's sovereignty against those who would like to deprive it of this sovereignty by delivering the country to the imperialists.

Over a long period of time and with utmost restraint and patience, the fraternal Communist Parties of the socialist countries took political measures to help the Czechoslovak people to halt the antisocialist forces' offensive in Czechoslovakia. And only after exhausting all such measures did they undertake to bring in armed forces.

The allied socialist countries' soldiers who are in Czechoslovakia are proving in deeds that they have no task other than to defend the socialist gains in that country. They are not interfering in the country's internal affairs, and they are waging a struggle not in words but in deeds for the principles of self-determination of Czechoslovakia's peoples, for their inalienable right to decide their destiny themselves after profound and careful consideration, without intimidation by counterrevolutionaries, without revisionist and nationalist demagoguery.

Those who speak of the "illegality" of the allied socialist countries' actions in Czechoslovakia forget that in a class society there is and can be no such thing as nonclass law. Laws and the norms of law are subordinated to the laws of the class struggle and the laws of social development. These laws are clearly formulated in the documents jointly adopted by the Communist and Workers' Parties.

The class approach to the matter cannot be discarded in the name of legalistic considerations. Whoever does so and forfeits the only correct, class-oriented criterion for evaluating legal norms begins to measure events with the yardsticks of bourgeois law. Such an approach to the question of sovereignty means, for example, that the world's progressive forces could not oppose the revival of neo-Nazism in the FRG, the butcheries of Franco and Salazar or the reactionary outrages of the "black colonels" in Greece, since these are the "internal affairs" of "sovereign states." It is typical that both the Saigon puppets and their American protectors concur completely in the notion that sovereignty forbids supporting the struggle of the progressive forces. After all, they shout from the housetops that the socialist states that are giving aid to the Vietnamese people in their struggle for independence and freedom are violating Vietnam's sovereignty. Genuine revolutionaries, as internationalists, cannot fail to support progressive forces in all countries in their just struggle for national and social liberation.

The interests of the socialist commonwealth

and the entire revolutionary movement and the interests of socialism in Czechoslovakia demand full exposure and political isolation of the reactionary forces in that country, consolidation of the working people and consistent fulfillment of the Moscow agreement between the Soviet and Czechoslovak leaders.

There is no doubt that the actions taken in Czechoslovakia by the five allied socialist countries in Czechoslovakia, actions aimed at defending the fundamental interests of the socialist commonwealth and primarily at defending Czechoslovakia's independence and sovereignty as a socialist state, will be increasingly supported by all who really value the interests of the present-day revolutionary movement, the peace and security of peoples, democracy and socialism.

11 GEARÓID Ó TUATHAIL AND JOHN AGNEW

"Geopolitics and Discourse: Practical Geopolitical Reasoning in American Foreign Policy"

from *Political Geography Quarterly* (1992)

The Cold War, Mary Kaldor recently noted, has always been a discourse, a conflict of words, "capitalism" versus "socialism" (Kaldor, 1990). Noting how Eastern Europeans always emphasize the power of words, Kaldor adds that the way we describe the world, the words we use, shape how we see the world and how we decide to act. Descriptions of the world involve geographical knowledge and Cold War discourse has had a regularized set of geographical descriptions by which it represented international politics in the post-war period. The simple story of a great struggle between a democratic "West" against a formidable and expansionist East has been the most influential and durable geopolitical script of this period. This story, which today appears outdated, was a story which played itself out not in Central Europe but in exotic "Third-World" locations, from the sands of the Ogaden in the Horn of Africa, to the mountains of El Salvador, the jungles of Vietnam and the valleys of Afghanistan. Of course, the plot was not always a simple one. It has been complex and nuanced, making the post-war world a dynamic, dramatic and sometimes ironic one – ironies such as Cuban troops guarding Gulf Oil facilities against black UNITA forces supported by a racist South African government. Yet the story was a compelling one which brought huge military-industrial complexes into existence on both sides of the "East–West" divide and rigidly disciplined the possibilities for alternative political practices throughout the world. All regional conflicts, up until very recently, were reduced to its terms and its logic. Now with this story's unravelling and its geography blurring, it is time to ask how did the Cold War in its geopolitical guise come into existence and work?

This paper is not an attempt directly to answer such questions. Rather it attempts to establish a conceptual basis for answering them. It seeks to outline a re-conceptualization of geopolitics in terms of discourse and apply this to the general case of American foreign policy. Geopolitics, some will argue, is, first and foremost, about practice and not discourse; it is about actions taken against other powers, about invasions, battles and the deployment of military force. Such practice is certainly geopolitical but it is only through discourse that the building up of a navy or the decision to invade a foreign country is made meaningful and justified. It is through discourse that leaders act, through the mobilization of certain simple geographical understandings that foreign-policy actions are explained and through ready-made geographically infused reasoning that wars are rendered meaningful. How we understand and constitute our social world is through the socially structured use of language (Franck and Weisband, 1971; Todorov, 1984). Political speeches and the like afford us a means of recovering the self-understandings of influential actors in world politics. They help us understand the social construction of worlds and the role of geographical knowledge in that social construction.

The paper is organized into two parts. The first part attempts to sketch a theory of geopolitics by employing the concept of discourse. Four suggestive theses on the implications of conceptualizing geopolitics in discursive terms are briefly outlined. The second part addresses the question of American geopolitics and provides an account of some consistent features of the practical geopolitical reasoning by which

American foreign policy has sought to write a geography of international politics. This latter part involves a detailed analysis of two of the most famous texts on the origins of the Cold War: George Kennan's "Long Telegram" of 1946 and his "Mr X" article in 1947. The irony of these influential geopolitical representations of the USSR is that they were not concrete geographical representations but overdetermined and ahistorical abstractions. It is the anti-geographical quality of geopolitical reasoning that this paper seeks to illustrate.

GEOPOLITICS AND DISCOURSE

Geopolitics, as many have noted, is a term which is notoriously difficult to define (Kristof, 1960). In conventional academic understanding geopolitics concerns the geography of international politics, particularly the relationship between the physical environment (location, resources, territory, etc.) and the conduct of foreign policy (Sprout and Sprout, 1960). Within the geopolitical tradition the term has a more precise history and meaning. A consistent historical feature of geopolitical writing, from its origins in the late nineteenth century to its modern use by Colin Gray and others, is the claim that geopolitics is a foil to idealism, ideology and human will. This claim is a long-standing one in the geopolitical tradition which from the beginning was opposed to the proposition that great leaders and humans will alone determine the course of history, politics and society. Rather, it was the natural environment and the geographical setting of a state which exercised the greatest influence on its destiny (Ratzel, 1969; Mackinder, 1890). Karl Haushofer argued that the study of Geopolitik demonstrated the "dependence of all political events on the enduring conditions of the physical environment" (Bassin, 1987: 120). In a 1931 radio address he remarked:

> geopolitics takes the place of political passion and development dictated by natural law reshapes the work of the arbitrary transgression of human will. The natural world, beaten back with sword or pitchfork, irrepressibly reasserts itself in the face of the earth. This is geopolitics! (Haushofer translated in Bassin, 1987: 120).

By its own understandings and terms, geopolitics is taken to be a domain of hard truths, material realities and irrepressible natural facts. Geopoliticians have traded on the supposed objective materialism of geopolitical analysis. According to Gray (1988: 93), "geopolitical analysis is impartial as between one or another political system or philosophy". It addresses the base of international politics, the permanent geopolitical realities around which the play of events in international politics unfolds. These geopolitical realities are held to be durable, physical determinants of foreign policy. Geography, in such a scheme, is held to be a non-discursive phenomenon: it is separate from the social, political and ideological dimensions of international politics.

The great irony of geopolitical writing, however, is that it was always a highly ideological and deeply politicized form of analysis. Geopolitical theory from Ratzel to Mackinder, Haushofer to Bowman, Spykman to Kissinger was never an objective and disinterested activity but an organic part of the political philosophy and ambitions of these very public intellectuals. While the forms of geopolitical writing have varied among these and other authors, the practice of producing geopolitical theory has a common theme: the production of knowledge to aid the practice of statecraft and further the power of the state.

Within political geography, the geopolitical tradition has long been opposed by a tradition of resistance to such reasoning. A central problem that has dogged such resistance is its lack of a coherent and comprehensive theory of geopolitical writing and its relationship to the broader spatial practices that characterize the operation of international politics. This paper proposes such a theory by re-conceptualizing the conventional meaning of geopolitics using the concept of discourse. Our foundational premise is the contention that geography is a social and historical discourse which is always intimately bound up with questions of politics and ideology (Ó Tuathail, 1989). Geography is never a natural, non-discursive phenomenon which is separate from ideology and outside politics. Rather, geography as a discourse is a form of power/knowledge itself (Foucault, 1980; Ó Tuathail, 1989).

Geopolitics, we wish to suggest, should be critically re-conceptualized as a discursive practice by which intellectuals of statecraft "spatialize" international politics in such a way as to represent it as a "world" characterized by particular types of places, peoples and dramas. In our understanding, the study of geopolitics is the study of the spatialization of international politics by core powers and hegemonic states. This definition needs careful explication.

The notion of discourse has become an important object of investigation in contemporary critical social science, particularly that which draws inspiration from the writings of the French philosopher Michel Foucault (MacDonell, 1986). Within the discipline of international relations, there has been a series of attempts to incorporate the notion of discourse into the study of the practices of international politics (Alker and Sylvan, 1986; Ashley, 1987; Shapiro, 1988; Der Derian and Shapiro, 1989). Dalby (1988, 1990a, 1990b) and Ó Tuathail (1989) have attempted to extend the concept into political geography. Discourses are best conceptualized as sets of capabilities people have, as sets of socio-cultural resources used by people in the construction of meaning about their world and their activities. It is NOT simply speech or written statements but the rules by which verbal speech and written statements are made meaningful. Discourses enable one to write, speak, listen and act meaningfully. They are a set of capabilities, an ensemble of rules by which readers/listeners and speakers/audiences are able to take what they hear and read and construct it into an organized, meaningful whole. Alker and Sylvan (1986) articulate the distinction this way:

> As backgrounds, discourses must be distinguished from the verbal productions which readers or listeners piece together. As we prefer to use the term people do not read or listen to a discourse: rather, they employ a discourse or discourses in the processes of reading or listening to a verbal production. Discourses do not present themselves as such; what we observe are people and verbal productions.

Discourses, like grammars, have a virtual and not an actual existence. They are not overarching constructs in the way that "structures" are sometimes represented. Rather, they are real sets of capabilities whose existence we infer from their realizations in activities, texts and speeches. Neither are they absolutely deterministic. Discourses enable. One can view these capabilities or rules as permitting a certain bounded field of possibilities and reasoning as the process by which certain possibilities are actualized. The various actualizations of possibilities have consequences for the further reproduction and transformation of discourse. The actualization of one possibility closes off previously existent possibilities and simultaneously opens up a new series of somewhat different possibilities. Discourses are never static but are constantly mutating and being modified by human practice. The study of geopolitics in discursive terms, therefore, is the study of the socio-cultural resources and rules by which geographies of international politics get written.[1]

The notion of "intellectuals of statecraft" refers to a whole community of state bureaucrats, leaders, foreign-policy experts and advisors throughout the world who comment upon, influence and conduct the activities of statecraft. Ever since the development of the modern state system in the sixteenth century there has been a community of intellectuals of statecraft. Up until the twentieth century this community was rather small and restricted, with most intellectuals also being practitioners of statecraft. In the twentieth century, however, this community has become quite extensive and internally specialized. Within the larger states at least, one can differentiate between types of intellectuals of statecraft on the basis of their institutional setting and style of reasoning. Within civil society there are "defense intellectuals" associated with particular defense contractors and weapons systems. There is also a specialized community of security intellectuals in various public think-tanks (e.g. the RAND Corporation, the Hoover Institute, the Georgetown Center for Strategic and International Studies) who write and comment upon international affairs and strategy (Cockburn, 1987; Dalby, 1990b). One finds a different form of intellectualizing from public intellectuals of statecraft such as Henry Kissinger or Zbigniew Brzezinski who, as former top governmental officials, command a wide audience for their opinions in national newspapers and foreign-policy journals. Within political society itself there are different gradations amongst the foreign-policy community from those who design, articulate and order foreign policy from the top to those actually charged with implementing particular foreign policies and practicing statecraft (whether diplomatic or military) on a daily basis. All can claim to be intellectuals of statecraft for they are constantly engaged in reasoning about statecraft though all may not have the function of intellectuals in the conventional sense, but rather in the sense of Gramsci's "organic" intellectuals (Gramsci, 1971).

We wish to propose four theses which follow from our preliminary observations on reasoning processes and intellectuals of statecraft. The first of these is that the study of geopolitics as we have defined it involves the comprehensive study of statecraft as a set of social practices. Geopolitics is not a discrete and relatively contained activity confined only to a small group of "wise men" who speak in the language of classical geopolitics. Simply to describe a foreign-policy problem is to engage in geopolitics, for one is implicitly and tacitly normalizing a particular world. One could describe geopolitical reasoning as the creation of the backdrop or setting upon which "international politics" takes place, but such would be a simplistic view. The creation of such a setting is itself part of world politics. This setting itself is more than a single backdrop but an active component part of the drama of world politics. To designate a place is not simply to define a location or setting. It is to open up a field of possible taxonomies and trigger a series of narratives, subjects and appropriate foreign-policy responses. Merely to designate an area as Islamic is to designate an implicit foreign policy (Said, 1978, 1981). Simply to describe a different or indeed the same place as "Western" (e.g. Egypt) is silently to operationalize a competing set of foreign-policy operators. Geopolitical reasoning begins at a very simple level and is a pervasive part of the practice of international politics. It is an innately political process of representation by which the intellectuals of statecraft designate a world and "fill" it with certain dramas, subjects, histories and dilemmas. All statespersons engage in the practice; it is one of the norms of the world political community.

Our second thesis is that most geopolitical reasoning in world politics is of a practical and not a formal type. Practical geopolitical reasoning is reasoning by means of consensual and unremarkable assumptions about places and their particular identities. This is the reasoning of practitioners of statecraft, of statespersons, politicians and military commanders. This is to be contrasted with the formal geopolitical reasoning of strategic thinkers and public intellectuals (such as those founding the "geopolitical tradition") who work in civil society and produce a highly codified system of ideas and principles to guide the conduct of statecraft. The latter forms of knowledge tend to have highly formalized rules of statement, description and debate. By contrast, practical geopolitical reasoning tends to be of a common-sense type which relies on the narratives and binary distinctions found in societal mythologies. In the case of colonial discourse, there are contrasts between white and non-white, civilized and backward, Western and non-Western, adult and child. The operation of such distinctions in European foreign policy during the age of empire is well known (Kiernan, 1969; Gates, 1985). US foreign policy towards the Philippines and Latin America during the latter half of the nineteenth century and the beginning of the twentieth century is also replete with such distinctions (Black, 1988; Hunt, 1987; Karnow, 1989). In Cold War discourse the contrast was, as Truman codified it in his famous Truman Doctrine statement of March 1947, between a way of life based upon the will of the majority and distinguished by free institutions, representative government, free elections, guarantees of individual liberty, freedom of speech and religion and freedom from political oppression versus a way of life based on the will of a minority forcibly imposed upon the majority. This latter way of life relied upon terror and oppression, a controlled press and radio, fixed elections and the suppression of personal freedoms. Such were the criteria by which places were to be judged and spatially divided into different geographical camps in the post-war period.

Our third thesis is that the study of geopolitical reasoning necessitates studying the production of geographical knowledge within a particular state and throughout the modern world system. Geographical knowledge is produced at a multiplicity of different sites throughout not only the nation-state, but the world political community. From the classroom to the living-room, the newspaper office to the film studio, the pulpit to the presidential office, geographical knowledge about a world is being produced, reproduced and modified. The challenge for the student of geopolitics is to understand how geographical knowledge is transformed into the reductive geopolitical reasoning of intellectuals of statecraft. How are places reduced to security commodities, to geographical abstractions which need to be "domesticated", controlled, invaded or bombed rather than understood in their complex reality? How, for example, did Truman metamorphose the situation in Greece in March 1947 – it was the site of a complex civil war at the time – into the Manichean terms of the Truman Doctrine? The answer we suspect is rather ironic given the common-sense meaning of geography as "place facts": geopolitical reasoning works by the active suppression of the complex geographical reality of places in favor of controllable geopolitical abstractions.

Our fourth thesis concerns the operation of geopolitical reasoning within the context of the modern world-system. Throughout the history of the modern world-system, intellectuals of statecraft from core states – particularly those states which are competing for hegemony – have disproportionate influence and power over how international political space is represented. A hegemonic world power, such as the United States in the immediate post-war period, is by definition a "rule-writer" for the world community. Concomitant with its material power is the power to represent world politics in certain ways. Those in power within the institutions of the hegemonic state become the deans of world politics, the administrators, regulators and geographers of international affairs. Their power is a power to constitute the terms of geopolitical world order, an ordering of international space which defines the central drama of international politics in particularistic ways. Thus not only can they represent in their own terms particular regional conflicts, whose causes may be quite localized (e.g. the Greek civil war), but they can help create conditions whereby peripheral and semi-peripheral states actively adopt and use the geopolitical reasoning of the hegemon. Examples of this range from the institutionalization of laws to suppress "Communism" in certain states (even though the state may not have an organized Communist movement; the laws are simply ways to suppress a broad range of dissent; e.g. the case of El Salvador) to the slavish parroting of approved Cold War discourse in international organizations and forums.

PRACTICAL GEOPOLITICAL REASONING IN AMERICAN FOREIGN POLICY

Given our re-conceptualization of geopolitics, any analysis of American geopolitics must necessarily be more than an analysis of the formal geopolitical reasoning of a series of "wise men" of strategy (Mahan, Spykman, Kissinger and others). American geopolitics involves the study of the different historical means by which US intellectuals of statecraft have spatialized international politics and represented it as a "world" characterized by particular types of places, peoples and dramas. Such is obviously a vast undertaking and we wish to make but three general observations on the contours of American geopolitical reasoning. Before doing so, however, it is important to note two factors about the American case. First, we must acknowledge the key role the Presidency plays in the assemblage of meaning about international politics within the United States (and internationally since the US became a world power). In ethnographic terms, the US President is the chief bricoleur of American political life, a combination of storyteller and tribal shaman. One of the great powers of the Presidency, invested by the sanctity, history and rituals associated with the institution – the fact that the media take their primary discursive cues from the White House – is the power to describe, represent, interpret and appropriate. It is a formidable power but not an absolute power, for the art of description and appropriation (e.g. President Reagan's representation of the Nicaraguan contras as the "moral equivalents of the founding fathers') must have resonances with the Congress, the established media and the American public. The generation of such resonances often requires the repetition and re-cycling of certain themes and images even though the socio-historical context of their use may have changed dramatically. One has the attempted production of continuity by the incorporation of "strategic terms" (Turton, 1984), "key metaphors" (Crocker, 1977) and "key symbols" (Herzfeld, 1982) into geopolitical reasoning. Behind all of these is the assumption of a power of appropriateness in the use of certain relatively fixed terms and phrases (Parkin, 1978).

Secondly, we must recognize that American involvement with world politics has followed a distinctive cultural logic or set of presuppositions and orientations, what Gramsci called "Americanismo" (De Grazia, 1984–85). In particular, economic freedom – in the form of "free" business activity and the political conditions necessary for this – has been a central element in American culture. This has given rise to an attempt to reconstruct foreign places in an American image. US foreign-policy experiences with Mexico, China, Central America, the Caribbean and the Philippines all bear witness to this fundamental feature of US foreign policy (Agnew, 1983; Karnow, 1989).

The first of our three observations on practical geopolitical reasoning in American foreign policy is that representations of "America" as a place are pervasively mythological. "America" is a place which is at once real, material and bounded (a territory with quiddity) yet also a mythological, imaginary and universal ideal with no specific spatial bounds. Ever since early modern times, North America and the Caribbean have had the transgressive aura of a place "beyond the line", as Dunn (1972: ch. 1) terms it, where might made right and the European treaties did not apply. By its own lore, the origins of the country are mythic and its location divine. In his famous pamphlet Common Sense, written in 1776 in support of the American rebellion, Thomas Paine wrote:

> This new world hath been the asylum for the persecuted lovers of civil and religious liberty from every part of Europe. Either have they fled, not from the tender embraces of the mother, but from the cruelty of the monster.... Everything that is right or natural pleads for separation. The blood of the slain, the weeping voice of nature cries 'TIS TIME TO PART. Even the distance at which the Almighty hath placed England and America, is a strong and natural proof, that the authority of the one, over the other, was never the design of Heaven. The time likewise at which the continent was discovered, adds weight to the argument, and the manner in which it was peopled increases the force of it. The reformation was preceded by the discovery of America, as if the Almighty graciously meant to open a sanctuary to the persecuted in future years, when home should afford neither friendship nor safety (Paine, 1969: 39, 4041).

The dramatic hyperbole of Paine's geopolitical reasoning is part of the mythological origins of the American state. In the popular imagination "America" was "discovered"; it was a new, empty, pristine place, a New World. Despite the obvious inadequacies of this view, such an imaginary geography can still be found in contemporary American political culture and in the articulation of US foreign policy. Speaking over 210 years later on 2 February 1988 in an address to the nation supporting the Nicaraguan contras, President Ronald Reagan remarked:

> My friends, I have often expressed my belief that the Almighty had a reason for placing this great and good land, the "New World", here between two vast oceans. Protected by the seas, we have enjoyed the blessings of peace – free for almost two centuries now from the tragedy of foreign aggression on our mainland. Help us to keep that precious gift secure. Help us to win support for those who struggle for the same freedoms we hold dear. In doing so, we will not just be helping them; we will be helping ourselves, our children, and all the peoples of the world. We will be demonstrating that America is still a beacon of hope, still a

> light unto the nations. Yes, a great opportunity to show that hope still burns bright in this land and over our continent, casting a glow across centuries, still guiding missions – to a future of peace and freedom (Reagan, 1988: 35).

The continuity between the two texts is evidence of the durability of particular narratives in American political discourse. It is a structuralist fallacy to think of this narrative as having a "deep structure" or a primordial set of binary oppositions – e.g. Old World: New World, despotism totalitarianism: freedom – to which everything else can be reduced. As a discourse its existence is virtual not actual and is assembled and re-assembled differently by presidents and other intellectuals of statecraft. Such discourse freely fuses fact with fiction and reality with the imaginary to produce a reasoning where neither is distinguishable from the other.[2] Both narratives read like primitive ethnographic tales: the origins of a tribe from the wanderings of persecuted members of other tribes, the flight from persecution, the chosen land, divine guidance, blessings, precious gifts, beacons and monsters. America's first leaders are known even today in American political culture as the "founding fathers".

Secondly, there is a tension between a universal omnipresent image of "America" and a different spatially-bounded image of the place. On one hand, American discourse consistently plays upon the unique geographical location of "America" yet simultaneously asserts that the principles of this "New World" are universal and not spatially confined there. The geography evoked in the American Declaration of Independence was not continental or hemispheral but universal. Its concern was with "the earth", the "Laws of Nature and of Nature's God", and all of "mankind". In this universalist vision, "America" is positioned as being equivalent with the strivings of a universal human nature. "The cause of America", Paine (1969: 23) proclaimed, "is in a great measure the cause of all mankind". The freedoms it struggles for are, in Reagan's terms, the freedoms desired by "all the peoples of the world". "America" is at once a territorially-defined state and a universal ideal, a place on the North American continent and a mythical homeland of freedom.

For the late eighteenth and most of the nineteenth century, the spatially-bounded sense of "America" was the one that predominated in US foreign-policy rhetoric. Even though the United States had closer economic, cultural and political ties with Europe than any other place, its foreign-policy rhetoric defined it as a separate and distinct sphere. "Europe", George Washington observed in his farewell address (1796), "has a set of primary interests which to us have none or a very remote relation. Hence she must be engaged in frequent controversies, the causes of which are essentially foreign to our concerns" (Richardson, 1905, vol. 1: 214). Washington's geopolitical reasoning was largely a negative one which defined the American sphere as extra-European (like Persia and Turkey) rather than a system complete and to itself. For others, notably Thomas Jefferson, Henry Clay and John Quincy Adams, there was a distinct "American system". Jefferson, writing in 1813 to the geographer Alexander von Humboldt on the five Spanish–American colonies in rebellion (which the US recognized in 1822; earlier recognition moves were defeated), noted:

> But in whatever government they end, they will be American governments, no longer to be involved in the never-ceasing broils of Europe. The European nations constitute a separate division of the globe; their localities make them a part of a distinct system; they have a set of interests of their own in which it is our business never to engage ourselves. America has a hemisphere to itself. It must have its separate system of interests; which must not be subordinated to those of Europe (Quoted in Whitaker, 1954: 29).

The "American system" was not, however, to be a multi-lateralist, pan-American affair or a counterpose to the Holy Alliance as Henry Clay had suggested in 1821. John Quincy Adams, who actively opposed such a policy, did not advocate isolationism so much as oppose any multi-lateral moves on the US's part (in concert with Great Britain or the South American republics). His position was unilateralist not isolationist. In 1820 he wrote to President Monroe:

> As to an American system, we have it; we constitute the whole of it; there is no commonality of interests or of principles between North and South America. Mr. Torres and Bolivar and O'Higgins talk about an American system as much as the Abbe Correa, but there is no basis for any such system (Quoted in Bemis, 1945: 367).

The unilateral declaration of what later became known as the Monroe Doctrine affirmed such a position, stating that the political system of the European powers is different from that of America. Therefore, the United States would "consider any attempt on their part to extend their system to any portion of this hemisphere as dangerous to our peace and safety". An "American hemisphere", of course, was an arbitrary social construct – for the United States can be located in many different hemispheres, depending on where one decides to center them (e.g. a Northern hemisphere, a so-called Western hemisphere or a predominantly land hemisphere: see Boggs, 1945). Such geopolitical reasoning was imaginary and the putative bonds of affinity between the Latin republics of South America and the white Anglo-Saxon republic of the North equally imaginary.

By the late nineteenth century, the increasing wealth and power of the US state, together with the scramble for colonies among the European powers, produced a foreign policy which subordinated the hemispheral identity of the United States to universalist themes and identities concerning race, civilization and Christianity. McKinley, acting under divine inspiration, saw it as the task of the United States to uplift and civilize the Philippines (while simultaneously preventing it from falling into the hands of commercial rivals France and Germany [Lafeber, 1963]) while Roosevelt's famous "corollary" of 1904 declared:

> Chronic wrongdoing or an impotence which results in a general loosening of the ties of civilized society, may in America, as elsewhere, ultimately require intervention by some civilized nation, and in the western Hemisphere the adherence of the United States to the Monroe Doctrine may force the United States, however reluctantly, in flagrant cases of wrongdoing or impotence, to the exercise of an international police power (Richardson, 1905, vol. IX: 7053).

The geopolitical reasoning by which domestic slavery and continental US expansionism worked – i.e. those concerning civilized versus uncivilized territories, superior and inferior races, adult and child identifications of peoples with white Anglo-Saxon males as the adults – were drawn upon to help write global political space. The United States was beginning to consider itself a "world power" with "principles" that were no longer qualified as contingently applicable to the "American hemisphere". McKinley and Theodore Roosevelt's racial script was followed by Woodrow Wilson's crusade for what he and US political culture took to be democracy. That Wilsonian internationalism did not succeed was partly due to the re-invigoration of the mythology that an isolationist "America" is the true and pure "America". Yet while the United States in the 1930s steered clear of political alliances with the rest of the world, its business enterprises continued their long-standing economic expansionism overseas. By the time of the Truman Doctrine, the US no longer conceptualized itself as *a* world power but as *the* world power. The geopolitical reasoning of Truman, as noted earlier, was abstract and universal. Containment had no clearly conceptualized geographical limitations. Its genuine space was the abstract universal isotropic plane wherein right does perpetual battle with wrong, liberty with totalitarianism and Americanism with the forces of un-Americanism.

A third feature of American discourse is the strong lines it draws between the space of the "Self" and the space of the "Other" (Todorov, 1984; Dalby, 1988, 1990a, 1990b). Like the cultural maps of many nations, American political discourse is given shape by a frontier which separates civilization from savagery in Turner's (1920) terms or an "Iron Curtain" marking the free world from the "evil empire". Robertson (1980: 92) notes:

> Frontiers and lines are powerful symbols for Americans. The moving frontier was never only a geographical line: it was a palpable barrier which separated the wilderness from civilization. It distinguished Americans, with their beliefs and their ideals, from savages and strangers, those "others" who could not be predicted or trusted. It divided

the American nation from other nations, and marked its independence.

While such a point is valid, one can overstate the uniquely American character of this practice. Early European experiences, particularly the Iberian reconquista against the "infidel" and the English colonial experience with "heathens" in Ireland, were factors in the formation of imperialism as a "way of life" in the United States (Meinig, 1986; Williams 1980). European discourses on colonialism, we have already noted, found their way into US foreign-policy practice not only in Theodore Roosevelt's time but even in determining the shape of the postwar world. The processes of geopolitical world ordering in US foreign policy in the late 1940s are worthy of some detailed examination. Taylor (1990) provides an account of the practical geopolitical reasoning of British intellectuals of statecraft (chiefly Churchill, Bevin and the British Foreign Office) during 1945. Let us consider the case of the two most famous American texts of that period, the "LongTelegram" and "Mr X" texts of George Kennan.

The figure of George Kennan looms large in the annals of American foreign policy for it was Kennan who helped codify and constitute central elements of what became Cold War discourse. Kennan himself was, as Stephanson (1989: 157) observes, a man of the North, one to whom the vast heterogeneous area of the Third World was "a foreign space, wholly lacking in allure and best left to its own no doubt tragic fate". The crucial division in the world for Kennan and the many others who made up the Atlanticist security community was that between the West and the East, between the world of maritime trading democracies and the Oriental world of xenophobic modern despotism. Trained at Princeton and in Germany and Estonia, Kennan developed something of an Old World *Weltanschauung* and brought this to bear in his early analyses of the USSR and world politics when working at the US Embassy in Moscow and later as Head of the Policy Planning Staff in Washington DC. In Kennan's two texts one can find at least three different strategies by which the USSR is represented. Each is worth exploring in detail.

The USSR as Oriental

Orientalism is premised, as Said (1978: 12) notes, on a primitive geopolitical awareness of the globe as composed of two unequal worlds, the Orient and the Occident. For Kennan and the Cold War discourse he helped codify, the USSR is part of the "Other" world, the Oriental world. In his famous "Long Telegram" Kennan describes the Soviet government as pervaded by an atmosphere of Oriental secretiveness and conspiracy. In the "Mr X" article published in *Foreign Affairs* in July 1947 he expounds on his thesis that the "political personality of Soviet power" is "the product of ideology and circumstances", the latter being the stamp of Russia's history and geography:

> The very teachings of Lenin himself require great caution and flexibility in the pursuit of communist purposes. Again, these precepts are fortified by the lessons of Russian history: of centuries of obscure battles between nomadic forces over the stretches of a vast unfortified plain. Here caution, circumspection, flexibility and deception are the valuable qualities, and their value finds natural appreciation in the Russian or oriental mind (Kennan, 1947: 574).

In an earlier passage, Kennan had noted the paranoia of Soviet leaders. "Their particular brand of fanaticism", he noted, "was too fierce and too jealous to envisage any permanent sharing of power". In a revealing sentence he then noted: "From the Russian–Asiatic world out of which they had emerged they carried with them a scepticism as to the possibilities of permanent and peaceful coexistence of rival forces" (Kennan, 1947: 570). Pietz (1988) notes that the Cold War discourse Kennan helped shape was "post-colonialist" in the sense that it drew upon and was assembled from many familiar and pervasive colonial discourses such as Orientalism and the putative primitiveness of non-Western regions and spaces. Totalitarianism, the theoretical anchor of Cold War discourse, came to be known as "nothing other than traditional Oriental despotism plus modern police technology" (Pietz, 1988: 58).[3]

The USSR as potential rapist

Another pre-existent source from which Cold War discourse and representations of the USSR were assembled was patriarchal mythology — particularly that concerning fables of female vulnerability, rape and guardianship. In the descriptions being constructed around the USSR and Communism at this time the image of penetration was frequently evoked.[4] The leaders of the USSR were a "frustrated" and "discontented" lot who "found in Marxist theory a highly convenient rationalization for their own instinctive desires" (Kennan, 1947: 569). Marxism was only a "fig leaf of moral and intellectual responsibility which cloaked essentially naked instinctive desires". These instinctive desires produced Soviet "aggressiveness" (another favorite Cold War description of the USSR) and "fluid and constant pressure to extend the limits of Russian police power which are together the natural and instinctive urges of Russian rulers" (Kennan, 1946: 54).

In the face of this instinctive behavior, the US needed to be aware that the USSR "cannot be charmed or talked out of existence" (Kennan, 1947: 576). The USSR was a wily and flexible power that would employ a variety of different "tactical maneuvers" (e.g. peaceful co-existence) to woo the West, particularly a vulnerable and psychologically-weakened Western Europe which was disposed to wishful thinking. Given this situation, the policy of the United States needed to be "that of a long-term, patient but firm and vigilant containment of Russian expansive tendencies" (Kennan, 1947: 575). The United States needed to act as the tough masculine guardian of Western Europe. If the policy of "adroit and vigilant application of counter-force at a series of constantly shifting geographical and political points, corresponding to the shifts and maneuvers of Soviet policy" was patiently followed by the United States, then the weaknesses of the Soviet Union itself would become apparent. Turning the sexual grid of intelligibility on the USSR itself, Kennan (1947: 578) wrote that as long as the deficiencies that characterize Soviet society are not corrected, "Russia will remain economically a vulnerable, and in a certain sense an impotent, nation, capable of exporting its enthusiasms and of radiating the strange charm of its primitive political vitality but unable to back up those articles of export by the real evidence of material power and prosperity". A testimony to the durability of this image is the rhetoric of the early Bush administration where Gorbachev's foreign policy was spoken of as a charm offensive aimed at the seduction of Western Europe.

The Red flood

In tandem with the patriarchal mythology described above, one also had the recurring representation of Soviet foreign policy and Communism as a flood. The image of the Red flood was a particularly powerful element in fascist mythology during the inter-war period where, as Theweleit (1987: 230) chronicles in Weimar Germany, the powerful metaphor "engenders a clearly ambivalent state of excitement. It is threatening but also attractive ...". Many different elements are at play here: situations and boundaries are fluid, solid ground becomes soft and swampy, barriers are breached, repressed instincts come bursting forth – water and sea as symbolic of the unconscious, the undisciplined id — and conditions are unrestrained, anarchic and dangerous. The response of the Freikorps, in Theweleit's account, is to act as firm, erect dams against this anarchic degeneration of society. With both feet securely planted on solid ground, they contained the Red flood and brought death to all that flowed. The very foundations of society, after all, were under attack. Switching to Kennan's Mr X article, we find the following graphic passage which defines the very nature of the Soviet threat to Western Europe:

> It's [the USSR's] political action is a fluid stream which moves constantly, wherever it is permitted to move, towards a given goal. Its main concern is to make sure that it has filled every nook and cranny available to it in the basin of world power. But if it finds unassailable barriers in its path, it accepts these philosophically and accommodates itself to them. The main thing is that there should always be pressure, unceasing constant pressure, towards the desired goal (Kennan, 1947: 575).

The image of the flood, which has also a sexual dimension (unrestrained, gushing desire, etc.), is critical, for it is by this means that the geography of containment becomes constituted. If the Soviet threat has the characteristics of a flood then one needs firm and vigilant containment along all of the Soviet border. Containment is thus constituted as a virtually global and not singularly Western European task. Effective containment in Western Europe, so the scenario goes, will lead to increasing Soviet pressure on the Middle East and Asia which eventually could result in the USSR spilling out into one or more of these regions. Such an image is easily reinforced by appropriate cartographic visuals featuring bleeding red maps of the USSR spreading outwards, or menacingly penatrating arrows busily trying to break out. The explanation of why US security managers instinctively read the North Korean invasion of South Korea as an act of Soviet expansionism certainly must address the power of such pre-existent images and scenarios. The formal geopolitical reasoning found in the different strategies of containment (Gaddis, 1982) rested, we suspect, on the flimsy foundations of widely shared practical geopolitical preconceptions.[5]

CONCLUSION

The Cold War as a discourse may have lost its credibility and meaning as a consequence of the events of 1989 but it is clear from the Gulf crisis [over the reflagging of Kuwaiti ships] that intellectuals of statecraft in the West at least, and the military-industrial complex behind them, will try to create a new set of enemies (the "irrational Third-World despot") in a re-structured world order. The reductive nature of the practical geopolitical reasoning used in the 1990–91 Gulf crisis by President Bush and Prime Minister Thatcher looks all too familiar. The character of foreign places and foreign enemies is represented as fixed. In 1947 when George Kennan declared that "there can be no appeal to common mental approaches" (1947: 574) in US dealings with the USSR he was effectively negating his own profession, namely diplomacy. The possibility of an open dialogue between the USSR and the United States was excluded a priori because the character of the USSR was already historically and geographically determined and thus effectively immutable. The irony of practical geopolitical representations of place is that in order to succeed they actually necessitate the abrogation of genuine geographical knowledge about the diversity and complexity of places as social entities. Describing the USSR then (or Iraq today) as Orientalist is a work of geographical abstractionism. A complex, diverse and heterogeneous social mosaic of places is hypostatized into a singular overdetermined and predictable actor. As a consequence therefore the United States was put in the ironic situation of being simultaneously tremendously geographically ignorant of the USSR (and today Iraq) yet fetishistically preoccupied with that state and its influence in world politics.

The global economic and political restructuring of the contemporary age has been both a consequence and a generator of changing geographical sensibilities. The marked "time–space compression" wrought by modern telecommunications and the globalization of capital, ideologies and culture has bound the fate of places more intimately together but has also opened up a series of possibilities for new types of subjectivities and new forms of political solidarity between places (Agnew and Corbridge, 1989). Globalization has enabled certain critical social movements to make connections between their struggles and the struggles of other critical social movements in very different places (see for example Kaldor and Falk, 1987; Walker, 1988). Contemporary geography, in deconstructing its own vocabulary and critically exploring the forms of practical geopolitical reasoning that circulate within states, can ally to these critical social movements. It can help create descriptions of the world based not on reductive geopolitical reasoning but on critical geographical knowledge.

REFERENCES

Agnew, J. A. (1983) An excess of "national exceptionalism": towards a new political geography of

American foreign policy. *Political Geography Quarterly* 2, 151–166.

Agnew, J. A. and Corbridge, S. (1989) The new geopolitics: the dynamics of geopolitical disorder. In *A World in Crisis? Geographical Perspectives* (R. J. Johnston and P. J. Taylor eds), pp. 266–288.

Alker, H. and Sylvan, D. (1986). Political discourse analysis. Paper presented at the Annual Meeting of the American Political Science Association, Washington DC, September.

Ashley, R. (1987) The geopolitics of geopolitical space: towards a critical social theory of international politics. *Alternatives* 14, 403–434.

Augelli, E. and Murphy, C. (1988) *America's Quest for Supremacy and the Third World: An Essay in Gramscian Analysis*. London: Pinter.

Bassin, M. (1987) Race contra space: the conflict between German *Geopolitik* and national socialism. *Political Geography Quarterly* 6, 115–134.

Baudrillard, J. (1988) *America*. New York: Verso.

Bemis, S. (1945) *John Quincy Adams and the Foundations of American Foreign Policy.* New York: Alfred A. Knopf.

Black, G. (1988) *The Good Neighbor: How the United States Wrote the History of Latin America*. New York: Pantheon.

Boggs, S. (1945) This hemisphere. *Department of State Bulletin* 6 May.

Cockburn, A. (1987) The defense intellectual: Edward N. Luttwak. *Grand Street* 6(3), 161–174.

Crabb, C. (1982) *The Doctrines of American Foreign Policy.* Baton Rouge: Louisiana State University Press.

Crocker, J. C. (1977) The social functions of rhetorical forms. In *The Social Use of Metaphor: Essays on the Anthropology of Rhetoric* (J. D. Sapir and J. C. Crocker eds), pp. 33–66. Philadelphia: University of Pennsylvania Press.

Dalby, S. (1988) Geopolitical discourse. The Soviet Union as Other. *Alternatives* 13, 415–422.

Dalby, S. (1990a) American security discourse: the persistence of geopolitics. *Political Geography Quarterly* 9, 171–188.

Dalby, S. (1990b) *Creating the Second Cold War: The Discourse of Politics*. London: Pinter.

Davis, M. (1986) *Prisoners of the American Dream*. London: Verso.

De Grazia, V. (1984–85) *Americanismo d'Esportazione*. Las Critica Sociologica 71–72, 5–22.

Der Derian, J. and Shapiro, M., Eds. (1989) *International/Intertextual Relations*. Lexington, Mass.: Lexington Books.

Dunn, R. S. (1972) *Sugar and Slaves: The Rise of the Planter Class in the English West Indies, 1634–1713*. New York: Norton.

Ertzold, T. and Gaddis, J. (1978) *Containment, Documents on American Foreign Policy and Strategy, 1945–1950*. New York: Columbia University Press.

Foucault, M. (1980) *Power/Knowledge*. New York: Pantheon.

Franck, T. M. and Weisband, E. (1971) *Word Politics: Verbal Strategy Among the Superpowers*. New York: Oxford University Press.

Gaddis, J. L. (1982) *Strategies of Containment*. New York: Oxford University Press.

Gates, H. L., Ed. (1985) "Race", writing and difference. *Critical Inquiry* 12 (1).

Gramsci. A. (1971) *Selections from the Prison Notebooks*. New York: International Publishers.

Gray, C. (1977) *The Geopolitics of the Nuclear Era.* Beverly Hills, CA: Sage Publications.

Gray, C. (1988) *The Geopolitics of Superpower.* Lexington: University of Kentucky Press.

Herzfeld, M. (1982) The etymology of excuses: aspects of rhetorical performance in Greece. *American Ethnologist* 9, 644–663.

Hunt, M. (1987) *Ideology and US Foreign Policy.* New Haven: Yale University Press.

Kaldor, M. (1990) After the Cold War. *New Left Review* 80, 25–37.

Kaldor, M. and Falk, R. (1987) *Dealignment: A New Foreign Policy Perspective*. New York: Basil Blackwell.

Karnow, S. (1989) *In Our Image: America's Empire in the Philippines*. New York: Random House.

Kennan, G. (1946) The "Long Telegram". In *Containment: Documents on American Foreign Policy and Strategy, 1945–1950* (T. Etzol and J. Gaddis eds), pp. 50–63. New York: Columbia University Press.

Kennan, G. ["Mr X"] (1947) The sources of Soviet conduct. *Foreign Affairs* 25, 566–582.

Kennan, G. (1967) *Memoirs 1925–1950*. Boston: Little Brown.

Kiernan, V. (1969) *The Lords of Human Kind.* Boston: Little Brown.

Kristoff, L. (1960) The origins and evolution of geopolitics. *Journal of Conflict Resolution* 4, 15–51.

Lafeber, W. (1963) *The New Empire: An Interpretation of American Expansionism, 1860–1898*. Ithaca, NY: Cornell University Press.

Luttwak, E. (1976) *The Grand Strategy of the Roman Empire*. Baltimore: John Hopkins Press.

MacDonell, D. (1986) *Theories of Discourse: An Introduction*. New York: Basil Blackwell.

Mackinder, H. (1890) The physical basis of political

geography. *The Scottish Geographical Magazine* 6, 78–84.

Meinig, D. (1986) *The Shaping of America. Volume 1: Atlantic America, 1492–1800*. New Haven: Yale University Press.

Ó Tuathail, G. (1989) Critical geopolitics: the social construction of place and space in the practice of statecraft. Unpublished PhD thesis, Syracuse University.

Paine, T. (1969) *The Essential Thomas Paine*. New York: Mentor.

Parker, G. (1985) *Western Geopolitical Thought in the Twentieth Century*. New York: St. Martin's Press.

Parkin, D. (1972) *The Cultural Definition of Political Response*, London: Academic Press.

Peitz, W. (1988). The "post-colonialism" of Cold War discourse. *Social Text* 19/20, 55–75.

Ratzel, F. (1969) The laws of the spatial growth of states. In *The Structure of Political Geography* (R. Kasperson and J. Minghi eds), pp. 17–29. Chicago: Aldine.

Reagan, R. (1988) Peace and Democracy for Nicaragua. Address to the nation on February 2, 1988. *Department of State Bulletin* 88(2133), 32–35.

Richardson, J. (1905). *A Compilation of Messages and Papers of the Presidents, 1789–1902*, 12 vols. Washington DC: Bureau of National Literature and Art.

Robertson, J. O. (1980) *American Myth, American Reality*. New York: Hill and Wang.

Said, E. (1978) *Orientalism*. New York: Vintage Books.

Said, E. (1981) *Covering Islam*. New York: Vintage Books.

Shapiro, M. (1988) *The Politics of Representation*. Madison: University of Wisconsin Press.

Sprout, H. and Sprout, M. (1960) Geography and international politics in an era of revolutionary change. *Journal of Conflict Resolution* 4, 145–161.

Stephanson, A. (1989) *George Kennan and the Art of Foreign Policy*. Boston: Harvard University Press.

Talbott, S. (1989) *The Master of the Game: Paul Nitze and the Nuclear Peace*. New York: Vintage Books.

Taylor, P. J. (1989) *Political Geography: World Economy, Nation-State and Locality*, 2nd edn. London: Longman.

Taylor, P. J. (1990) *Britain and the Cold War: 1945 as Geopolitical Transition*. London: Pinter.

Theweleit, K. (1987) *Male Fantasies. Volume 1: Women, Floods, Bodies, History*. Minneapolis: University of Minnesota Press.

Todorov, T. (1984). *The Conquest of America: The Question of the Other*, trans. by R. Howard. New York: Harper Torchbooks.

Turner, F. J. (1920) *The Frontier of American History*. New York: Henry Holt.

Turton, A. (1984) Limits of ideological domination and the formation of social consciousness. In *History and Peasant Consciousness in South-east Asia* (A. Turton and S. Tanabe eds), pp. 19–73. Osaka: National Museum of Ethnology, Senri Enthological Studies, no. 13.

Walker, R. (1988) *One World, Many Worlds: Struggles for a Just World Peace*. Boulder, CO: Lynne Rienner.

Whitaker, A. (1954) *The Western Hemisphere Idea: Its Rise and Decline*. Ithaca: Cornell University Press.

Williams, W. A. (1980) *Empire as a Way of Life*. Oxford: Oxford University Press.

Yannas, P. (1989) Containment disourse and the making of "Greece". Paper presented at the joint annual ISA and BISA Conference, London, March 28–April 1.

NOTES

1 In attempting to use Foucault and critical international-relations theories in political geography, there is a tendency to speak loosely of the "discourse of geopolitics" or "geopolitical discourse". Such phrases can be unhelpful, for they suggest that geopolitics is a dicrete discourse itself. This is not our contention. We prefer to use the term "geopolitical reasoning" to describe the spatialization of international politics that results from the employment of discourses in foreign-policy practice.

2 Jean Baudrillard (1988: 7) has termed America "the only remaining primitive society", a society of ferocious ritualism and hyperbolic primitivism that has "far outstripped its own moral, social or ecological rationale". For a discussion of the political and economic realities of living in American mythology, see Davis (1986).

3 Kennan's successor as Head of the Policy Planning Staff was Paul Nitze. In urging that the US develop the H-Bomb or "Super", as it was known in security discourse, Nitze argued that the "threat to Western Europe seemed to me singularly like that which Islam had posed centuries before, with its combination of ideological zeal and fighting power" (Nitze, quoted in Talbott, 1989: 52). The influence of a classical

education on intellectuals of statecraft (see Luttwak, 1976) with its narratives of fights between civilization and barbarian hordes, seems worthy of further exploration. Inquiry in this area may help explain the appeal of Mackinder's ideas to elements of the security community in this period.

4 In Volume One of his memoirs Kennan (1967), who by this time had supposedly repudiated many of his earlier conceptions of the USSR, nevertheless repeatedly returns to the image of penetration in discussions of Soviet power.

5 There are a series of other strategies by which the USSR is represented in the early Cold War discourse codified by Kennan and numerous others. The writing of territory and states in organic terms prompted a medicalization of certain regions (e.g. Western Europe as a weak patient needing aid against disease) and the use of psychological terms to describe the Other (e.g. the USSR as a paranoid personality). See Yannas (1989).

Cartoon 5 Ronald Reagan's speech icons (by Jack Ohman)
President Ronald Reagan was known as the "great communicator." His formulaic evocation of the myths of "national exceptionalism" (the flag, "founding fathers" and apple pie), his definition of America in terms of military strength (aircraft carriers, missiles and tanks) and enemies (Ayatollah Khomeini and the Soviet "evil empire"), and his "Christian values" and "morning in America" talk are all evident in this parody of his 1987 "State of the Union" address to Congress.
Source:

12 POLICY STATEMENT OF THE COMMITTEE ON THE PRESENT DANGER

"Common Sense and the Common Danger"

from *Alerting America: The Papers of the Committee on the Present Danger* (1984)

I

Our country is in a period of danger, and the danger is increasing. Unless decisive steps are taken to alert the nation, and to change the course of its policy, our economic and military capacity will become inadequate to assure peace with security.

The threats we face are more subtle and indirect than was once the case. As a result, awareness of danger has diminished in the United States, in the democratic countries with which we are naturally and necessarily allied, and in the developing world.

There is still time for effective action to ensure the security and prosperity of the nation in peace, through peaceful deterrence and concerted alliance diplomacy. A conscious effort of political will is needed to restore the strength and coherence of our foreign policy; to revive the solidarity of our alliances; to build constructive relations of cooperation with other nations whose interests parallel our own – and on that sound basis to seek reliable conditions of peace with the Soviet Union, rather than an illusory détente.

Only on such a footing can we and the other democratic industrialized nations, acting together, work with the developing nations to create a just and progressive world economy – the necessary condition of our own prosperity and that of the developing nations and Communist nations as well. In that framework, we shall be better able to promote human rights, and to help deal with the great and emerging problems of food, energy, population, and the environment.

II

The principal threat to our nation, to world peace, and to the cause of human freedom is the Soviet drive for dominance based upon an unparalleled military buildup.

The Soviet Union has not altered its long-held goal of a world dominated from a single center – Moscow. It continues, with notable persistence, to take advantage of every opportunity to expand its political and military influence throughout the world: in Europe; in the Middle East and Africa; in Asia; even in Latin America; in all the seas.

The scope and sophistication of the Soviet campaign have been increased in recent years, and its tempo quickened. It encourages every divisive tendency within and among the developed states and between the developed and the underdeveloped world. Simultaneously, the Soviet Union has been acquiring a network of positions including naval and air bases in the Southern Hemisphere which support its drive for dominance in the Middle East, the Indian Ocean, Africa, and the South Atlantic.

For more than a decade, the Soviet Union has been enlarging and improving both its strategic and its conventional military forces far more rapidly than the United States and its allies. Soviet military power and its rate of growth cannot be explained or justified by considerations of self-defense. The Soviet Union is consciously seeking what its spokesmen call "visible preponderance" for the Soviet sphere. Such preponderance, they explain, will permit the Soviet Union "to transform the conditions of world politics" and determine the direction of its development.

The process of Soviet expansion and the worldwide deployment of its military power threaten our interest in the political independence of our friends and allies, their and our fair access to raw materials, the freedom of the seas, and in avoiding a preponderance of adversary power.

These interests can be threatened not only by direct attack, but also by envelopment and indirect aggression. The defense of the Middle East, for example, is vital to the defense of Western Europe and Japan. In the Middle East the Soviet Union opposes those just settlements between Israel and its Arab neighbors which are critical to the future of the area. Similarly, we and much of the rest of the world are threatened by renewed coercion through a second round of Soviet-encouraged oil embargoes.

III

Soviet expansionism threatens to destroy the world balance of forces on which the survival of freedom depends. If we see the world as it is, and restore our will, our strength and our self-confidence, we shall find resources and friends enough to counter that threat. There is a crucial moral difference between the two superpowers in their character and objectives. The United States – imperfect as it is – is essential to the hopes of those countries which desire to develop their societies in their own ways, free of coercion.

To sustain an effective foreign policy, economic strength, military strength, and a commitment to leadership are essential. We must restore an allied defense posture capable of deterrence at each significant level and in those theaters vital to our interests. The goal of our strategic forces should be to prevent the use of, or the credible threat to use, strategic weapons in world politics; that of our conventional forces, to prevent other forms of aggression directed against our interests. Without a stable balance of forces in the world and policies of collective defense based upon it, no other objective of our foreign policy is attainable.

As a percentage of Gross National Product, US defense spending is lower than at any time in twenty-five years. For the United States to be free, secure and influential, higher levels of spending are now required for our ready land, sea, and air forces, our strategic deterrent, and, above all, the continuing modernization of those forces through research and development. The increased level of spending required is well within our means so long as we insist on all feasible efficiency in our defense spending. We must also expect our allies to bear their fair share of the burden of defense.

From a strong foundation, we can pursue a positive and confident diplomacy, addressed to the full array of our economic, political and social interests in world politics. It is only on this basis that we can expect successfully to negotiate hardheaded and verifiable agreements to control and reduce armaments.

If we continue to drift, we shall become second best to the Soviet Union in overall military strength; our alliances will weaken; our promising rapprochement with China could be reversed. Then we could find ourselves isolated in a hostile world, facing the unremitting pressures of Soviet policy backed by an overwhelming preponderance of power. Our national survival itself would be in peril, and we should face, one after another, bitter choices between war and acquiescence under pressure.

IV

We are Independents, Republicans and Democrats who share the belief that foreign and national security policies should be based upon fundamental considerations of the nation's future and well being, not that of one faction or party.

We have faith in the maturity, good sense and fortitude of our people. But public opinion must be informed before it can reach considered judgments and make them effective in our democratic system. Time, weariness, and the tragic experience of Vietnam have weakened the bipartisan consensus which sustained our foreign policy between 1940 and the mid-1960s. We must build a fresh consensus to expand the opportunities and diminish the dangers of a world in flux.

We have therefore established the Committee

on the Present Danger to help promote a better understanding of the main problems confronting our foreign policy, based on a disciplined effort to gather the facts and a sustained discussion of their significance for our national security and survival.

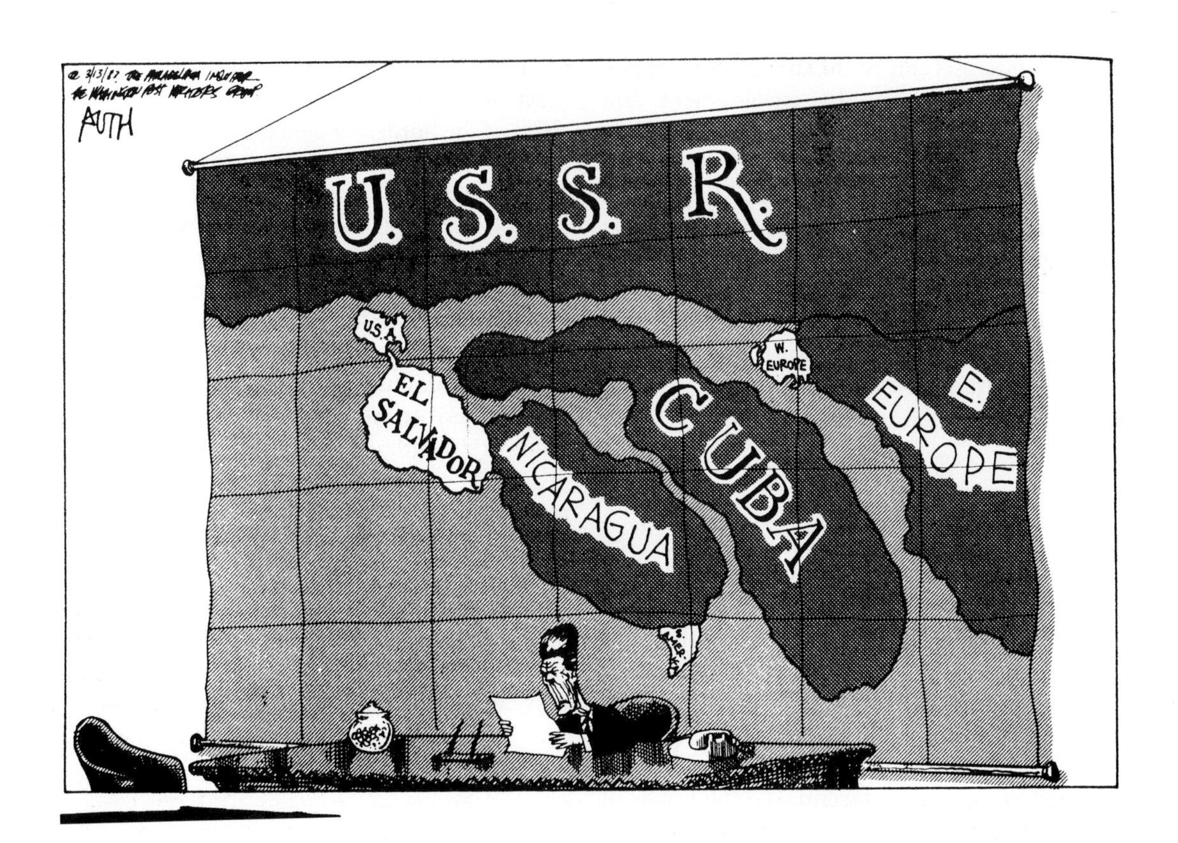

Cartoon 6 Ronald Reagan's mental map (by Auth)
Ronald Reagan's revival of the simple-minded rhetoric of the "Soviet threat" in the early 1980s justified his huge military budgets and wars of aggression in Central America against the Sandanistas and leftist guerrillas in El Salvador. His black and white, "us" versus "them" rhetoric and systematic exaggeration of the threat as all pervading is satirized here by Tony Auth.
Source: AUTH, Universal Press Syndicate

13 END COMMITTEE

"Appeal for European Nuclear Disarmament (END)"

from *Protest and Survive* (1980)

We are entering the most dangerous decade in human history. A third world war is not merely possible, but increasingly likely. Economic and social difficulties in advanced industrial countries, crisis, militarism and war in the third world compound the political tensions that fuel a demented arms race. In Europe, the main geographical stage for the East–West confrontation, new generations of ever more deadly nuclear weapons are appearing.

For at least twenty-five years, the forces of both the North Atlantic and the Warsaw alliance have each had sufficient nuclear weapons to annihilate their opponents, and at the same time to endanger the very basis of civilized life. But with each passing year, competition in nuclear armaments has multiplied their numbers, increasing the probability of some devastating accident or miscalculation.

As each side tries to prove its readiness to use nuclear weapons, in order to prevent their use by the other side, new, more "usable" nuclear weapons are designed and the idea of "limited" nuclear war is made to sound more and more plausible. So much so that this paradoxical process can logically only lead to the actual use of nuclear weapons.

Neither of the major powers is now in any moral position to influence smaller countries to forgo the acquisition of nuclear armament. The increasing spread of nuclear reactors and the growth of the industry that installs them, reinforce the likelihood of worldwide proliferation of nuclear weapons, thereby multiplying the risks of nuclear exchanges.

Over the years, public opinion has pressed for nuclear disarmament and detente between the contending military blocs. This pressure has failed. An increasing proportion of world resources is expended on weapons, even though mutual extermination is already amply guaranteed. This economic burden, in both East and West, contributes to growing social and political strain, setting in motion a vicious circle in which the arms race feeds upon the instability of the world economy and vice versa: a deathly dialectic.

We are now in great danger. Generations have been born beneath the shadow of nuclear war, and have become habituated to the threat. Concern has given way to apathy. Meanwhile, in a world living always under menace, fear extends through both halves of the European continent. The powers of the military and of internal security forces are enlarged, limitations are placed upon free exchanges of ideas and between persons, and civil rights of independent minded individuals are threatened, in the West as well as the East.

We do not wish to apportion guilt between the political and military leaders of East and West. Guilt lies squarely upon both parties. Both parties have adopted menacing postures and committed aggressive actions in different parts of the world.

The remedy lies in our own hands. We must act together to free the entire territory of Europe, from Poland to Portugal, from nuclear weapons, air and submarine bases, and from all institutions engaged in research into or manufacture of nuclear weapons. We ask the two super powers to withdraw all nuclear weapons from European territory. In particular, we ask the Soviet Union to halt production of SS 20 medium-range missiles and we ask the United States not to implement the decision to develop cruise missiles and Pershing I missiles for deployment in Westem Europe. We also urge the ratification of the SALT II agreement, as a necessary step towards the renewal of effective negotiations on general and complete disarmament.

At the same time, we must defend and extend the right of all citizens, East or West, to take part in this common movement and to engage in every kind of exchange. We appeal to our friends in Europe, of every faith and persuasion, to consider urgently the ways in which we can work together for these common objectives. We envisage a European-wide campaign, in which every kind of exchange takes place; in which representatives of different nations and opinions confer and co-ordinate their activities; and in which less formal exchanges, between universities, churches, women's organizations, trade unions, youth organizations, professional groups and individuals, take place with the object of promoting a common object: to free all of Europe from nuclear weapons.

We must commence to act as if a united, neutral and pacific Europe already exists. We must learn to be loyal, not to "East" or "West", but to each other, and we must disregard the prohibitions and limitations imposed by any national state.

It will be the responsibility of the people of each nation to agitate for the expulsion of nuclear weapons and bases from European soil and territorial waters, and to decide upon its own means and strategy, concerning its own territory. These will differ from one country to another, and we do not suggest that any single strategy should be imposed. But this must be part of a trans-continental movement in which every kind of exchange takes place.

We must resist any attempt by the statesmen of East or West to manipulate this movement to their own advantage. We offer no advantage to either NATO or the Warsaw Alliance. Our objectives must be to free Europe from confrontation, to enforce detente between the United States and the Soviet Union, and, ultimately, to dissolve both great power alliances.

In appealing to fellow-Europeans, we are not turning our backs on the world. In working for the peace of Europe we are working for the peace of the world. Twice in this century Europe has disgraced its claims to civilization by engendering world war. This time we must repay our debts to the world by engendering peace.

This appeal will achieve nothing if it is not supported by determined and inventive action, to win more people to support it. We need to mount an irresistible pressure for a Europe free of nuclear weapons.

We do not wish to impose any uniformity on the movement nor to pre-empt the consultations and decisions of those many organizations already exercising their influence for disarmament and peace. But the situation is urgent. The dangers steadily advance. We invite your support for this common objective, and we shall welcome both your help and advice.

14 MIKHAIL GORBACHEV

"New Political Thinking"

from *Perestroika: New Thinking for Our Country and the World* (1988)

The fundamental principle of the new political outlook is very simple: *nuclear war cannot be a means of achieving political, economic, ideological or any other goals*. This conclusion is truly revolutionary, for it means discarding the traditional notions of war and peace. It is the political function of war that has always been a justification for war, a "rational" explanation. Nuclear war is senseless; it is irrational. There would be neither winners nor losers in a global nuclear conflict: world civilization would inevitably perish. It is a suicide, rather than a war in the conventional sense of the word.

But military technology has developed to such an extent that even a non-nuclear war would now be comparable with a nuclear war in its destructive effect. That is why it is logical to include in our category of nuclear wars this "variant" of an armed clash between major powers as well.

Thereby, an altogether different situation has emerged. A way of thinking and a way of acting, based on the use of force in world politics, have formed over centuries, even millennia. It seems they have taken root as something unshakable. Today, they have lost all reasonable grounds. Clausewitz's dictum that war is the continuation of policy only by different means, which was classical in his time has grown hopelessly out of date. It now belongs to the libraries. For the first time in history, basing international politics on moral and ethical norms that are common to all humankind, as well as humanizing interstate relations, has become a vital requirement.

A new dialectic of strength and security follows from the impossibility of a military – that is, nuclear – solution to international differences. Security can no longer be assured by military means – neither by the use of arms or deterrence, nor by continued perfection of the "sword" and the "shield." Attempts to achieve military superiority are preposterous. Now such attempts are being made in space. It is an astonishing anachronism which persists due to the inflated role played by militarists in politics. From the security point of view the arms race has become an absurdity because its very logic leads to the destabilization of international relations and eventually to a nuclear conflict. Diverting huge resources from other priorities, the arms race is lowering the level of security, impairing it. It is in itself an enemy of peace. The only way to security is through political decisions and disarmament. In our age genuine and equal security can be guaranteed by constantly lowering the level of the strategic balance from which nuclear and other weapons of mass destruction should be completely eliminated.

Perhaps this frightens some people. "What is to be done with the military-industrial complex then?" they ask. The jobs and wages of so many people are involved. This issue was specially analyzed in one of the most recent works of Nobel Prize laureate V. Leontyev, and he has proved that the militarists' arguments do not hold water, from an economic standpoint. This is what I think: to begin with, each job in the military-industrial complex costs two or three times more than one in a civilian industry. Three jobs could be created instead. Secondly, even today sectors of the military economy are connected with the civilian economy, doing much for the latter. So, this is a starting point for utilizing their possibilities for peaceful purposes. Thirdly, the USSR and the USA could come up with large joint programs, pooling our resources and our scientific and intellectual potentials in order to solve the most diverse problems for the benefit of humankind.

The new political outlook calls for the recognition of one more simple axiom: security is indivisible. It is either equal security for all or none at all. The only solid foundation for security is the recognition of the interests of all peoples and countries and of their equality in international affairs. The security of each nation should be coupled with the security for all members of the world community. Would it, for instance, be in the interest of the United States if the Soviet Union found itself in a situation whereby it considered it had less security than the USA? Or would we benefit by a reverse situation? I can say firmly that we would not like this. So, adversaries must become partners and start looking jointly for a way to achieve universal security.

We can see the first signs of new thinking in many countries, in different strata of society. And this is only natural, because it is the way of mutually advantageous agreements and reciprocal compromises on the basis of the supreme common interest – preventing a nuclear catastrophe. Consequently, there should be no striving for security for oneself at the expense of others.

The new outlooks influence equally strongly the character of military doctrines. Those should be strictly the doctrines of defense. And this is connected with such new or comparatively new notions as the reasonable sufficiency of armaments, non-aggressive defense, the elimination of imbalance and asymmetries in various types of armed forces, separation of the offensive forces of the two blocs, and so on and so forth.

Universal security in our time rests on the recognition of the right of every nation to choose its own path of social development, on the renunciation of interference in the domestic affairs of other states, on respect for others in combination with an objective self-critical view of one's own society. A nation may choose either capitalism or socialism. This is its sovereign right. Nations cannot and should not pattern their life either after the United States or the Soviet Union. Hence, political positions should be devoid of ideological intolerance.

Ideological differences should not be transferred to the sphere of interstate relations, nor should foreign policy be subordinate to them, for ideologies may be poles apart, whereas the interest of survival and prevention of war stand universal and supreme. On a par with the nuclear threat, the new political mode of thinking considers the solution of other global problems, including those of economic development and ecology, as an indispensable condition for assuring a lasting and just peace. To think in a new way also means to see a direct link between disarmament and development.

We stand for the internationalization of the efforts to turn disarmament into a factor of development. In a message to the International Conference on this subject in New York in late August 1987, I wrote: "The implementation of the basic principle 'disarmament for development' can and must rally mankind, and facilitate the formation of a global consciousness." The Delhi Declaration on Principles for a Nuclear-Weapon-Free and Non-Violent World, which was signed by Prime Minister Rajiv Gandhi of the Republic of India and myself in November 1986, contains words which I'd like to cite here as well:

> In the nuclear age, humanity must evolve a new mode of political thought, a new concept of the world that would provide reliable guarantees for humanity's survival. People want to live in a safer and a more just world. Humanity deserves a better fate than being a hostage to nuclear terror and despair. It is necessary to change the existing world situation and to build a nuclear-weapon-free world, free of violence and hatred, fear and suspicion.

There are serious signs that the new way of thinking is taking shape, that people are coming to understand what brink the world has approached. But this process is a very difficult one. And the most difficult thing is to ensure that this understanding is reflected in the actions of the policy-makers, in their minds. But I believe that the new political mentality will force its way through, for it was born of the realities of our time.

Cartoon 7 Cold Warriors anonymous (by Wuerker)
The "addiction" of both the superpowers to geopolitical interventionism during the Cold War is satirized by Matt Wuerker. Here he represents Gorbachev's "new political thinking" as an admission of guilt to a councilor. President Bush, however, is represented as still "in denial" about the US's addiction to geopolitical interventionism.
Source: M. Wuerker

PART 3

New World Order Geopolitics

PART 3

INTRODUCTION

Gearóid Ó Tuathail

In 1988, the Committee on the Present Danger geopolitician Colin Gray began his book *The Geopolitics of Superpower* with the following declaration:

> For as far into the future as can be claimed contemporarily relevant, the Soviet Union is going to remain *the* source of danger – narrowly to American national security, more broadly (and quite literally) to the exercise of the values of Western civilization (1988: 1–2, emphasis his).

One year later, the Berlin Wall was torn down by enthusiastic Germans while the velvet and violent revolutions of 1989 overthrew the communist dictatorships of Eastern Europe. Three years later, the Soviet Union itself collapsed as an imperial structure. Defined as an antagonistic relationship of competition between two superpowers, the Cold War was at an end as one of the competing superpowers collapsed under the weight of its own contradictions.

The Cold War, however (as argued in the previous section), was never simply an antagonistic relationship but a system of geopolitical control with an elaborate complex of state institutions, military forces, economic interests, political coalitions, cultural values, and intellectuals of statecraft on each side. While the Soviet complex began to disintegrate, the Western complex of ideology, institutions and intellectuals remained coherent and in place. Describing the end of the Cold War as akin to a race where two horses were running around a track, the historian Bruce Cumings (1991) remarked that one (the USSR/Warsaw Pact) broke its leg while the other (the USA/NATO) kept on running regardless.

The evident disintegration of the Soviet Empire, nevertheless, provoked a serious crisis of meaning in world politics, for the Cold War geopolitical narratives that had mapped out global spatial strategy for intellectuals and practitioners of geopolitics since 1947 were no longer useful or credible. For the institutions and intellectuals of the Cold War national security state, the end of the Cold War was experienced as a condition of geopolitical vertigo, a state of confusion where the old nostrums of the Cold War were redundant and new ones had not yet been invented, issued and approved.

The need to definitively map this condition of chaos, confusion and geopolitical flux was an institutional imperative for the Cold War bureaucracies of the West for, with the collapse of the Soviet Union, their official geopolitical *raison d'être* disappeared. Once unquestioned necessities, the Pentagon, Central Intelligence Agency and plethora of geopolitical think-tanks and other military-industrial-knowledge institutions that had handsomely lived off the Cold War for so long, were now suddenly exposed as bloated bureaucracies built on exaggeration and hyperbole. Most damning of all was that none had predicted the collapse of communism, the implosion of the Soviet Union and the demise of the Cold War. All had aggrandized themselves for years by peddling simpleminded discourses of danger about the Soviet Union. Now the collapse of the Soviet Empire had revealed not only the bankruptcy of communism but the intellectual bankruptcy of the West's own institutions of geopolitical expertise. Cold War geopoliticians like Colin Gray were revealed as "experts" (like the famous emperor) without clothes. Institutional complexes of power/

knowledge like the Central Intelligence Agency and disciplinary complexes of power/knowledge like Sovietology (the study of the Soviet Union) and security studies (the study of the national security needs of the West) were so blinkered by Cold War ideology that they missed the actual collapse of the Soviet Empire.

As one might expect, however, the intellectuals and institutions associated with the national security state were slow to acknowledge their intelligence failures. Rather they moved swiftly to re-legitimate themselves by making the very formlessness of the post-Cold War world order a threat. Led by a president in George Bush who had spent his professional life fighting the Cold War (he was a former Director of the CIA among other things), the US national security state painfully acknowledged the end of the Cold War and found new legitimation in the disorder of the post-Cold War geopolitical landscape. Counciling against the dangers of "euphoria," the Bush administration pronounced "uncertainty," "unpredictability," "instability" and "chaos" as the new threats (Ó Tuathail, 1992). In responding to the Iraqi invasion of the tiny state of Kuwait, it found the reason and rationale to define a "new world order" with America at its center with a "unique responsibility" to do the "hard work" of bringing freedom to the rest of the globe (Reading 18).

The national exceptionalism and triumphalism that characterized President Bush's declaration of a "new world order" during the Gulf War of 1990–91 was already evident as early as the summer of 1989 in US strategic and political culture. A striking example is Francis Fukuyama's essay "The End of History?" (Reading 15) and the admiration it received from many in the Western media. Fukuyama worked at the time as deputy director of policy planning in the Bush administration, having previously worked in the Reagan administration and at the Rand Corporation, a Cold War think-tank based in Santa Monica, California, that was home to conservative intellectuals of statecraft specializing in the technostrategy of nuclear war. While Fukuyama's position recalled the memory of George Kennan, his essay received attention because of its timeliness, sensationalist thesis (making it appealing to the media) and the neoconservative network of friends and contacts around the journal *The National Interest*, its original place of publication. Funded by the Harry Bradley Foundation, a prominent neoconservative organization, the John M. Olin Foundation, a wealthy manufacturer who made his fortune largely in munitions, and the Smith Richardson Foundation, yet another organization committed to funding neoconservative journals and intellectuals, the journal *The National Interest* was first established by Irving Kristol, the dean of American neoconservatives, in 1985 as a publishing outlet for intellectuals of statecraft on the right of the political spectrum (Atlas, 1989). Fukuyama's essay was first presented in the summer of 1989 issue as a major philosophical statement with only prominent neoconservative intellectuals responding to its claims.

Fukuyama's provocative thesis is, as the critic Christopher Hitchens described it, "self-congratulation raised to the status of philosophy" (Atlas, 1989: 42). Drawing upon the writings of the early nineteenth-century German philosopher Georg Wilheim Friedrich Hegel as interpreted in the 1930s by the Paris-based Russian emigré philosopher Alexandre Kojève, Fukuyama claims that we are now witnessing the end of History (capital H) as a struggle over ideas and principles. Claiming that Hegel had proclaimed the end of History in 1806 with the triumph of Napoleon (over the Prussian monarchy at the Battle of Jena) and the universal principles of the French revolution (claims contested by Hegel scholars), Fukuyama uses Kojève's argument that Western Europe and the United States represent the "universal homogeneous state" that Hegel first identified in the Napoleonic state to assert that the West is the culmination of all historical progress. In Fukuyama's vulgarization of Hegel's more complex and subtle arguments, the "end of History" is that point at which humanity has actualized the universal truths first expressed in the French revolution, the "principles of liberty and equality." History does not literally end because most states are still struggling to reach and actualize these universal truths. However, a few vanguard states have reached and actualized these universal truths. Unlike most states, the "universal homogeneous state" has reached the pinnacle of historical evolution. It is "homogeneous" because "all prior contradictions" (like geopolitical or class divisions, for example) "are resolved and all human needs are satisfied. There is no struggle or conflict over 'large' issues, and consequently no need for generals and statesmen; what remains is primarily economic activity."

In making this sweeping argument, Fukuyama ignores some inconvenient details and simplifies others. That the Napoleonic state, a dictatorship and empire after all, represented the "principles of liberty and equality" is highly questionable. That the principles of "liberty and equality" necessarily translate into Western-style political democracy and capitalist neoliberal economics is also questionable. Fukuyama, however, uses these general principles as idealized self-understandings to justify the status quo. The actual existing political democracy and capitalist neoliberalism of Western Europe and North America represent for him "the vanguard of civilization," the culmination of human history.

Fukuyama later elaborated his thesis in greater detail in a subsequent book (Fukuyama, 1992). Many critics have challenged his use of both Hegel and Kojève, pointing out that Fukuyama has them making arguments that they do not actually make (Ryan, 1992). For our purposes, Fukuyama's article is important because it is an early neoconservative attempt to re-make Cold War discourse in the light of the imminent collapse of communism in Europe. Like geopoliticians in the past, Fukuyama adopts a divine view of the globe and pronounces from upon high about the meaning of world politics. In keeping with the imperial hubris found in many of the earlier texts we have examined, Fukuyama considers his own state and the West more generally as the consummation of history, the fulfillment of human historical destiny. All other states are supposedly struggling to attain the pinnacle of perfection the West has reached. The West, in short, is best. Because it had reached the "end of History" the West is "post-historical," whereas the rest of the world is still struggling in the "historical."

This conceptual divide between the West and the rest, the post-historical, and the historical is how Fukuyama re-maps the Cold War's First, Second and Third Worlds. In Fukuyama's conceptual map the so-called "Third World" is largely irrelevant: "For our purposes, it matters very little what strange thoughts occur to people in Albania or Burkino Faso, for we are interested in what one could in some sense call the common ideological heritage of mankind." The Second World of communism is where Fukuyama detects major ideological transformation. Despite being a nominally Marxist state, Fukuyama claims that China can no longer "act as a beacon of illiberal forces around the world." The Soviet Union is also no longer an alternative to liberalism for Gorbachev's "democratization and decentralization principles are highly subversive of the fundamental precepts of both Marxism and Leninism." The ideological transformations in both states lead Fukuyama to proclaim the passing of Marxism–Leninism as a "living ideology of world historical significance."

It is important to note that Fukuyama's re-working of the Western Cold War discourse of capitalism versus communism, the free world versus totalitarianism, into a new conceptual map characterized by a divide between the West versus the rest, the post-historical versus the historical, maintains the former discourse's antipathy for geography. Just as Cold War geopolitics displaced geographical specificity with categorical universals, so also does Fukuyama's vulgarized Hegelian schema. Places across the globe are read not in terms of their geographical particularity but in terms of sweeping, abstract and universal Western philosophical categories. Once again, geographical uniqueness is overridden by idealized universals, this time the divide between the "historical" and the "post-historical." The end of History thesis certainly does not mark the beginning of geography in geopolitical discourse.

As a map of meaning designed to make sense of world politics in the early 1990s, Fukuyama's scheme is flawed in two significant ways. First, it is a remarkably ethnocentric schema which fails to acknowledge the serious problems –

what Fukuyama after Hegel would call "contradictions" – that beset Western states. This triumphalist complacency is a function both of the imprecise concepts Fukuyama uses – principally the notions of the "universal homogeneous state" and "liberalism" which are self-idealized and imaginary concepts rather than historical ones – and those he does not use – principally capitalism but also militarism, racism and patriarchy. The "universal homogeneous state" in Fukuyama stretches historically from Napoleon to NATO, including along the way such radically different states like the United States, South Korea, Japan and even, after agricultural reform, China! To categorize certain states as "liberal" does not really tell us very much about the specific geographical structure of states and the contradictions of their particular historical versions of liberalism, its compromise with nationalism, militarism and patriarchal white supremacy in different states. Western states are far from being universally liberal or homogeneous.

In ignoring capitalism and suggesting a receding of the class issue, Fukuyama is ignoring the turbulent "creative destruction" wrought by the globalization of capitalism in recent decades and the marked increase in income inequality across the West, particularly in the United Kingdom and the United States. To assume that the Western modernity represented by the European Union and the United States represents a stable culmination of civilization bereft of serious "contradictions" (like class, race, identity, environment and globalization issues) is dangerously naive. The claim that "the egalitarianism of modern America represents the essential achievement of the classless society envisioned by Marx" (part of his dismissal of the questions of class and race) reveals Fukuyama's preference for idealized self-images and ideological self-deceptions rather than empirical evidence and concrete historical realities. Such claims, of course, are perfect legitimations for the privileges of the already powerful.

Second, Fukuyama's assumption that the declining appeal of Marxist–Leninist ideology and the supposed spread of the liberalism of the "universal homogeneous state" (what he calls "post-historical consciousness") will lead to the receding of international conflict between states and the growing "Common Marketization" of world politics is unduly optimistic. Again, the problem is that Fukuyama's abstract philosophical narrative leads him to sweeping conclusions that elide the messy and complex territoriality of world politics. While Marxist–Leninist ideology may be in decline, anti-Western and anti-capitalist ideologies are far from dead. Furthermore, these are likely to flourish as the dislocations caused by capitalist globalization are contested and resisted by more and more political movements (as is currently happening in places from Eastern Europe and Russia to Egypt, India and Saudi Arabia). Finally, one need not fall back on essentialist realist or neorealist assumptions about the interstate system to argue that states are complex entities motivated by a variety of forces and ideologies which sometimes, indeed often, lead them into conflict with each other. Fukuyama's assumption that international life for those who have reached the end of history (Western Europe and North America) is far more preoccupied with economics than with politics and strategy echoes the common fallacious assumption that capitalist democratic states are pacific and not bellicose. This remarkable statement ignores the whole history of Cold War militarism and the fact that Western Europe was one of the most militarized places on the planet by the late 1980s. The contrast Fukuyama offers is a false one, for preoccupation with economics can also be preoccupation with politics, strategy and military power, a fact that was underscored by the US's intervention to protect the West's supply of oil after the Iraqi invasion of Kuwait in August 1990.

Not all establishment readings of the potential post-Cold War world order were as triumphalist as Fukuyama's 1989 essay. As the Cold War antagonism between the superpowers wound down, the sense of Western "victory" was tempered for many by an awareness of the limits and indeed the economic cost – in deficit spending and imperial overstretch – of that victory. To certain economic nationalists in the United States, if the Cold War had a winner it was neither the Soviet Union nor the United States but Japan, a resurgent geo-economic power by the early 1990s that many claimed would eventually overtake the United States and become the world's "number one" (Vogel, 1979). Echoing Fukuyama's identification of the salience of economics in the new world order, yet disagreeing with his reasoning and conclusions, Edward Luttwak argued a year later in *The National Interest* that the waning of the Cold War marked a shift from geopolitics to geo-economics in world politics. Generalizing in a somewhat glib manner, he claimed that "[e]veryone, it appears, now agrees that the methods of commerce are displacing military methods" in world politics.

While not so simple as Luttwak suggests, concern with the economic basis of political and military strength was particularly pronounced after the collapse of the Soviet Union as America's official enemy, the Eastern Other that had helped define the Western Self for over four decades. As a consequence of both the globalization of the American economy and the Reagan administration's military buildup of the 1980s, the United States in the early 1990s had record budget and trade deficits. America's largest trade deficit was with Japan and, for some at least, this geo-economic superpower with a huge trade surplus was America's new enemy, a resourceful and inscrutable Eastern Other that revealed an economically enfeebled Western Self. Rather than adopt the traditional Cold War conception of Japan as a Western power "just like us," these geo-economic intellectuals argued for a "revisionist" conception of Japan that recognized how the Japanese state was different from Western states. Some revisionist intellectuals talked about the "economic colonization" of the United States by Japanese transnationals (Frantz and Collins, 1989) while others, like the journalist James Fallows, echoed George Kennan in calling for the "containment" of Japan (Fallows, 1989). These arguments registered themselves in political and popular culture at the time in books like *The Coming War With Japan* (Friedman and Lebard, 1991), *In the Shadow of the Rising Sun* (Dietrich, 1991) and novels (and later a Hollywood movie) like Michael Crichton's *Rising Sun*.

Things, however, were not as simple as Luttwak's argument about a transition from geopolitics to geo-economics suggests. A quintessential neoconservative Cold War intellectual with strong ties to defense contractors and strategic studies institutes, Luttwak seemed an unlikely convert to arguments that, for some more liberal geo-economic intellectuals at least (like James Fallows), required the United States to re-think its massive military spending policies. Yet, upon closer examination, Luttwak's reasoning is merely an extension of the essentialist realist assumptions that had underpinned and legitimated Cold War militarism. Essentialist because they posit an absolute truth about states without regard to history, these realist assumptions held, as Luttwak put it, that states as

> spatial entities structured to jealously delimit their own territories, to assert their exclusive control within them, and variously to attempt to influence events beyond their borders ... are inherently inclined to strive for relative advantage against like entities on the international scene, even if only by means other than force.

As bureaucracies, states are, Luttwak claimed, "impelled by the bureaucratic urges of role-preservation and role-enhancement to acquire a 'geo-economic' substitute for their decaying geopolitical role." Conflict between states, as a consequence, is inevitable, though with the waning of the Cold War this conflict is more and more likely to be geo-economic rather than geopolitical in nature.

Though appealing in its simplicity, Luttwak's thesis, which he subsequently expanded into a book (Luttwak, 1993), is flawed both by its conceptualization of a transition from geopolitics to geo-economics and in its reliance on ahistorical and unjustified realist assumptions about the nature of states. Like Fukuyama's earlier opposition between economics and politics/strategy, Luttwak's opposition between geopolitics and geo-economics mischaracterizes a more complex reality. For a start, Cold War geopolitics was also about geo-economics, the policy of Cold War militarism being closely associated with an international Pax Americana and the power of a domestic military-industrial complex (Cox, 1987; Sherry,

1995). The end of the Cold War did not mark the end of geopolitics per se, merely the end of Cold War geopolitics (except in US foriegn policy towards Cuba). The globalization of the US economy and the increasing power of transnational corporations were not something opposed by political leaders of the US state but actively encouraged by them. Both the Reagan and Bush administrations (and indeed the subsequent Clinton administrations) subscribed to the discourse of transnational liberalism (neoliberalism), a belief in the extension of the principles of free trade and deregulation worldwide (Agnew and Corbridge, 1995). Globalization, for them, was a good and positive development, a sum-sum game where everyone wins. The hegemony of transnational liberalism in the US state belies the ahistorical argument of Luttwak about the zero-sum mentality of states. While states are impelled by bureaucratic urges of role-preservation and role-enhancement, this does not "naturally" mean that they will therefore act geo-economically. In any case, in a world of deterritorializing economies where "who is us," as Robert Reich (1991) notes, is the apposite question (i.e. the nominal nationality of corporations has been exceeded by their functional transnationality), the meaning of acting geo-economically is unclear. Geopolitics and geo-economics are not opposites but concepts entwined in each other (as the US–Japan debate over a new jet fighter, the FSX, revealed; Ó Tuathail, 1992).

The event that provoked the official attempt to delimit a new world order was, as we have already noted, the Iraqi invasion of Kuwait. Cleansed of its associations earlier in the twentieth century with Adolf Hitler's plans for re-making the map of world politics, the idea of a new world order was championed first by Soviet President Mikhail Gorbachev who envisioned it as a world order beyond Soviet–American antagonism that would be characterized by interstate cooperation through the United Nations to address transnational threats to the planet, like environmental degradation and the proliferation of nuclear weapons. Meeting with Gorbachev in Helsinki to assure the Soviet leader's cooperation in his plan to oust Iraq's military from Kuwait, Bush appropriated the idea to conceptualize a new era "free from the threat of terror, stronger in the pursuit of justice, and more secure in the quest for peace, an era in which the nations of the world, East and West, North and South, can prosper and live in harmony" (Reading 17). In practical terms, the new world order for Bush was a world where the United States, in alliance with those who were willing to follow, did the ordering. Any change in the status quo geopolitical order unfavorable to the United States and the interests of "the West," such as the Iraqi invasion of Kuwait, was considered unlawful aggression that "would not stand." By contrast, any change in the existing geopolitical order initiated by the United States, such as the US military invasion of Panama, or favorable to it, such as the collapse of the Soviet Union, was acceptable and necessary. Central to Bush's vision of a new world order were the myths of American national exceptionalism. Echoing the imperial hubris of old, Bush believed that America's interests were universal interests, the interests and aspirations of everyone. America was humanity's best hope. In his 1991 State of the Union speech (Reading 18), Bush declared that "the hopes of humanity turn to us. We are Americans. We have a unique responsibility to do the hard work of freedom." Given this re-cycling of longstanding nationalist exceptionalist themes and from the evidence of the Gulf War – a war fought to guarantee Western access to cheap petroleum and restore an antidemocratic but pro-Western monarchy – Bush's new world order sounded a lot like the old Cold War world order. It was different, however, for this time America was the sole remaining superpower, a power without the check of any serious rival superpower.

A critical perspective on the meaning of the Gulf War is provided by Timothy Luke in his essay "The Discipline of Security Studies and the Codes of Containment: Learning from Kuwait" (Reading 19). By "codes" Luke means code words like "communism" and "totalitarianism," the strategic code words of the discourse of Cold War geopolitics. Luke makes the important argument that though classic Cold War discourse is fading,

> Cold War-style reasoning continues to dominate US strategic thought inasmuch as the premise of containment, directed against any threatening evil otherness now rather than simply communism, and balance-of-power politics, tied to the correlation of forces in particular regional competitions for primacy, underpins Washington's response to foreign crises (1991: 317).

In a much more sophisticated way than Luttwak, Luke describes how economic, cultural and political globalization, and the move to a more informational and transnational form of corporate capitalism, is transforming such traditional anchoring principles of world politics as state sovereignty, territorial integrity and place-bound community. Power, Luke notes, is no longer bound to place but "also often flows more placelessly beneath, behind, between, and beyond boundaries set into space as new senses of artificial location become very fluid or mobile, defined by shifting connections into the networks of information carrying these flows" (1991: 319). Luke is not suggesting a simple transition from geopolitics to chronopolitics, from traditional spatial sovereignty to the pace and speed of informational exchange. Rather, both space and speed become an entwined hybrid – (s)pace – as the territoriality of the state system is overlain with the networks of global telecommunications and the global webs of transnational corporate capitalism (Castells, 1996). In the condition of postmodernity, the "real" becomes "hyperreal," or more real than the real itself, as it is informationalized and televisualized.

Luke argues that the story of the Gulf War is a uniquely suggestive and revealing instance of the tendencies he identifies and describes. Though displaced from their territory, the Kuwaiti royal family was able to protect most of their wealth from the invading Iraqis (for it was a portfolio of non-Kuwaiti based transnational investments, what Luke calls "hyperreal estate") and wage an "air war" or informational war on the airwaves of the Western media (with the aid of a well-connected Washington public relations firm) to persuade the Western powers, principally the United States, to go to war for them so they could return as the pro-Western but anti-democratic rulers of Kuwait (MacArthur, 1992).

Luke also demonstrates how President George Bush conducted an "air war" of his own in the Western media, hyping the Kuwaiti invasion – a place most Americans could not find on a map but which was well known to the former oil man Bush – into a global crisis that was the moral equivalent of World War II. This materialist petrowar was fought as a televisual retrowar, a re-run of the quintessential "good war" in which brave allied soldiers battled the dark evil forces of Hitler/Saddam Hussein. Geopolitical discourse became akin to a movie script which featured Iraq as the "expansionist equivalent of Nazi Germany," while "Kuwait assumed the role of the totalitarian empire's weak helpless victim, like Czechoslovakia. . . ." As Luke notes, fighting for, and liberating Kuwait, therefore, "becomes equivalent to invading and taking back Europe from fascism. From the victory, an entirely new world order will be born, based on the notion of collective defense against aggression and flexible containment of current-day, or would-be Saddams . . ." (1991: 329–330).

From the perspective of the West's military-industrial complex, the Gulf War was a perfect opportunity to re-legitimate itself and re-define the fluid post-Cold War world as one where "rogue states" (like Iraq, Iran, Libya and Syria) and "nuclear outlaws" menaced the security of the Western world (Reading 20). Bush used the Gulf War to tout the great success of weapons systems like the Patriot anti-ballistic missile system, even though later it was revealed that the Patriot did not destroy any Iraqi SCUD missiles but did cause considerable damage on the ground. In his State of the Union speech, Bush promoted the favorite program of American militarists, the Strategic Defense Initiative, while simultaneously claiming, somewhat hollowly – given his enthusiasm and that of many Americans for the Gulf War – that the American "nation does not glory in war." With the Gulf War as a public relations spectacular, the Pentagon and other Cold War bureaucracies were able to justify a new Rogue Doctrine as their post-Cold War military posture.

The nature of this doctrine is examined by

peace studies scholar Michael Klare, who argues that it is "a product of a determined Pentagon effort to create a new foreign threat to justify military spending in the wake of the Cold War" (Reading 20). Central to the rogue state doctrine is the argument that the US military must maintain an ability to fight two major regional wars (Desert Storm size wars) at the same time. This "two war" standard is the strategic linchpin of US defense strategy, a strategy that justifies the persistence of the existent service structure (despite obvious duplication) and maintains defense spending at high levels despite the fact that the Gulf War demonstrated the hollowness of the threat from rogue states. The "two war" standard, in short, is designed to deter military cutbacks rather than real geopolitical threats. Like the "Soviet threat" before it, it rests on hyperbole and exaggeration rather than on credible and serious strategic intelligence.

While the Cold War military-industrial complex and "society of security" – a society organized around discourses of danger and committed to the containment of the Otherness these discourses imagine – managed to re-legitimate itself with some adjustments to a world order without a clear superpower enemy, the meaning and coherence of "the West" as an identity remained in doubt as a consequence of the passing of the Cold War. Certainly the disappearance of the external Soviet Other that helped define an imagined and highly mythologized internal Western Self for four decades contributed to this crisis of meaning, but so also did the marked globalization of corporate identity and economic activity from the late 1960s onwards. In addition, multicultural voices, the perspectives of women, racial and ethnic minorities, within Western states were beginning to challenge the privileged positions of authority historically enjoyed by conservative white males within these states, especially in the military and in foreign policy analysis and practice.

For some neoconservatives, the rising tenor of the creed of "multiculturalism" – a creed that challenged the West's exceptionalist myths and its limited application of the principles of liberty and equality – threatened to balkanize "the West" as an identity from within. Instead of reading it as an attempt to deepen democracy and equality, neoconservatives saw multiculturalism as a threatening "de-Westernization" of their own narrow mythic visions of the West. Seeking to defend their traditional privileges in the name of defending "the West," some neoconservative intellectuals such as Samuel Huntington cast the whole post-Cold War world order as a cultural war between different civilizational groups. Articulated first in *Foreign Affairs*, the journal of the Council on Foreign Relations, Huntington wrote of a worldwide "clash of civilizations" which ultimately pitted "the West against the Rest," a clash that was not simply a geographical clash between a Western "here" (the United States and Europe) and a Rest "over there" (all lands beyond these regions) but more profoundly a clash between an "us" (recognized and represented by neoconservative white males like Huntington) and a "them" (all others), a clash that was also internal to Western states and other "torn states" across the globe (Reading 21).

Like geopoliticians before him, Huntington adopts an Olympian view and declares that the

> great divisions among humankind and the dominating source of conflict [in the new world order] will be cultural. Nation states will remain the most powerful actors in world politics, but the principal conflicts in global politics will occur between nations and groups of different civilizations. The clash of civilizations will dominate world politics (1993: 22).

From his lofty vantage point, Huntington declares the existence of seven, possibly eight, essential civilizational blocs, all of which are in conflict. Using the same ahistorical and totalizing reasoning used in Cold War discourse, Huntington claims that these civilizations are primordial and "stretch back deep into history." Identity – "what are you?" – is a cultural given that cannot be changed. Civilizational identities and conflicts between them, he claims, lead to "civilization rallying" as groups and states belonging to one civilization that become involved in a war with peoples from another civilization rally to the support of their own "kin and country."

The most fundamental of all civilizational

clashes, the "central axis of world politics," for Huntington is the conflict between "the West and the Rest." In ideological terms, this is merely the continuation of the Cold War as a civilizational struggle by a different name, though this time Japan is ambiguously Western while a Confucian–Islamic network of "weapon states" (another name for rogue states) are the new dangerous Otherness against which "the West" must act. Echoing Cold War-style discourses of danger, Huntington argues that this new enemy is relentlessly building up its military and threatening "Western interests, values and power" across the globe.

Huntington's "clash of civilizations" thesis gave rise to considerable debate within the US strategic establishment (Ajami, 1993; Rubenstein and Crocker, 1994). Like Fukuyama and Luttwak, Huntington subsequently produced a book on his thesis (Huntington, 1996). A critical perspective on Huntington's thesis is provided in my own essay, "Samuel Huntington and the 'Civilizing' of Global Space" (Reading 22). This essay contextualizes Huntington as a neoconservative intellectual of statecraft and seeks to challenge the ahistorical and essentialist reasoning he employs to map the new world order as a civilizational tableaux. Rather than civilizations actually existing in the way Huntington imagines them, I argue that Huntington's "civilizing of global space," that is his "taming" of global space by declaring it composed of a series of essential civilizations, is a neoconservative attempt to re-legitimate a "society of [national] security" and its imperial style of strategic reasoning after the Cold War. Like other establishment intellectuals of statecraft, Huntington tries to "civilize" the turbulence and chaos of the new world (dis)order by proclaiming a map of world politics that shores up "the West" around conservative values, resuscitates Cold War bureaucracies by delineating danger Otherness, and re-defines new enemies to mobilize against both abroad (like "Islamic fundamentalists") and within (immigrants and multiculturalists).

While world politics certainly entered a new era with the collapse of the Berlin Wall in 1989, the interpretation and meaning of the new world (dis)order that followed has been fiercely contested. The Cold War may be over but Cold War militarism, intellectuals of statecraft and geopolitical styles of reasoning persist. The nature of the new world (dis)order remains essentially contested and so should geopolitical discourse as an "expert" system of power/knowledge seeking to delimit and define a map of that world order in ways that serve existing power structures and systems of authority.

Across the globe, forces of transformation and change are engaging traditional structures of power, from class hierarchies and bureaucratic privileges within states to racial inequalities and patriarchal supremacy within cultures. Represented by some as a clash between cosmopolitanism and nationalism, global flows and local fundamentalism, Jihad versus McWorld, these global/local transformations are complex and not reducible to formulaic terms and oppositions (Barber, 1995; Herod *et al.*, 1997; Kofman and Youngs, 1996). Refusing the rush to essentialize change and delimit it as a kind of geographical antagonism is the best intellectual defense against the nostrums of geopoliticians seeking to "sell" us a world where the dramas are simple, the identities pure and the antagonisms clear. The geographical heterogeneity and hybridity of the world is always much messier than our geopolitical maps of it.

REFERENCES AND FURTHER READING

On Fukuyama and the end of the Cold War

Atlas, J. (1989) "What is Fukuyama Saying?" *New York Times Magazine*, October 22.

Cumings, B. (1991) "Trilateralism and the new world order," *World Policy Journal*, 8: 195–226.

Fukuyama, F. (1992) *The End of History and the Last Man*, New York: Free Press.

Gray, C. (1988) *The Geopolitics of Superpower*, Lexington, Kentucky: University of Kentucky Press.

Hobsbawm, E. (1994) *The Age of Extremes: A History of the World, 1914–1991*, New York: Vintage.

Ryan, A. (1992) "Professor Hegel Goes to Washington," *New York Times Book Review*, March 26.

Ó Tuathail, G. (1992) "The Bush Administration and

the 'End' of the Cold War," *Geoforum*, 23: 437–452.

On Luttwak, geo-economics and Japan as threat

Crichton, M. (1992) *The Rising Sun*, New York: Ballantine Books.
Dietrich, W. (1991) *In the Shadow of the Rising Sun*, University Park: Pennsylvania State University Press.
Fallows, J. (1989) "Containing Japan," *The Atlantic Monthly*, May, 40–54.
Frantz, D. and Collins, C. (1989) *Selling Out*, Chicago: Contemporary Books.
Friedman, G. and Lebard, M. (1991) *The Coming War with Japan*, New York: St Martin's Press.
Luttwak, E. (1993) *The Endangered American Dream*, New York: Simon and Schuster.
Ó Tuathail, G. (1992) "Pearl Harbor Without Bombs: A Critical Geopolitics of the US–Japan 'FSX' Debate," *Environment and Planning A*, 24: 975–994.
——(1993) "Japan as Threat: Geo-economic Discourses on the USA–Japan relationship in US civil society, 1987–1991," in *The Political Geography of the New World Order*, ed. Colin Williams, London: Belhaven.
——(1996) *Critical Geopolitics*, London: Routledge.
Vogel, E. (1979) *Japan as Number One*, New York: Harper and Row.

On globalization and the society of security

Agnew, J. and Corbridge, S. (1995) *Mastering Space: Hegemony, Territory and International Political Economy*, London: Routledge.
Barber, B. (1995) *Jihad Versus McWorld*, New York: Ballantine.
Campbell, D. (1992) *Writing Security: United States Foreign Policy and the Politics of Identity*, Minneapolis: University of Minnesota Press.
Castells, M. (1996) *The Rise of the Network Society*. Oxford: Blackwell.
Cox, R. (1987) *Production, Power and World Order*, New York: Columbia University Press.
Herod, A., Ó Tuathail, G. and Roberts, S. (1998) *An Unruly World? Globalization, Governance and Geography*, London: Routledge.
Kofman, E. and Youngs, G. (1996) *Globalization: Theory and Practice*, London: Pinter.
Luke, T. (1991) "The Discipline of Security Studies and the Codes of Containment: Learning From Kuwait," *Alternatives*, 16: 315–344.
Reich, R. (1991) *The Work Of Nations*, New York: Knopf.

On the Gulf War

Brescheeth, H. and Yuval-Davis, N. (1991) *The Gulf War and the New World Order*, London: Zed.
DerDerian, J. (1992) *Antidiplomacy: Spies, Terror, Speed, and War*, Oxford: Blackwell.
Kellner, D. (1992) *The Persian Gulf TV War*, Boulder: Westview.
MacArthur, J. (1992) *Second Front: Censorship and Propoganda in the Gulf War*, New York: Hill and Wang.
Peters, C. (ed.) (1992) *Collateral Damage: The "New World Order" at Home and Abroad*, Boston: South End Press.
Sifry, M. and Cerf, C. (eds) (1991) *The Gulf War Reader*, New York: Random House.
Smith, J.E. (1992) *George Bush's War*, New York: Henry Holt.

On Huntington, neoconservatives and cultural wars

Ajami, F. (1993) "The Summoning," *Foreign Affairs*, 72: 41–53.
Huntington, S. (1996) *The Clash of Civilizations and the Remaking of World Order*, New York: Simon and Schuster.
McNeill, W. (1997) "Decline of the West?" *The New York Review of Books*, January 9: 18–22.
Rubenstein, R. and Crocker, J. (1994) "Challenging Huntington," *Foreign Policy*, 96: 113–128.
Sherry, M. (1995) *In the Shadow of War*, New Haven: Yale University Press.

Cartoon 8 Workers of the world (by Wuerker)
Turning an old communist slogan on its head, Matt Wuerker represents the 1989 revolutions in Eastern Europe as revolts against communist icons in favor of Western consumer icons. The workers of the world have become consumers of world commodities. The pot-belly on Ronald McDonald, however, indicates that the utopia of mass consumption is not what it seems but has ugly consequences, namely unhealthy and environmentally wasteful habits of consumption.
Source: M. Wuerker

15

FRANCIS FUKUYAMA

"The End of History?"

from *The National Interest* (1989)

In watching the flow of events over the past decade or so, it is hard to avoid the feeling that something very fundamental has happened in world history. The past year has seen a flood of articles commemorating the end of the Cold War, and the fact that "peace" seems to be breaking out in many regions of the world. Most of these analyses lack any larger conceptual framework for distinguishing between what is essential and what is contingent or accidental in world history, and are predictably superficial. If Mr. Gorbachev were ousted from the Kremlin or a new Ayatollah proclaimed the millennium from a desolate Middle Eastern capital, these same commentators would scramble to announce the rebirth of a new era of conflict.

And yet, all of these people sense dimly that there is some larger process at work, a process that gives coherence and order to the daily headlines.The twentieth century saw the developed world descend into a paroxysm of ideological violence, as liberalism contended first with the remnants of absolutism, then bolshevism and fascism, and finally an updated Marxism that threatened to lead to the ultimate apocalypse of nuclear war. But the century that began full of self-confidence in the ultimate triumph of Western liberal democracy seems at its close to be returning full circle to where it started: not to an "end of ideology" or a convergence between capitalism and socialism, as earlier predicted, but to an unabashed victory of economic and political liberalism.

The triumph of the West, of the Western idea, is evident first of all in the total exhaustion of viable systematic alternatives to Western liberalism. In the past decade, there have been unmistakable changes in the intellectual climate of the world's two largest communist countries, and the beginnings of significant reform movements in both. But this phenomenon extends beyond high politics and it can be seen also in the ineluctable spread of consumerist Western culture in such diverse contexts as the peasants' markets and color television sets now omnipresent throughout China, the cooperative restaurants and clothing stores opened in the past year in Moscow, the Beethoven piped into Japanese department stores, and the rock music enjoyed alike in Prague, Rangoon, and Tehran.

What we may be witnessing is not just the end of the Cold War, or the passing of a particular period of postwar history, but the end of history as such: that is, the end point of mankind's ideological evolution and the universalization of Western liberal democracy as the final form of human government. This is not to say that there will no longer be events to fill the pages of *Foreign Affairs'* yearly summaries of international relations, for the victory of liberalism has occurred primarily in the realm of ideas or consciousness and is as yet incomplete in the real or material world. But there are powerful reasons for believing that it is the ideal that will govern the material world *in the long run*. To understand how this is so, we must first consider some theoretical issues concerning the nature of historical change.

The notion of the end of history is not an original one. Its best known propagator was Karl Marx, who believed that the direction of historical development was a purposeful one determined by the interplay of material forces, and would come to an end only with the achievement of a communist utopia that would finally resolve all prior contradictions. But the concept of history as a dialectical process with a beginning, a middle, and an end was borrowed by Marx from his great German predecessor, Georg Wilhelm Friedrich Hegel.

For better or worse, much of Hegel's historicism has become part of our contemporary intellectual baggage. The notion that mankind has progressed through a series of primitive stages of consciousness on his path to the present, and that these stages corresponded to concrete forms of social organization, such as tribal, slave-owning, theocratic, and finally democratic-egalitarian societies, has become inseparable from the modern understanding of man. Hegel was the first philosopher to speak the language of modern social science, insofar as man for him was the product of his concrete historical and social environment and not, as earlier natural right theorists would have it, a collection of more or less fixed "natural" attributes. The mastery and transformation of man's natural environment through the application of science and technology was originally not a Marxist concept, but a Hegelian one. Unlike later historicists whose historical relativism degenerated into relativism *tout court*, however, Hegel believed that history culminated in an absolute moment – a moment in which a final, rational form of society and state became victorious.

It is Hegel's misfortune to be known now primarily as Marx's precursor, and it is our misfortune that few of us are familiar with Hegel's work from direct study, but only as it has been filtered through the distorting lens of Marxism. In France, however, there has been an effort to save Hegel from his Marxist interpreters and to resurrect him as the philosopher who most correctly speaks to our time. Among those modern French interpreters of Hegel, the greatest was certainly Alexandre Kojève, a brilliant Russian emigre who taught a highly influential series of seminars in Paris in the 1930s at the *Ecole Practique des Hautes Etudes*. While largely unknown in the United States, Kojève had a major impact on the intellectual life of the continent. Among his students ranged such future luminaries as Jean-Paul Sartre on the Left and Raymond Aron on the Right; postwar existentialism borrowed many of its basic categories from Hegel via Kojève.

Kojève sought to resurrect the Hegel of the *Phenomenology of Mind*, the Hegel who proclaimed history to be at an end in 1806. For as early as this Hegel saw in Napoleon's defeat of the Prussian monarchy at the Battle of Jena the victory of the ideals of the French Revolution, and the imminent universalization of the state incorporating the principles of liberty and equality. Kojève, far from rejecting Hegel in light of the turbulent events of the next century and a half, insisted that the latter had been essentially correct. The Battle of Jena marked the end of history because it was at that point that the *vanguard* of humanity (a term quite familiar to Marxists) actualized the principles of the French Revolution. While there was considerable work to be done after 1806 – abolishing slavery and the slave trade, extending the franchise to workers, women, blacks, and other racial minorities, etc. – the basic *principles* of the liberal democratic state could not be improved upon. The two world wars in this century and their attendant revolutions and upheavals simply had the effect of extending those principles spatially, such that the various provinces of human civilization were brought up to the level of its most advanced outposts, and of forcing those societies in Europe and North America at the vanguard of civilization to implement their liberalism more fully.

The state that emerges at the end of history is liberal insofar as it recognizes and protects through a system of law man's universal right to freedom, and democratic insofar as it exists only with the consent of the governed. For Kojève, this so-called "universal homogenous state" found real-life embodiment in the countries of postwar Western Europe – precisely those flabby, prosperous, self-satisfied, inward-looking, weak-willed states whose grandest project was nothing more heroic than the creation of the Common Market. But this was only to be expected. For human history and the conflict that characterized it was based on the existence of "contradictions": primitive man's quest for mutual recognition, the dialectic of the master and slave, the transformation and mastery of nature, the struggle for the universal recognition of rights, and the dichotomy between proletarian and capitalist. But in the universal homogenous state, all prior contradictions are resolved and all human needs are satisfied. There is no struggle or conflict over "large" issues, and consequently no need for generals or statesmen; what remains is primarily economic activity. And indeed, Kojève's life was consistent with his teaching. Believing that there was no more work for philosophers as well, since Hegel (correctly understood) had already achieved absolute knowledge, Kojève left teaching after the war and spent the remainder of his life working as a bureaucrat in the European Economic Community, until his death in 1968.

To his contemporaries at mid-century, Kojève's proclamation of the end of history must have seemed like the typical eccentric solipsism of a French intellectual, coming as it did on the heels of World War II and at the very height of the Cold War. To comprehend how Kojève could have been so audacious as to assert that history has ended, we must first of all understand the meaning of Hegelian idealism.

For Hegel, the contradictions that drive history exist first of all in the realm of human consciousness, i.e. on the level of ideas – not the trivial election year proposals of American politicians, but ideas in the sense of large unifying world views that might best be understood under the rubric of ideology. Ideology in this sense is not restricted to the secular and explicit political doctrines we usually associate with the term, but can include religion, culture, and the complex of moral values underlying any society as well. Hegel's view of the relationship between the ideal and the real or material worlds was an extremely complicated one, beginning with the fact that for him the distinction between the two was only apparent. He did not believe that the real world conformed or could be made to conform to ideological preconceptions of philosophy professors in any simple minded way, or that the "material" world could not impinge on the ideal. Indeed, Hegel the professor was temporarily thrown out of work as a result of a very material event, the Battle of Jena. But while Hegel's writing and thinking could be stopped by a bullet from the material world, the hand on the trigger of the gun was motivated in turn by the ideas of liberty and equality that had driven the French Revolution.

For Hegel, all human behavior in the material world, and hence all human history, is rooted in a prior state of consciousness – an idea similar to the one expressed by John Maynard Keynes when he said that the views of men of affairs were usually derived from defunct economists and academic scribblers of earlier generations. This consciousness may not be explicit and self-aware, as are modern political doctrines, but may rather take the form of religion or simple cultural or moral habits. And yet this realm of consciousness *in the long run* necessarily becomes manifest in the material world, indeed creates the material world in its own image. Consciousness is cause and not effect, and can develop autonomously from the material world; hence the real subtext underlying the apparent jumble of current events is the history of ideology.

Failure to understand that the roots of economic behavior lie in the realm of consciousness and culture leads to the common mistake of attributing material causes to phenomena that are essentially ideal in nature. For example, it is common place in the West to interpret the reform movements first in China and most recently in the Soviet Union as the victory of the material over the ideal – that is, a recognition that ideological incentives could not replace material ones in stimulating a highly productive modern economy, and that if one wanted to prosper one had to appeal to baser forms of self-interest. But the deep defects of socialist economies were evident thirty or forty years ago to any one who chose to look. Why was it that these countries moved away from central planning only in the 1980s? The answer must be found in the consciousness of the elites and leaders ruling them, who decided to opt for the "Protestant" life of wealth and risk over the "Catholic" path of poverty and security. That change was in no way made inevitable by the material conditions in which either country found itself on the eve of the reform, but instead came about as the result of the victory of one idea over another.

For Kojève, as for all good Hegelians, understanding the underlying processes of history requires understanding developments in the realm of consciousness or ideas, since consciousness will ultimately remake the material world in its own image. To say that history ended in 1806 meant that mankind's ideological evolution ended in the ideals of the French or American Revolutions: while particular regimes in the real world might not implement these ideals fully, their theoretical truth is absolute and could not be improved upon. Hence it did not matter to Kojève that the consciousness of the postwar generation of Europeans had not been universalized throughout the world; if ideological development had in fact ended, the homogenous state would eventually become victorious throughout the material world.

I have neither the space nor, frankly, the ability to defend in depth Hegel's radical idealist perspective. The issue is not whether Hegel's system was right, but whether his perspective might uncover the problematic nature of many materialist explanations we often take for granted. This is not to deny the role of material factors as such. To a literal minded idealist, human society can be built around any arbitrary set of principles regardless of their relationship to the material world. And in fact men have proven themselves able to endure the most extreme material hardships in the name of ideas that exist in the realm of the spirit alone, be it the divinity of cows or the nature of the Holy Trinity.

But while man's very perception of the material world is shaped by his historical consciousness of it, the material world can clearly affect in return the viability of a particular state of consciousness. In particular, the spectacular abundance of advanced liberal economies and the infinitely diverse consumer culture made possible by them seem to both foster and preserve liberalism in the political sphere. I want to avoid the materialist determinism that says that liberal economics inevitably produces liberal politics, because I believe that both economics and politics presuppose an autonomous prior state of consciousness that makes them possible. But that state of consciousness that permits the growth of liberalism seems to stabilize in the way one would expect at the end of history if it is underwritten by the abundance of a modern freemarket economy. We might summarize the content of the universal homogenous state as liberal democracy in the political sphere combined with easy access to VCRs and stereos in the economic.

Have we in fact reached the end of history? Are there, in other words, any fundamental "contradictions" in human life that cannot be resolved in the context of modern liberalism, that would be resolvable by an alternative political-economic structure? If we accept the idealist premises laid out above, we must seek an answer to this question in the realm of ideology and consciousness. Our task is not to answer exhaustively the challenges to liberalism promoted by every crackpot messiah around the world, but only those that are embodied in important social or political forces and movements, and which are therefore part of world history. For our purposes, it matters very little what strange thoughts occur to people in Albania or Burkina Faso, for we are interested in what one could in some sense call the common ideological heritage of mankind.

In the past century, there have been two major challenges to liberalism, those of fascism and of communism. The former saw the political weakness, materialism, anomie, and lack of community of the West as fundamental contradictions in liberal societies that could only be resolved by a strong state that forged a new "people" on the basis of national exclusiveness. Fascism was destroyed as a living ideology by World War II. This was a defeat, of course, on a very material level, but it amounted to a defeat of the idea as well. What destroyed fascism as an idea was not universal moral revulsion against it, since plenty of people were willing to endorse the idea as long as it seemed the wave of the future, but its lack of success. After the war, it seemed to most people that German fascism as well as its other European and Asian variants were bound to self-destruct. There was no material reason why new fascist movements could not have sprung up again after the war in other locales, but for the fact that expansionist ultranationalism, with its promise of unending conflict leading to disastrous military defeat, had completely lost its appeal. The ruins of the

Reich chancellory as well as the atomic bombs dropped on Hiroshima and Nagasaki killed this ideology on the level of consciousness as well as materially, and all of the proto-fascist movements spawned by the German and Japanese examples like the Peronist movement in Argentina or Subhas Chandra Bose's Indian National Army withered after the war.

The ideological challenge mounted by the other great alternative to liberalism, communism, was far more serious. Marx, speaking Hegel's language, asserted that liberal society contained a fundamental contradiction that could not be resolved within its context, that between capital and labor, and this contradiction has constituted the chief accusation against liberalism ever since. But surely, the class issue has actually been successfully resolved in the West. As Kojève (among others) noted, the egalitarianism of modern America represents the essential achievement of the classless society envisioned by Marx. This is not to say that there are not rich people and poor people in the United States, or that the gap between them has not grown in recent years. But the root causes of economic inequality do not have to do with the underlying legal and social structure of our society, which remains fundamentally egalitarian and moderately redistributionist, so much as with the cultural and social characteristics of the groups that make it up, which are in turn the historical legacy of premodern conditions. Thus black poverty in the United States is not the inherent product of liberalism, but is rather the "legacy of slavery and racism" which persisted long after the formal abolition of slavery.

As a result of the receding of the class issue, the appeal of communism in the developed Western world, it is safe to say, is lower today than any time since the end of World War I. This can be measured in any number of ways: in the declining membership and electoral pull of the major European communist parties, and their overtly revisionist programs; in the corresponding electoral success of conservative parties from Britain and Germany to the United States and Japan, which are unabashedly pro-market and anti-statist; and in an intellectual climate whose most "advanced" members no longer believe that bourgeois society is something that ultimately needs to be overcome. This is not to say that the opinions of progressive intellectuals in Western countries are not deeply pathological in any number of ways. But those who believe that the future must inevitably be socialist tend to be very old, or very marginal to the real political discourse of their societies.

One may argue that the socialist alternative was never terribly plausible for the North Atlantic world, and was sustained for the last several decades primarily by its success outside of this region. But it is precisely in the non-European world that one is most struck by the occurrence of major ideological transformations. Surely the most remarkable changes have occurred in Asia. Due to the strength and adaptability of the indigenous cultures there, Asia became a battleground for a variety of imported Western ideologies early in this century. Liberalism in Asia was a very weak reed in the period after World War I; it is easy today to forget how gloomy Asia's political future looked as recently as ten or fifteen years ago. It is easy to forget as well how momentous the outcome of Asian ideological struggles seemed for world political development as a whole.

The first Asian alternative to liberalism to be decisively defeated was the fascist one represented by Imperial Japan. Japanese fascism (like its German version) was defeated by the force of American arms in the Pacific war, and liberal democracy was imposed on Japan by a victorious United States. Western capitalism and political liberalism when transplanted to Japan were adapted and transformed by the Japanese in such a way as to be scarcely recognizable. Many Americans are now aware that Japanese industrial organization is very different from that prevailing in the United States or Europe, and it is questionable what relationship the factional maneuvering that takes place with the governing Liberal Democratic Party bears to democracy. Nonetheless, the very fact that the essential elements of economic and political liberalism have been so successfully grafted onto uniquely Japanese traditions and institutions guarantees their survival in the long run. More important is the contribution that Japan has made in turn to world history by following in the footsteps of the United States to create a truly universal consumer culture that has become both a symbol and an underpinning of the universal homogenous state. V.S. Naipaul travelling in Khomeini's Iran shortly after the revolution noted the omnipresent signs advertising the products of Sony, Hitachi, and JVC, whose appeal remained virtually irresistible and gave the lie to the regime's pretensions of restoring a state based on the rule of the *Shariah*. Desire for access to the consumer culture, created in large measure by Japan, has played a crucial role in fostering the spread of economic liberalism throughout Asia, and hence in promoting political liberalism as well.

The economic success of the other newly industrializing countries (NICs) in Asia following on the example of Japan is by now a familiar story. What is important from a Hegelian standpoint is that political liberalism has been following economic liberalism, more slowly than many had hoped but with seeming inevitability. Here again we see the victory of the idea of the universal homogenous state. South Korea had developed into a modern, urbanized society with an increasingly large and well-educated middle class that could not possibly be isolated from the larger democratic trends around them. Under these circumstances it seemed intolerable to a large part of this population that it should be ruled by an anachronistic military regime while Japan, only a decade or so ahead in economic terms, had parliamentary institutions for over forty years. Even the former socialist regime in Burma, which for so many decades existed in dismal isolation from the larger trends dominating Asia, was buffeted in the past year by pressures to liberalize both its economy and political system. It is said that unhappiness with strongman Ne Win began when a senior Burmese officer went to Singapore for medical treatment and broke down crying when he saw how far socialist Burma had been left behind by its ASEAN neighbors.

But the power of the liberal idea would seem much less impressive if it had not infected the largest and oldest culture in Asia, China. The simple existence of communist China created an alternative pole of ideological attraction, and as such constituted a threat to liberalism. But the past fifteen years have seen an almost total discrediting of Marxism–Leninism as an economic system. Beginning with the famous third plenum of the Tenth Central Committee in 1978, the Chinese Communist party set about decollectivizing agriculture for the 800 million Chinese who still lived in the countryside. The role of the state in agriculture was reduced to that of a tax collector, while production of consumer goods was sharply increased in order to give peasants a taste of the universal homogenous state and thereby an incentive to work. The reform doubled Chinese grain output in only five years, and in the process created for Deng Xiao-ping a solid political base from which he was able to extend the reform to other parts of the economy. Economic statistics do not begin to describe the dynamism, initiative, and openness evident in China since the reform began.

China could not now be described in any way as a liberal democracy. At present, no more than 20 per cent of its economy has been marketized, and most importantly it continues to be ruled by a self-appointed Communist party which has given no hint of wanting to devolve power. Deng has made none of Gorbachev's promises regarding democratization of the political system and there is no Chinese equivalent of *glasnost*. The Chinese leadership has in fact been much more circumspect in criticizing Mao and Maoism than Gorbachev with respect to Brezhnev and Stalin, and the regime continues to pay lip service to Marxism–Leninism as its ideological underpinning. But anyone familiar with the outlook and behavior of the new technocratic elite now governing China knows that Marxism and ideological principle have become virtually irrelevant as guides to policy, and that bourgeois consumerism has a real meaning in that country for the first time since the revolution. The various slowdowns in the pace of reform, the campaigns against "spiritual pollution" and crackdowns on political dissent are more properly seen as tactical adjustments made in the process of managing what is an extraordinarily difficult political transition. By ducking the question of political reform while putting the economy on a new footing, Deng has managed to avoid the breakdown of authority that has accompanied Gorbachev's *perestroika*. Yet the pull of the liberal idea continues to be very strong as economic power devolves and the economy becomes more open to the outside world. There are currently over 20,000 Chinese students studying in the US and other Western countries, almost all of them the children of the Chinese elite. It is hard to believe that when they return home to run the country they will be content for China to be the only country in Asia unaffected by the larger democratizing trend. The student demonstrations in Beijing that broke out first in December 1986 and recurred recently on the occasion of Hu Yao-bang's death were only the beginning of what will inevitably be mounting pressure for change in the political system as well.

What is important about China from the standpoint of world history is not the present state of the reform or even its future prospects. The central issue is the fact that the People's Republic of China can no longer act as a beacon for illiberal forces around the world, whether they be guerrillas in some Asian jungle or middle class students in Paris. Maoism, rather than being the pattern for Asia's future, became an anachronism, and it was the mainland Chinese who in fact were decisively influenced by the prosperity and dynamism of their overseas co-ethnics – the ironic ultimate victory of Taiwan. Important as these changes in China have been, however, it is developments in the Soviet Union – the original "homeland of the world proletariat" – that have put the final nail in the coffin of the Marxist–Leninist alternative to liberal democracy.

What has happened in the four years since Gorbachev's coming to power is a revolutionary assault on the most fundamental institutions and principles of Stalinism, and their replacement by other principles which do not amount to liberalism per se but whose only connecting thread is liberalism. This is most evident in the economic sphere, where the reform economists around Gorbachev have become steadily more radical in their support for free markets, to the point where some like Nikolai Shmelev do not mind being compared in public to Milton Friedman. . . .

[. . .]

The Soviet Union could in no way be described as a liberal or democratic country now, nor do I think that it is terribly likely that *perestroika* will succeed such that the label will be thinkable any time in the near future. But at the end of history it is not necessary that all societies become successful liberal societies, merely that they end their ideological pretensions of representing different and higher forms of human society. And in this respect I believe that something very important has happened in the Soviet Union in the past few years: the criticisms of the Soviet system sanctioned by Gorbachev have been so thorough and devastating that there is very little chance of going back to either Stalinism or Brezhnevism in any simple way. Gorbachev has finally permitted people to say what they had privately understood for many years, namely, that the magical incantations of Marxism–Leninism were nonsense, that Soviet socialism was not superior to the West in any respect but was in fact a monumental failure. The conservative opposition in the USSR, consisting both of simple workers afraid of unemployment and inflation and of party officials fearful of losing their jobs and privileges, is outspoken and may be strong enough to force Gorbachev's ouster in the next few years. But what both groups desire is tradition, order, and authority; they manifest no deep commitment to Marxism–Leninism, except insofar as they have invested much of their own lives in it. For authority to be restored in the Soviet Union after Gorbachev's demolition work, it must be on the basis of some new and vigorous ideology which has not yet appeared on the horizon.

If we admit for the moment that the fascist and communist challenges to liberalism are dead, are there any other ideological competitors left? Or put another way, are there contradictions in liberal society beyond that of class that are not resolvable? Two possibilities suggest themselves, those of religion and nationalism.

The rise of religious fundamentalism in recent years within the Christian, Jewish, and Muslim traditions has been widely noted. One is inclined to say that the revival of religion in some way attests to a broad unhappiness with the impersonality and spiritual vacuity of liberal consumerist societies. Yet while the emptiness at the core of liberalism is most certainly a defect in the ideology – indeed, a flaw that one does not need the perspective of religion to recognize – it is not at all clear that it is remediable through politics. Modern liberalism itself was historically a consequence of the weakness of religiously-based societies which, failing to agree on the nature of the good life, could not provide even the minimal preconditions of peace and stability. In the contemporary world only Islam has offered a theocratic state as a political alternative to both liberalism and communism. But the doctrine has little appeal for non-Muslims, and it is hard to believe that the movement will take on any universal significance. Other less organized religious impulses have been successfully satisfied within the sphere of personal life that is permitted in liberal societies.

The other major "contradiction" potentially unresolvable by liberalism is the one posed by nationalism and other forms of racial and ethnic consciousness. It is certainly true that a very large degree of conflict since the Battle of Jena has had its roots in nationalism. Two cataclysmic world wars in this century have been spawned by the nationalism of the developed world in various guises, and if those passions have been muted to a certain extent in postwar Europe, they are still extremely powerful in the Third World. Nationalism has been a threat to liberalism historically in Germany, and continues to be one in isolated parts of "post-historical" Europe like Northern Ireland.

But it is not clear that nationalism represents an irreconcilable contradiction in the heart of liberalism. In the first place, nationalism is not one single phenomenon but several, ranging from mild cultural nostalgia to the highly organized and elaborately articulated doctrine of National Socialism. Only systematic nationalisms of the latter sort can qualify as a formal ideology on the level of liberalism or communism. The vast majority of the world's nationalist movements do not have a political program beyond the negative desire of independence from some other group or people, and do not offer anything like a comprehensive agenda for socio-economic organization. As such, they are compatible with doctrines and ideologies that do offer such agendas. While they may constitute a source of conflict for liberal societies, this conflict does not arise from liberalism itself so much as from the fact that the liberalism in question is incomplete. Certainly a great deal of the world's ethnic and nationalist tension can be explained in terms of peoples who are forced to live in unrepresentative political systems that they have not chosen.

While it is impossible to rule out the sudden appearance of new ideologies or previously unrecognized contradictions in liberal societies, then, the present world seems to confirm that the fundamental principles of socio-political organization have not advanced terribly far since 1806. Many of the wars and revolutions fought since that time have been undertaken in the name of ideologies which claimed to be more advanced than liberalism, but whose pretensions were ultimately unmasked by history. In the meantime, they have helped to spread the universal homogenous state to the point where it could have a significant effect on the overall character of international relations.

What are the implications of the end of history for international relations? Clearly, the vast bulk of the Third World remains very much mired in history, and will be a terrain of conflict for many years to come. But let us focus for the time being on the larger and more developed states of the world who, after all, account for the greater part of world politics. Russia and China are not likely to join the developed nations of the West as liberal societies any time in the foreseeable future, but suppose for a moment that Marxism–Leninism ceases to be a factor driving the foreign policies of these states – a prospect which, if not yet here, the last few years have made a real possibility. How will the overall characteristics of a de-ideologized world differ from those of the one with which we are familiar at such a hypothetical juncture?

The most common answer is – not very much. For there is a very widespread belief among many observers of international relations that underneath the skin of ideology is a hard core of great power national interest that guarantees a fairly high level of competition and conflict between nations. Indeed, according to one academically popular school of international relations theory, conflict inheres in the international system as such, and to understand the prospects for conflict one must look at the shape of the system – for example, whether it is bipolar or multipolar – rather than at the specific character of the nations and regimes that constitute it. This school in effect applies a Hobbesian view of politics to international relations, and assumes that aggression and insecurity are universal characteristics of human societies rather than the product of specific historical circumstances.

Believers in this line of thought take the relations that existed between the participants in the classical nineteenth-century European balance of power as a model for what a de-ideologized contemporary world would look like. Charles Krauthammer, for example, recently explained that if as a result of Gorbachev's reforms the USSR is shorn of Marxist–Leninist ideology, its behavior will revert to that of nineteenth-century imperial Russia. While he finds this more reassuring than the threat posed by a communist Russia, he implies that there will still be a substantial degree of competition and conflict in the international system, just as there was say between Russia and Britain or Wilhelmine Germany in the last century. This is, of course, a convenient point of view for people who want to admit that something major is changing in the Soviet Union, but do not want to accept responsibility for recommending the radical policy redirection implicit in such a view. But is it true?

In fact, the notion that ideology is a superstructure imposed on a substratum of permanent great power interest is a highly questionable proposition. For the way in which any state defines its national interest is not universal but rests on some kind of prior ideological basis, just as we saw that economic behavior is determined by a prior state of consciousness. In this century, states have adopted highly articulated doctrines with explicit foreign policy agendas legitimizing expansionism, like Marxism–Leninism or National Socialism.

The expansionist and competitive behavior of nineteenth-century European states rested on no less ideal a basis; it just so happened that the ideology driving it was less explicit than the doctrines of the twentieth century. For one thing, most "liberal" European societies were illiberal insofar as they believed in the legitimacy of imperialism, that is, the right of one nation to rule over other nations without regard for the wishes of the ruled. The justifications for imperialism varied from nation to nation, from a crude belief in the legitimacy of force, particularly when applied to non-Europeans, to the White Man's Burden and Europe's Christianizing mission, to the desire to give people of color access to the culture of Rabelais and Moliere. But whatever the particular ideological basis, every "developed" country believed in the acceptability of higher civilizations ruling lower ones – including, incidentally, the United States with regard to the Philippines. This led to a drive for pure territorial aggrandizement in the latter half of the century and played no small role in causing the Great War.

The radical and deformed outgrowth of nineteenth-century imperialism was German fascism, an ideology which justified Germany's right not only to rule over non-European peoples, but over *all* non-German ones. But in retrospect it seems that Hitler represented a diseased bypath in the general course of European development, and since his fiery defeat, the legitimacy of any kind of territorial aggrandizement has been thoroughly discredited. Since World War II, European nationalism has been defanged and shorn of any real relevance to foreign policy, with the consequence that the nineteenth-century model of great power behavior has become a serious anachronism. The most extreme form of nationalism that any Western European state has mustered since 1945 has been Gaullism, whose self-assertion has been confined largely to the realm of nuisance politics and culture. International life for the part of the world that has reached the end of history is far more preoccupied with economics than with politics or strategy.

The developed states of the West do maintain defense establishments and in the post-war period have competed vigorously for influence to meet a worldwide communist threat. This behavior has been driven, however, by an external threat from states that possess overtly expansionist ideologies, and would not exist in their absence. To take the "neo-realist" theory seriously, one would have to believe that "natural" competitive behavior would reassert itself among the OECD states were Russia and China to disappear from the face of the earth. That is, West Germany and France would arm themselves against each other as they did in the 1930s, Australia and New Zealand would send military advisers to block each others' advances in Africa, and the US–Canadian border would become fortified. Such a prospect is, of course, ludicrous: minus Marxist–Leninist ideology, we are far more likely to see the "Common Marketization" of world politics than the disintegration of the EEC into nineteenth-century competitiveness. Indeed, as our experience in dealing with Europe on matters such as terrorism or Libya prove, they are much further gone than we down the road that denies the legitimacy of the use of force in international politics, even in self-defense.

The automatic assumption that Russia shorn of its expansionist communist ideology should pick up where the czars left off just prior to the Bolshevik Revolution is therefore a curious one. It assumes that the evolution of human consciousness has stood still in the meantime, and that the Soviets, while picking up currently fashionable ideas in the realm of economics, will return to foreign policy views a century out of date in the rest of Europe. This is certainly not what happened to China after it began its reform process. Chinese competitiveness and expansionism on the world scene have virtually disappeared: Beijing no longer sponsors Maoist insurgencies or tries to cultivate influence in distant African countries as it did in the 1960s. This is not to say that there are not troublesome aspects to contemporary Chinese foreign policy, such as the reckless sale of ballistic missile technology in the Middle East; and the PRC continues to manifest traditional great power behavior in its sponsorship of the Khmer Rouge against Vietnam. But the former is explained by commercial motives and the latter is a vestige of earlier ideologically-based rivalries. The new China far more resembles Gaullist France than pre-World War I Germany.

The real question for the future, however, is the degree to which Soviet elites have assimilated the consciousness of the universal homogenous state that is post-Hitler Europe. From their writings and from my own personal contacts with them, there is no question in my mind that the liberal Soviet intelligentsia rallying around Gorbachev has arrived at the end-of-history view in a remarkably short time, due in no small measure to the contacts they have had since the Brezhnev era with the larger European civilization around them. "New political thinking," the general rubric for their views, describes a world dominated by economic concerns, in which there are no ideological grounds for major conflict between nations, and in which, consequently, the use of military force becomes less legitimate. As Foreign Minister Shevardnadze put it in mid-1988:

> The struggle between two opposing systems is no longer a determining tendency of the present day era. At the modern stage, the ability to build up material wealth at an accelerated rate on the basis of front-ranking science and high-level techniques and technology, and to distribute it fairly, and through joint efforts to restore and protect the resources necessary for mankind's survival acquires decisive importance.

The post-historical consciousness represented by "new thinking" is only one possible future for the Soviet Union, however. There has always been a very strong current of great Russian chauvinism in the Soviet Union, which has found freer expression since the advent of *glasnost*. It may be possible to return to traditional Marxism–Leninism for a while as a simple rallying point for those who want to restore the authority that Gorbachev has dissipated. But as in Poland, Marxism–Leninism is dead as a mobilizing ideology: under its banner people cannot be made to work harder, and its adherents have lost confidence in themselves. Unlike the propagators of traditional Marxism–Leninism, however, ultra nationalists in the USSR believe in their Slavophile cause passionately, and one gets the sense that the fascist alternative is not one that has played itself out entirely there.

The Soviet Union, then, is at a fork in the

road: it can start down the path that was staked out by Western Europe forty-five years ago, a path that most of Asia has followed, or it can realize its own uniqueness and remain stuck in history. The choice it makes will be highly important for us, given the Soviet Union's size and military strength, for that power will continue to preoccupy us and slow our realization that we have already emerged on the other side of history.

The passing of Marxism–Leninism first from China and then from the Soviet Union will mean its death as a living ideology of world historical significance. For while there may be some isolated true believers left in places like Managua, Pyongyang, or Cambridge, Massachussets, the fact that there is not a single large state in which it is a going concern undermines completely its pretensions to being in the vanguard of human history. And the death of this ideology means the growing "Common Marketization" of international relations, and the diminution of the likelihood of large-scale conflict between states.

This does not by any means imply the end of international conflict *per se*. For the world at that point would be divided between a part that was historical and a part that was post-historical. Conflict between states still in history, and between those states and those at the end of history, would still be possible.There would be still be a high and perhaps rising level of ethnic and nationalist violence, since those are impulses completely played out, even in parts of the post-historical world. Palestinians and Kurds, Sikhs and Tamils, Irish Catholics and Walloons, Armenians and Azeris, will continue to have their unresolved grievances. This implies that terrorism and wars of national liberation will continue to be an important item on the national agenda. But large-scale conflict must involve large states still caught in the grip of history, and they are what appear to be passing from the scene.

The end of history will be a very sad time. The struggle for recognition, the willingness to risk one's own life for a purely ideological struggle that called forth daring, courage, imagination, and idealism, will be replaced by economic calculation, the endless solving of technical problems, environmental concerns, and the satisfaction of sophisticated consumer demands. In the post-historical period there will be neither art nor philosophy, just the perpetual caretaking of the museum of human history. I can feel in myself, and see in others around me, a powerful nostalgia for the time when history existed. Such nostalgia, in fact, will continue to fuel competition and conflict even in the post-historical world for some time to come. Even though I recognize its inevitabilty, I have the most ambivalent feelings for the civilization that has been created in Europe since 1945, with its North Atlantic and Asian offshoots. Perhaps this very prospect of centuries of boredom at the end of history will serve to get history started once again.

16 EDWARD N. LUTTWAK

"From Geopolitics to Geo-Economics: Logic of Conflict, Grammar of Commerce"

***The National Interest* (1990)**

Except for those unfortunate parts of the world where armed confrontations or civil strife persist for purely regional or internal reasons, the waning of the Cold War is steadily reducing the importance of military power in world affairs.

True, in the central strategic arena, where Soviet power finally encountered the de facto coalition of Americans, Europeans, Japanese, and Chinese, existing military forces have diminished very little so far. Nevertheless, as a Soviet–Western war becomes ever more implausible, the ability to threaten or reassure is equally devalued (and by the same token, of course, there is no longer a unifying threat to sustain the coalition against all divisive impulses). Either way, the deference that armed strength could evoke in the dealings of governments over all matters – notably including economic questions – has greatly declined, and seems set to decline further. Everyone, it appears, now agrees that the methods of commerce are displacing military methods – with disposable capital in lieu of firepower, civilian innovation in lieu of military-technical advancement, and market penetration in lieu of garrisons and bases. But these are all tools, not purposes; what purposes will they serve?

If the players left in the field by the waning importance of military power were purely economic entities – labor-sellers, entrepreneurs, corporations – then only the logic of commerce would govern world affairs. Instead of World Politics, the intersecting web of power relationships on the international scene, we would simply have World Business, a myriad economic interactions spanning the globe. In some cases, the logic of commerce would result in fierce competition. In others, the same logic would lead to alliances between economic entities in any location to capitalize ventures, vertically integrate, horizontally co-develop, co-produce, or co-market goods and services. But competitively or cooperatively, the action on all sides would always unfold without regard to frontiers. If that were to happen, not only military methods but the logic of conflict itself – which is adversarial, zero-sum, and paradoxical – would be displaced. This, or something very much like it, is in fact what many seem to have in mind when they speak of a new global interdependence and its beneficial consequences.

But things are not quite that simple. The international scene is still primarily occupied by states and blocs of states that extract revenues, regulate economic as well as other activities for various purposes, pay out benefits, offer services, provide infrastructures, and – of increasing importance – finance or otherwise sponsor the development of new technologies and new products. As territorial entities, spatially rather than functionally defined, states cannot follow a commercial logic that would ignore their own boundaries. What logic then do they follow?

- Do they seek to collect as much in revenues as their fiscal codes prescribe – or are they content to let other states or blocs of states tax away what they themselves could obtain? Since the former is the reality (that is, a zero-sum situation in which the gain of one is the loss of another), here the ruling logic is the logic of conflict.
- Do they regulate economic activities to achieve disinterestedly transnational purposes – or do they seek to maximize outcomes within their own boundaries, even if this means that the outcomes are suboptimal elsewhere? Since the

latter is the predominant, if not exclusive, reality, economic regulation is as much a tool of statecraft as military defenses ever were. Hence, insofar as external repercussions are considered, the logic of state regulation is in part the logic of conflict. As such, its attributes include the typically warlike use of secrecy and deception for the sake of surprise (as, for example, when product standards are first defined in secret consultations with domestic producers, long before their public enunciation).

- Do states and blocs of states pay out benefits and offer services transnationally – or (fractional aid allocations apart) do they strive to restrict such advantages to their own residents? Likewise, do they design infrastructures to maximize their transnational utility – or do they aim for domestically optimal and appropriately competitive configurations, regardless of how others are affected? Since the latter is the reality, the logic of state action is again in part the logic of conflict. (The competitive building of huge international airports in adjacent, minuscule, Persian Gulf sheikhdoms is an extreme example of such behavior, but such conduct is not uncommon in milder forms.)
- Finally, do states and blocs of states promote technological innovation for its own sake – or do they seek thereby to maximize benefits within their own boundaries? Since the latter is the reality, the logic of conflict applies. (Three obvious examples are the obstacles that long delayed the introduction of Concorde flights into US airports, Japanese barriers against US supercomputers and telecommunications, and the development of rival High Definition Television formats.)

As this is how things are, it follows that – even if we leave aside the persistence of armed confrontations in unfortunate parts of the world and wholly disregard what remains of the Cold War – World Politics is still not about to give way to World Business, i.e. the free interaction of commerce governed only by its own nonterritorial logic.

Instead, what is going to happen – and what we are already witnessing – is a much less complete transformation of state action represented by the emergence of "Geo-economics." This neologism is the best term I can think of to describe the admixture of the logic of conflict with the methods of commerce – or, as Clausewitz would have written, the logic of war in the grammar of commerce.

THE NATURE OF THE BEAST

With states and blocs of states still in existence, it could not be otherwise. As spatial entities structured to jealously delimit their own territories, to assert their exclusive control within them, and variously to attempt to influence events beyond their borders, states are inherently inclined to strive for relative advantage against like entities on the international scene, even if only by means other than force.

Moreover, states are subject to the internal impulses of their own bureaucracies, whose officials compete to achieve whatever goals define bureaucratic success, including goals in the international economic arena that may as easily be conflictual as competitive or cooperative. Actually much more than that is happening: *As bureaucracies writ large, states are themselves impelled by the bureaucratic urges of role-preservation and role-enhancement to acquire a "geo-economic" substitute for their decaying geopolitical role.*

There is also a far more familiar phenomenon at work: the instrumentalization of the state by economic interest groups that seek to manipulate its activities on the international scene for their own purposes, often by requiring adversarial "geo-economic" stances. No sphere of state action is immune: fiscal policy can be profitably used so as to place imports at a disadvantage; regulations, benefits, services, and infrastructures can all be configured to favor domestic interests in various ways; and, of course, the provision of state funds for domestic technological development is inherently discriminatory against unassisted foreign competitors.

The incidence of both adversarial bureaucratic impulses and adversarial manipulations of the state by interest groups will vary greatly from country to country. But fundamentally, states will tend to act "geo-economically" simply because of what they are: spatially-defined entities structured to outdo each other on the world scene. For all the other functions that states have acquired as providers of individual benefits, assorted services, and varied infrastructures, their raison d'etre and the ethos that sustains them still derive from their chronologically first function: to provide security from foes without (as well as outlaws within).

Relatively few states have had to fight to exist, but all states exist to fight – or at least they are structured as if that were their dominant function. Even though most of the existing 160-odd independent states have never fought any external wars, and most of those that have fought have not done so for generations, the governing structures of the modern state are still heavily marked by conflictual priorities, the need to prepare for, or to wage, interstate conflict. In how many major countries does the Minister for Telecommunications, or Energy, or Trade outrank the Defense Minister? Only – appropriately enough – in Japan, where Defense (*Boecho*) is a *Cho* or lesser department (translated as agency), as opposed to a *Sho* or ministry, as in Tsusansho, the Ministry of Trade. The *Boecho*'s head, while a minister, does not hold cabinet rank.

It is true, of course, that, under whatever name, 'geo-economics' has always been an important aspect of international life. In the past, however, the outdoing of others in the realm of commerce was overshadowed by strategic priorities and strategic modalities. Externally, if the logic of conflict dictated the necessity for cooperation against a common enemy while, in contrast, the logic of commerce dictated competition, the preservation of the alliance was almost always given priority. (That indeed is how all the commercial quarrels between the United States and Western Europe over frozen chickens, microchips, beef, and the rest – and between the United States and Japan – from textiles in the 1960s to supercomputers in the 1980s – were so easily contained during the past decades of acute Soviet–Western confrontation. As soon as commercial quarrels became noisy enough to attract the attention of political leaders on both sides, they were promptly suppressed by those leaders – often by paying off all parties – before they could damage political relations and thus threaten the imperative of strategic cooperation.) Internally, insofar as national cohesion was sustained against divisive social and economic tensions by the unifying urgencies of external antagonisms, it was armed conflict or the threat of it – not commercial animosities – that best served to unite nations.

Now, however, as the relevance of military threats and military alliances wanes, geo-economic priorities and modalities are becoming dominant in state action. Trade quarrels may still be contained by the fear of the economic consequence of an action–reaction cycle of punitive measures, but they will no longer simply be suppressed by political interventions on both sides, urgently motivated by the strategic imperative of preserving alliance cooperation against a common enemy. And if internal cohesion has to be preserved by a unifying threat, that threat must now be economic. Such a reordering of modalities is already fully manifest in the expressed attitudes of other Europeans to the new undivided Germany, and even more so in American attitudes toward Japan. Gorbachev's redirection of Soviet foreign policy had barely started when Japan began to be promoted to the role of the internally unifying Chief Enemy, judging by the evidence of opinion polls, media treatments, advertisements, and congressional pronouncements.

Should we conclude from all this that the world is regressing to a new age of mercantilism? Is that what "geo-economics" identifies, quite redundantly? Not so. The goal of mercantilism was to maximize gold stocks, whereas the goal of geo-economics (aggrandizement of the state aside) could only be to provide the best possible employment for the largest proportion of the population. In the past, moreover, when commercial quarrels evolved into political quarrels, they could become military confrontations almost automatically; and in turn military confrontations could readily lead to war.

In other words, mercantilism was *a subordinated modality*, limited and governed by the ever-present possibility that the loser in the mercantilist (or simply commercial) competition would switch to the grammar of war. Spain might decree that all trade to and from its American colonies could only travel in Spanish bottoms through Spanish ports, but British and

Dutch armed merchantmen could still convey profitable cargoes to disloyal colonists in defiance of Spanish sloops; and, with war declared, privateers could seize outright the even more profitable cargoes bound for Spain. Likewise, the Dutch sent their frigates into the Thames to reply to the mercantilist legislation of the British Parliament that prohibited their cabotage, just as much earlier the Portuguese had sunk Arab ships with which they could not compete in the India trade.

"Geo-economics," on the other hand, is emerging in a world where there is *no superior modality*. Import-restricted supercomputers cannot be forcibly delivered by airborne assault to banks or universities in need of them, nor can competition in the world automobile market be assisted by the sinking of export car ferries on the high seas. That force has lost the role it once had in the age of mercantilism – as an *admissible* adjunct to economic competition – is obvious enough. But of course the decay of the military grammar of geopolitics is far more pervasive than this, even if it is by no means universal.

Students of international relations may still be taught to admire the classic forms of realpolitik, with its structure of anticipatory calculations premised on the feasibility of war. But for some decades now the dominant elites of the greatest powers have ceased to consider war as a practical solution for military confrontations between them, because non-nuclear fighting would only be inconclusively interrupted by the fear of nuclear war, while the latter is self inhibiting. (In accordance with the always paradoxical logic of conflict, the application of the fusion technique meant that nuclear weapons exceeded the culminating point of utility, becoming less useful as they became more efficient.)

For exactly the same reason, military confrontations were themselves still considered very much worth pursuing – and rightly so, for war was thereby precluded throughout the decades of Soviet–Western antagonism. More recently, however, the dominant elites of the greatest powers appear to have concluded that military confrontations between them are only dissuasive of threats that are themselves most implausible. It is that new belief that has caused the decisive devaluation of military strength as an instrument of statecraft in the direct relations of the greatest powers.

Hence, while the methods of mercantilism could always be dominated by the methods of war, in the new "geo-economic" era not only the causes but also the instruments of conflict must be economic. If commercial quarrels do lead to political clashes, as they are now much more likely to do with the waning of the imperatives of geopolitics, those political clashes must be fought out with the weapons of commerce: the more or less disguised restriction of imports, the more or less concealed subsidization of exports, the funding of competitive technology projects, the support of selected forms of education, the provision of competitive infrastructures, and more.

PLAYING THE NEW GAME

The discussion so far has focused on the actual and prospective role of states and, by implication, of blocs of states engaged in "geo-economic" conduct. But what happens on the world economic scene will not of course be defined by such conduct; indeed the role of "geo-economics" in the doings and undoings of the world economy should be far smaller than the role of geopolitics in world politics as a whole.

First, the propensity of states to act geo-economically will vary greatly, even more than their propensity to act geopolitically. For reasons historical and institutional, or doctrinal and political, some states will maintain a strictly *laissez faire* attitude, simply refusing to act "geo-economically." Both the very prosperous and the very poor might be in that category, just as both Switzerland and Burma have long been geopolitically inactive. In other cases, the desirable scope of geo-economic activism by the state is already becoming a focal point of political debate and partisan controversy: witness the current Democratic–Republican dispute on "industrial policy" in the United States. In still other cases, such as that of France, the dominant elites that long insisted on a very ambitious degree of geopolitical activism (ambitious, that is, in terms of the resources available) are now easily shifting their emphasis to demand much more geo-economic activism from the French state. And then, of course, there are the states – Japan most notably – whose geo-economic propensities are not in question.

Second, there is the much more important limitation that states and blocs of states acting "geo-economically" must do so within an arena that is not exclusively theirs, in which they coexist with private economic operators large and small, from individuals to the largest multinational corporations. While states occupy virtually all of the world's political space, they occupy only a fraction of the total economic space, and global political economic trends such as privatization are reducing that fraction even further. (On the other hand, the role of states is increasing precisely in the economic sectors whose importance is itself increasing, sectors defined by the commercial application of the most advanced technologies.)

Of the different forms of coexistence between geo-economically active states and private economic operators, there is no end. Coexistence can be passive and disregarded, as in the relationship (or lack of it) between the state and the myriad small, localized service businesses. With neither wanting anything from the other – except for the taxes that the fiscal authorities demand – the two can simply coexist without interacting or communicating.

At the opposite extreme, there is the intense positive interaction between politically weighty businesses in need of state support on the world economic scene, and the bureaucracies or politicians that they seek to manipulate for their own purposes. Or, going the other way, there is the equally intense and equally positive interaction that occurs when states seek to guide large companies for their own geo-economic purposes, or even select them as their "chosen instrument" (a specialized form of coexistence that dates back at least to the seventeenth-century East India companies, Dutch and Danish as well as, most famously, British).

Even more common, no doubt, are the cases of reciprocal manipulation, most notably in the remarkably uniform dealings of the largest international oil companies – whether American, British, or French – with their respective (and otherwise very different) state authorities. In each case, the state has been both user and used, and the companies both instruments and instrumentalizers.

Negative state–private sector interactions are not likely to be common, but they could be very important when they do occur. Geo-economically active states that oppose rival foreign states will also obviously oppose private foreign companies that are the chosen instruments of those rivals, as well as private foreign companies that simply have the misfortune to stand in the way. An era of intense "geo-economic" activity might thus become an era of unprecedented risk for important private companies in important sectors. If they invest Y million of their funds to develop X technology, they may find themselves irremediably overtaken by the X project of country Z, funded by the taxpayer in the amount of 2Y million, or 20Y million for that matter. Or private companies may find themselves competing with foreign undercutters determined to drive them out of business, and amply funded for that purpose by their state authorities. As public funding for such purposes is likely to be concealed, a victim company may enter a market quite unaware of its fatal disadvantage. In such diverse ways the international economy will be pervasively affected by that fraction of its life that is geo-economic rather than simply economic in character (just as in the past the geopolitical activity of the few greatest powers decisively conditioned the politics of the many).

Perhaps the pan-Western trade accords of the era of armed confrontation with the Soviet Union – based on the original General Agreement on Tariffs and Trade – may survive without the original impulse that created them, and may serve to inhibit the overt use of tariffs and quotas as the geo-economic equivalent of fortified lines. And that inheritance of imposed amity may also dissuade the hostile use of all other "geo-economic" weapons, from deliberate regulatory impediments to customs-house conspiracies aimed at rejecting imports covertly – the commercial equivalents of the ambushes of war. But that still leaves room for far more important weapons: the competitive development of commercially important new technologies, the predatory financing of their sales during their embryonic stage, and the manipulation of the standards that condition their use – the geo-economic equivalents of the offensive campaigns of war.

Today, there is a palpably increasing tension between the inherently conflictual nature of states (and blocs of states) and the intellectual recognition of many of their leaders and citizens that while war is a zero-sum encounter by nature, commercial relations need not be and indeed rarely have been. The outcome of that tension within the principal countries and blocs will determine the degree to which we will live in a geo-economic world.

17 PRESIDENT GEORGE BUSH

"Toward a New World Order"

from *Public Papers of the Presidents of the United States* (1991)

We gather here tonight, witness to events in the Persian Gulf as significant as they are tragic. In the early morning hours of August 2, following negotiations and promises by Iraq's dictator Saddam Hussein not to use force, a powerful Iraqi army invaded its trusting and much weaker neighbor, Kuwait. Within 3 days, 120,000 Iraqi troops with 850 tanks had poured into Kuwait and moved south to threaten Saudi Arabia. It was then that I decided to check that aggression.

At this moment, our brave service men and women stand watch in that distant desert and on distant seas, side-by-side with the forces of more than 20 other nations. They are some of the finest men and women of the United States of America, and they're doing one terrific job. These valiant Americans were ready at a moment's notice to leave their spouses and their children, to serve on the front line halfway around the world. They remind us who keeps America strong, they do. . . .

[. . .]

Tonight, I want to talk to you about what's at stake – what we must do together to defend civilized values around the world and maintain our economic strength at home.

THE OBJECTIVES AND GOALS

Our objectives in the Persian Gulf are clear; our goals defined and familiar.

- Iraq must withdraw from Kuwait completely, immediately, and without condition.
- Kuwait's legitimate government must be restored.
- The security and stability of the Persian Gulf must be assured.
- American citizens abroad must be protected.

These goals are not ours alone. They have been endorsed by the UN Security Council five times in as many weeks. Most countries share our concern for principle, and many have a stake in the stability of the Persian Gulf. This is not, as Saddam Hussein would have it, the United States against Iraq. It is Iraq against the world.

As you know, I have just returned from a very productive meeting with Soviet President Gorbachev. I am pleased that we are working together to build a new relationship. In Helsinki, our joint statement affirmed to the world our shared resolve to counter Iraq's threat to peace. Let me quote:

> We are united in the belief that Iraq's aggression must not be tolerated. No peaceful international order is possible if larger states can devour their smaller neighbors.

Clearly, no longer can a dictator count on East–West confrontation to stymie concerted UN action against aggression. A new partnership of nations has begun.

A HISTORIC PERIOD OF COOPERATION

We stand today at a unique and extraordinary moment. The crisis in the Persian Gulf, as grave as it is, also offers a rare opportunity to move toward a historic period of cooperation. Out of these troubled times, our fifth objective – a new world order – can emerge; a new era – freer from the threat of terror, stronger in the pursuit of justice, and more secure in the quest for peace, an era in which the nations of the world, East and West, North and South, can prosper and live in harmony.

A hundred generations have searched for this elusive path to peace, while a thousand wars raged across the span of human endeavor. Today, that new world is struggling to be born, a world quite different from the one we have known, a world where the rule of law supplants the rule of the jungle, a world in which nations recognize the shared responsibility for freedom and justice, a world where the strong respect the rights of the weak.

This is the vision that I shared with President Gorbachev in Helsinki. He and other leaders from Europe, the Gulf, and around the world understand that how we manage this crisis today could shape the future for generations to come.

The test we face is great – and so are the stakes. This is the first assault on the new world that we seek, the first test of our mettle. Had we not responded to this first provocation with clarity of purpose, if we do not continue to demonstrate our determination, it would be a signal to actual and potential despots around the world.

America and the world must defend common vital interests. And we will. America and the world must support the rule of law. And we will. America and the world must stand up to aggression. And we will. And one thing more; in the pursuit of these goals, America will not be intimidated.

Vital issues of principle are at stake. Saddam Hussein is literally trying to wipe a country off the face of the earth. We do not exaggerate. Nor do we exaggerate when we say Saddam Hussein will fail.

Vital economic interests are at risk as well. Iraq itself controls some 10 per cent of the world's proven oil reserves. Iraq plus Kuwait controls twice that. An Iraq permitted to swallow Kuwait would have the economic and military power, as well as the arrogance, to intimidate and coerce its neighbors – neighbors that control the lion's share of the world's remaining oil reserves. We cannot permit a resource so vital to be dominated by one so ruthless. And we won't.

Recent events have surely proven that there is no substitute for American leadership. In the face of tyranny, let no one doubt American credibility and reliability. Let no one doubt our staying power. We will stand by our friends. One way or another, the leader of Iraq must learn this fundamental truth.

THE INTERNATIONAL RESPONSE AND OBLIGATION

From the outset, acting hand-in-hand with others, we have sought to fashion the broadest possible international response to Iraq's aggression. The level of world cooperation and condemnation of Iraq is unprecedented. Armed forces from countries spanning four continents are there at the request of King Fahd of Saudi Arabia to deter and, if need be, to defend against attack. Muslims and non-Muslims, Arabs and non-Arabs, soldiers from many nations stand shoulder-to-shoulder, resolute against Saddam Hussein's ambitions.

We can now point to five UN Security Council resolutions that condemn Iraq's aggression. They call for Iraq's immediate and unconditional withdrawal, the restoration of Kuwait's legitimate government, and categorically reject Iraq's cynical and self-serving attempt to annex Kuwait.

Finally, the United Nations has demanded the release of all foreign nationals held hostage against their will and in contravention of international law. It is a mockery of human decency to call these people "guests." They are hostages, and the whole world knows it.

[British] Prime Minister Margaret Thatcher, a dependable ally, said it all: "We do not bargain over hostages. We will not stoop to the level of using human beings as bargaining chips – ever." Of course, our hearts go out to the hostages and to their families. But our policy cannot change. And it will not change. America and the world policy cannot change. And it will not change. America and the world will not be blackmailed by this ruthless policy.

We are now in sight of a United Nations that performs as envisioned by its founders. We owe much to the outstanding leadership of Secretary General Javier Perez de Cuellar. The United Nations is backing up its words with action. The Security Council has imposed mandatory economic sanctions on Iraq, designed to force Iraq to relinquish the spoils of its illegal conquest. The Security Council has also taken the decisive step of authorizing the use of all means necessary to ensure compliance with these sanctions.

Together with our friends and allies, ships of the US Navy are today patroling Mideast waters. They have already intercepted more than 700 ships to enforce the sanctions. Three regional leaders I spoke with just yesterday told me that these sanctions are working. Iraq is feeling the heat.

We continue to hope that Iraq's leaders will recalculate just what their aggression has cost them. They are cut off from world trade, unable to sell their oil. And only a tiny fraction of goods gets through.

The communique with President Gorbachev made mention of what happens when the embargo is so effective that children of Iraq literally need milk or the sick truly need medicine. Then, under strict international supervision that guarantees the proper destination, food will be permitted.

At home, the material cost of our leadership can be steep. That is why Secretary of State Baker and Treasury Secretary Brady have met with many world leaders to underscore that the burden of this collective effort must be shared. We are prepared to do our share and more to help carry that load; we insist that others do their share as well.

The response of most of our friends and allies has been good. To help defray costs, the leaders of Saudi Arabia, Kuwait, and the United Arab Emirates have pledged to provide our deployed troops with all the food and fuel they need. Generous assistance will also be provided to stalwart front line nations, such as Turkey and Egypt.

I am also heartened to report that this international response extends to the neediest victims of this conflict – those refugees. For our part, we have contributed $28 million for relief efforts. This is but a portion of what is needed. I commend, in particular, Saudi Arabia, Japan, and several European nations which have joined us in this purely humanitarian effort.

There's an energy-related cost to be borne as well. Oil-producing nations are already replacing lost Iraqi and Kuwaiti output. More than half of what was lost has been made up. And we're getting superb cooperation. If producers, including the United States, continue steps to expand oil and gas production, we can stabilize prices and guarantee against hardship. Additionally, we and several of our allies always have the option to extract oil from our strategic petroleum reserves if conditions warrant. As I have pointed out before, conservation efforts are essential to keep our energy needs as low as possible. We must then take advantage of our energy sources across the board – coal, natural gas, hydro, and nuclear. Our failure to do these things has made us more dependent on foreign oil than ever before. Finally, let no one even contemplate profiteering from this crisis. We will not have it.

I cannot predict just how long it will take to convince Iraq to withdraw from Kuwait. Sanctions will take time to have their full intended effect. We will continue to review all options with our allies. But let it be clear: we will not let this aggression stand.

Our interest, our involvement in the gulf is not transitory. It predated Saddam Hussein's aggression and will survive it. Long after all our troops come home – and we all hope it is soon, very soon – there will be a lasting role for the United States in assisting the nations of the Persian Gulf. Our role then – to deter future aggression. Our role is to help our friends in their own self-defense, and, something else, to curb the proliferation of chemical, biological, ballistic missile, and, above all, nuclear technologies.

Let me also make clear that the United States has no quarrel with the Iraqi people. Our quarrel is with Iraq's dictator and with his aggression. Iraq will not be permitted to annex Kuwait. That is not a threat; that is not a boast; that is just the way it is going to be.

PUTTING OUR ECONOMIC HOUSE IN ORDER

Our ability to function effectively as a great power abroad depends on how we conduct ourselves at home. Our economy, our armed forces, our energy dependence, and our cohesion all determine whether we can help our friends and stand up to our foes.

For America to lead, America must remain strong and vital. Our world leadership and domestic strength are mutual and reinforcing, a woven piece, strongly bound as Old Glory. To revitalize our leadership, our leadership capacity, we must address our budget deficit – not after election day or next year, but now.

Higher oil prices slow our growth, and higher defense costs would only make our fiscal deficit problem worse. That deficit was already greater than it should have been – a projected $232 billion for the coming year. It must – it will – be reduced.

To my friends in Congress, together we must act this very month – before the next fiscal year begins on October 1st – to get America's economic house in order. The Gulf situation helps us realize we are more economically vulnerable than we ever should be. Americans must never again enter any crisis economic or military – with an excessive dependence on foreign oil and an excessive burden of federal debt....

MEETING RESPONSIBILITIES ABROAD

In the final analysis, our ability to meet our responsibilities abroad depends upon political will and consensus at home. This is never easy in democracies, for we govern only with the consent of the governed. Although free people in a free society are bound to have their differences, Americans traditionally come together in times of adversity and challenge.

Once again, Americans have stepped forward to share a tearful good-bye with their families before leaving for a strange and distant shore. At this very moment, they serve together with Arabs, Europeans, Asians, and Africans in defense of principle and the dream of a new world order. That is why they sweat and toil in the sand and the heat and the sun.

If they can come together under such adversity; if old adversaries like the Soviet Union and the United States can work in common cause; then surely we who are so fortunate to be in this great chamber – Democrats, Republicans, liberals, conservatives – can come together to fulfill our responsibilities here.

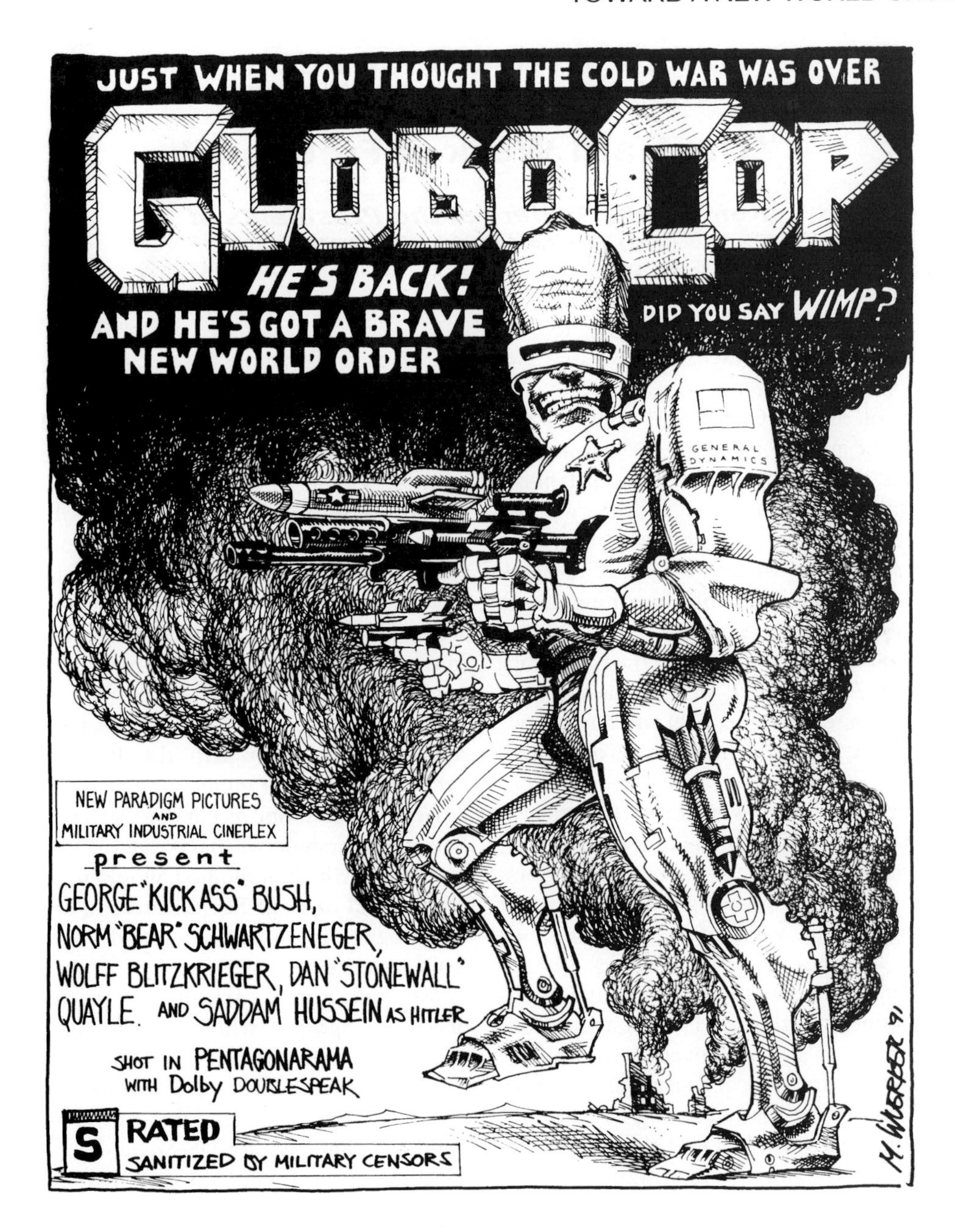

Cartoon 9 Globocop (by Wuerker)
The Gulf War and George Bush's "new world order" are represented as techno-fantasy and spectator-militarist entertainment in this cartoon by Matt Wuerker. The Cold War as a geopolitical production is over and "Military Industrial Cineplex" are selling a new geopolitical "movie" with militarist heroes, non-Western enemies, and lots of special effects.
Source: M. Wuerker

18 PRESIDENT GEORGE BUSH

"The Hard Work of Freedom"

from *Public Papers of the Presidents of the United States* (1992)

Members of the US Congress, I come to this house of the people to speak to you and all Americans, certain that we stand at a defining hour. Halfway around the world, we are engaged in a great struggle in the skies and on the seas and sands. We know why we're there. We are Americans, part of something larger than ourselves.

For two centuries, we've done the hard work of freedom. And, tonight, we lead the world in facing down a threat to decency and humanity.

What is at stake is more than one small country. It is a big idea: a new world order where diverse nations are drawn together in common cause to achieve the universal aspirations of mankind – peace and security, freedom, and the rule of law. Such is a world worthy of our struggle and worthy of our children's future.

The community of nations has resolutely gathered to condemn and repel lawless aggression. Saddam Hussein's unprovoked invasion, his ruthless, systematic rape of a peaceful neighbor, violated everything the community of nations holds dear. The world has said this aggression would not stand – and it will not stand.

Together, we have resisted the trap of appeasement, cynicism, and isolation that gives temptation to tyrants The world has answered Saddam's invasion with 12 UN resolutions, starting with a demand for Iraq's immediate and unconditional withdrawal and backed up by forces from 28 countries of 6 continents. With few exceptions, the world now stands as one.

The end of the Cold War has been a victory for all humanity. A year and a half ago, in Germany, I said that our goal was a Europe whole and free. Tonight, Germany is united. Europe has become whole and free, and America's leadership was instrumental in making it possible.

Our relationship with the Soviet Union is important, not only to us but to the world. That relationship has helped to shape these and other historic changes. But like many other nations, we have been deeply concerned by the violence in the Baltics, and we have communicated that concern to the Soviet leadership.

The principle that has guided us is simple: our objective is to help the Baltic peoples achieve their aspirations, not to punish the Soviet Union. In our recent discussions with the Soviet leadership, we have been given representations, which, if fulfilled, would result in the withdrawal of some Soviet forces, a reopening of dialogue with the republics, and a move away from violence.

We will watch carefully as the situation develops. And we will maintain our contact with the Soviet leadership to encourage continued commitment to democratization and reform. If it is possible, I want to continue to build a lasting basis for US–Soviet cooperation, for a more peaceful future for all mankind.

The triumph of democratic ideas in Eastern Europe and Latin America – and the continuing struggle for freedom elsewhere all around the world – all confirm the wisdom of our nation's founders. Tonight, we work to achieve another victory, a victory over tyranny and savage aggression.

We in this union enter the last decade of the 20th century thankful for our blessings, steadfast in our purpose, aware of our difficulties, and responsive to our duties at home and around the world.

For two centuries, America has served the world as an inspiring example of freedom and democracy. For generations, America has led the struggle to preserve and extend the blessings of liberty. And, today, in a rapidly changing world, American leadership is indispensable. Americans know that leadership brings burdens and sacrifices. But we also know why the hopes of humanity turn to us. We are Americans. We have a unique responsibility to do the hard work of freedom. And when we do, freedom works.

The conviction and courage we see in the Persian Gulf today is simply the American character in action. The indomitable spirit that is contributing to this victory for world peace and justice is the same spirit that gives us the power and the potential to meet our toughest challenges at home.

We are resolute and resourceful. If we can selflessly confront the evil for the sake of good in a land so far away, then surely we can make this land all that it should be. . . .

[. . .]

This nation was founded by leaders who understood that power belongs in the hands of people. And they planned for the future. And so must we, here and all around the world. As Americans, we know there are times when we must step forward and accept our responsibility to lead the world away from the dark chaos of dictators, toward the brighter promise of a better day. Almost 50 years ago, we began a long struggle against aggressive totalitarianism. Now we face another defining hour for America and the world.

There is no one more devoted, more committed to the hard work of freedom, than every soldier and sailor, every marine, airman, and Coast Guardsman, every man and woman now serving in the Persian Gulf

. . . What a wonderful, fitting tribute to them. Each of them has volunteered – volunteered to provide for this nation's defense, and now they bravely struggle, to earn for America, for the world, and for future generations, a just and lasting peace. Our commitment to them must be the equal of their commitment to their country. They are truly America's finest.

The war in the Gulf is not a war we wanted. We worked hard to avoid war. For more than 5 months, we, along with the Arab League, the European Community, and the United Nations, tried every diplomatic avenue. UN Secretary General Perez de Cuellar, Presidents Gorbachev [of the Soviet Union], Mitterrand [of France], Ozal [of Turkey], Mubarak [of Egypt], and Benjedid [of Algeria], Kings Fahd [of Saudi Arabia] and Hassan [of Morocco], Prime Ministers Major [of the United Kingdom] and Andreotti [of Italy] – just to name a few – all worked for a solution. But time and again, Saddam Hussein flatly rejected the path of diplomacy and peace.

The world well knows how this conflict began and when. It began on August 2nd, when Saddam invaded and sacked a small, defenseless neighbor. And I am certain of how it will end. So that peace can prevail, we will prevail.

Tonight, I am pleased to report that we are on course. Iraq's capacity to sustain war is being destroyed. Our investment, our training, our planning – all are paying off. Time will not be Saddam's salvation. Our purpose in the Persian Gulf remains constant: to drive Iraq out of Kuwait, to restore Kuwait's legitimate government, and to ensure the stability and security of this critical region.

Let me make clear what I mean by the region's stability and security. We do not seek the destruction of Iraq, its culture, or its people. Rather, we seek an Iraq that uses its great resources not to destroy, not to serve the ambitions of a tyrant, but to build a better life for itself and its neighbors. We seek a Persian Gulf where conflict is no longer the rule, where the strong are neither tempted nor able to intimidate the weak.

Most Americans know instinctively why we are in the Gulf. They know we had to stop Saddam now, not later. They know that this brutal dictator will do anything, will use any weapon, will commit any outrage, no matter how many innocents must suffer. They know we must make sure that control of the world's oil resources does not fall into his hands, only to finance further aggression. They know that we need to build a new, enduring peace based not on arms races and confrontation, but on shared principles and the rule of law. And we all realize that our responsibility to be the catalyst for peace in the region does not end with the successful conclusion of this war.

Democracy brings the undeniable value of thoughtful dissent, and we have heard some dissenting voices here at home: some - a handful - reckless; most responsible. But the fact that all voices have the right to speak out is one of the reasons we've been united in purpose and principle for 200 years.

Our progress in this great struggle is the result of years of vigilance and a steadfast commitment to a strong defense. Now, with remarkable technological advances like the Patriot missile, we can defend against ballistic missile attacks aimed at innocent civilians.

Looking forward, I have directed that the SDI [Strategic Defense Initiative] program be refocused on providing protection from limited ballistic missile strikes – whatever their source. Let us pursue an SDI program that can deal with any future threat to the United States, to our forces overseas, and to our friends and allies.

The quality of American technology, thanks to the American worker, has enabled us to successfully deal with difficult military conditions and help minimize precious loss of life. We have given our men and women the very best, and they deserve it. . . .

We will succeed in the Gulf. And, when we do, the world community will have sent an enduring warning to any dictator or despot, present or future, who contemplates outlaw aggression. The world can, therefore, seize this opportunity to fulfill the long-held promise of a new world order – where brutality will go unrewarded and aggression will meet collective resistance.

Yes, the United States bears a major share of leadership in this effort. Among the nations of the world, only the United States of America has had both the moral standing and the means to back it up. We are the only nation on this Earth that could assemble the forces of peace. This is the burden of leadership and the strength that has made America the beacon of freedom in a searching world.

This nation has never found glory in war. Our people have never wanted to abandon the blessings of home and work for distant lands and deadly conflict. If we fight in anger, it is only because we have to fight at all. And all of us yearn for a world where we will never have to fight again.

Each of us will measure, within ourselves, the value of this great struggle. Any cost in lives, any cost, is beyond our power to measure. But the cost of closing our eyes to aggression is beyond mankind's power to imagine. This we do know: our cause is just, our cause is moral, our cause is right.

Let future generations understand the burden and the blessings of freedom. Let them say we stood where duty required us to stand. Let them know that together we affirmed America, and the world, as a community of conscience.

The winds of change are with us now. The forces of freedom are together and united. And we move toward the next century, more confident than ever that we have the will at home and abroad to do what must be done – the hard work of freedom.

May God bless the United States of America.

19 TIMOTHY W. LUKE

"The Discipline of Security Studies and the Codes of Containment: Learning From Kuwait"

from *Alternatives: Social Transformation and Humane Governance* (1991)

WHAT HAPPENS AFTER THE END OF HISTORY?

For nearly five decades, the various disciplines of national security studies have been dedicated to defining and applying the complex codes of containment at the various economic, military, political, social, and strategic fronts of the struggle between, as these rhetorics of power framed it, capitalism and communism, the West and the East, democracy and totalitarianism, the United States and the Soviet Union. From Yalta to Malta, the frozen tundras of bloc politics provided a peculiarly fixed terrain, which the disciplinary readings of national security studies could somewhat reliably map with their anti-Communist/antitotalitarian codes of containment. During 1989–1991, however, tremendous changes, working from above and from below, have upended the fields of reference and zones of difference that once anchored the disciplinary reach of security studies to the strategic projects of Cold War-era containment. With the velvet and violent revolutions in Eastern Europe as well as perestroika in the Soviet Union, these frozen terrains of Cold War combat are melting into far more mushy, if not totally fluid, expanses of almost inchoate confusion.

The complex national security apparatus in the United States gained full articulation from 1945 to 1947 as the United States recognized that its vast economic resources, conventional military capabilities, and nuclear monopoly could be used to enscribe a new kind of transnational order upon Europe and Asia against the resistant designs of a much less capable, but equally expansionistic, Stalinist state socialism. To contain the Soviet Union, and restrain German unification, the NATO alliance against the USSR and Warsaw Pact provided a fixed frame of international conflict and competition for more than four decades. The Cold War, in large part, was an elaborate "strategy of global enscription that was both extensive and intensive in its disciplinary effects: the great scope of anticommunism as a discourse of danger was matched by its impact on the details of everyday life in the United States."[1] In fact, these effects were felt even in the darkest corners of the far-flung territories under the shadows of the nuclear umbrella providing extended deterrence beyond the United States.[2] However, things have been changing. The postwar division of Germany, which rested at the heart of these complex undertakings, ended in 1989–1990 as Berliners on both sides of the wall tore down this key physical and political barrier along the East–West frontier, and other Germans rapidly reunified the Democratic and Federal Republics under the guidelines of Bonn's liberal democratic constitution. Almost simultaneously, Czechoslovakia, Poland, Hungary, Romania, and Bulgaria also repudiated their involvement in Cold War bloc politics by overthrowing their ruling Communist parties in 1989–1990 and nullifying the Warsaw Pact in 1991. Finally, at the November 1990 Paris summit, Presidents Gorbachev and Bush along with other assembled chief executives of the major European states declared that the Cold War, and hence its traditional containment-driven conflicts, was dead and gone. Afterwards, President Bush jetted off to visit US troops in Saudi Arabia, and President Gorbachev flew back home to his battles to somehow keep the Soviet Union together as a "union" that is "soviet." Unlike the Soviet Union of old Cold War containment narratives, the Soviet Union now is not posing any serious ideological, political, or cultural threat to the West as "communism." Instead, it today is the resentful recipient of emergency European food relief, the eager customer of US fast-food franchises, and the uneasy scene of a Lebanon-like global media deathwatch as the new "sick man" of Europe.

These changes are extremely problematic inasmuch as the United States organized its national security, discursively and operationally, for nearly fifty years around four goals: (1) resisting a confident, expansionist, Communist Soviet Union anywhere in the world, especially in Germany and Western Europe; (2) keeping Germany from being unified without Washington's (and Moscow's) approval; (3) promoting the eventual liberation of peoples and states under Soviet occupation in Central and Eastern Europe; and (4) maintaining a nuclear deterrence structure capable of checking the Soviet Union from initiating World War III and/or expanding farther outside of 1939 (or 1941) national borders. One can cite a series of events sometime after Chernobyl and before Kuwait that have neutralized most of this discourse's primitive assumptions. The strategic codes of containment policy once did drive the West's resistance against the East in accord with the discursive demands of these basic goals, but now "anti-communism," as a discourse of danger, largely has run out of gas. Some see this as "the end of history."[3] Actually, it merely appears to be the end of Cold War history.

Without these guiding principles of Cold War conflict, then, what happens to the discipline of national security studies and the codes of containment in the United States and its fragmenting Cold War Western bloc? This article attempts to address these issues in reconsidering some of the larger still unexplained tendencies exposed by the recent war against Iraq over Kuwait. Cold War-style reasoning continues to dominate US strategic thought inasmuch as the premise of containment, directed against any threatening evil otherness now rather than simply communism, and balance-of-power politics, tied to the correlation of forces in particular regional competitions for primacy, underpins Washington's responses to foreign crises. Consequently, in the Kuwaiti conflict, one might find traces of new models of containment, new types of alliance, and new kinds of conflict to legitimate the disciplinary demands of contemporary national security as the United States faces the post-Cold War era of the 1990s and beyond. Wars frequently compress social changes into brief intense bursts of rapid transformation as well as perhaps heightening the range of critical insights that might be made about the implications of these war-induced changes. The Gulf War of 1991 is no exception to this rule.

FROM THE FLOW OF POWER IN SPACE TO THE POWER OF FLOWS APACE?

Beginning with the debates in the 1950s and 1960s about "technological society" or "post-industrial society," critical discourses of social analysis have remarked upon the many apparently new qualities of modern industrial society.[4] These transformations are still not completely understood, but they seem to be altering the most basic composition of the nation-state and the essential arrangements of the contemporary world-system of nation-states, transnational corporate commerce, and supranational ideopolitical blocs. Often these transformations are discussed as aspects of postmodernism.[5] One of the most pervasive influences, however, driving these shifts in structure and substance appears to be the "informationalization" of the social means of production, consumption, administration, and destruction during and after the 1950s and 1960s, when the global impact of mass telecommunications, electronic computerization, cybernetic automation, rapid transportation, and flexible accumulation began to be experienced more broadly.[6]

At this juncture, what provisional observations can be made about informationalization? Power, time, and space appear to take much different forms in informational society than those once found in industrial social relations with their perspectival sense of location, hierarchy, and organization. The organizational logic of preinformational society is anchored to places, as power draws boundaries around space; erects monetary, military, and managerial borders around space; and exercises a monopolistic writ of sovereignty within these delimited expanses by exerting, guiding, or directing its effects from point to point or place to place within space.[7] The stability, security, and sovereignty of state power, then, most often have been stated and comprehended in essentially spatial terms through geopolitical discourses of expansion, military defense, or economic development. Panoptic surveillance from the center and top of this space by state agencies works to normalize activities within it to suit the monetary, military, and managerial agendas of its state structures' leadership.

Security of place is assured by guarding against intrusions from competing, and usually contiguous, state apparatuses that seek to penetrate or annex more space to enact their economic, political, and cultural/ administrative agendas. Sovereignty follows from an almost mythic power of geographic authority, writing and drawing lines of identity and antagonism on the Earth. States, in turn, are those legitimate monopolies charged with enscribing, discursively and coercively, writs of difference – in money, religion, markets, ideology, and militaries – from what transpires within and without the geopolitical spaces framed by international borders. As Campbell claims, "The presence of sovereign states in an anarchic realm is a spatial conception that privileges a geopolitical reading of global politics."[8] By endogenizing various disciplines of monopolistic order inside, and exogenizing diverse practices of free-for-all anarchistic conflict outside, those borders defining each nation-state's place on the planet's terrain, the fictive practices of political self-rule, or national sovereignty, define themselves spatially against the landscapes (as national religions, coinage, and armies set limits of their power), cloudscapes (as aircraft made airspace a significant concern), and seascapes (as naval cannon made territorial waters more defensible) of the Earth in a conjunctive, centralizing hierarchical order. Defending borders, controlling airspace, and patrolling offshore waters all are regarded legitimately as essential practices for drawing, defining, and then disciplining the various places of national territory that contain the social activities differentiating this nation-state from that nation-state. As Walker noted, "The principle of state sovereignty suggests a spatial demarcation between those places in which the attainment of universal principles might be possible and those in which they are not."[9]

Always essentially fictive constructs of linear space in real time, nations, states, territories, or possessions are also discursive fields of state authorship enscribed upon individuals and groups, whose attributes and behaviors are continuously remanufactured by the coercive gaze or normalizing hand of state power. Informationalization, however, alters these power dynamics by generating new organizational logics nested in flexible accumulation's rapid and intense flows of ideas, goods, symbols, people, images, and money on a global scale, which are disjunctive and fragmenting, anarchical and disordered. Of course, a "transnational" flow of goods, capital, people, and ideas has existed for centuries; it certainly antedates even the rise of modern nation-states. However, this flow, at least until the late 1950s or early 1960s, tended to move more slowly, less, and more narrowly than the rush of products, ideas, persons, and money that developed with jet transportation, electronic telecommunication, massive decolonization, and extensive computerization after 1960. Hence, it appears that these greater intensities, rates, densities, levels, and velocities in the flow have quantitatively transmuted it into something qualitatively new and different. Today's global marketplace is very unlike the medieval spice trade, the Renaissance market in old manuscripts, or early modern intraimperial trade in slaves, raw materials, and hard specie. Rather than acceding to a privileged geopolitical (or geoeconomic) reading of global power, therefore, it might generate many different grammars for a less well-understood chronopolitical (or chronoeconomic) reading of planetary political processes.[10] Power today also often flows more placelessly beneath, behind, between, and beyond boundaries set into space as new senses of artificial location become very fluid or more mobile, defined by shifting connections into the networks of information carrying these flows.[11] Geopolitical barriers are articulated as cartographic traces, memberships in military pacts, and diverse denominational codes in national monetary currencies. Informational flows rarely are stymied for long by such barriers; indeed, cross-border flows of money, influence, and knowledge are heavily eroding such notions of geopolitical borders, while perhaps producing their own new kinds of barriers.[12]

Tightly bounded ethnogeographic settings are augmented, if not often almost entirely supplanted, by new complex cultural activities on continuously flowing mediascapes of transnational scope and content. The flow is an ephemerally existing configuration of particular images, symbols, and meanings about power, money, and value – channeled through transnational corporations, scientific communities, banks, and telecommunication networks – in a continually shifting sign flux at this or that point, setting its own terms of access, collaboration, and service through multiplex streams of code. Codes create new spaces, new times, and new powers in the operations of the modes of information. Information communicates its effects not merely by conveying content like cargo; it also "forms" by informing as something variable, adaptive, and multiplex. To become informational is to continually be in-formed and in-forming, and in-formation as codings and decodings are formed-in, forming-in, and formations-in communicative exchanges. With such interoperative dynamics, there is both a presence and an absence of borders, space, and process-in-time. Power, taken as coding capabilities and symbolic competence, determines access to these in forming cyberspaces; delineates in-formed monetary, military, and managerial connections; and defines oligopolistic formations-in interoperation within these manifold streams by containing, modulating, amplifying, or resisting their effects at various levels, rates, or dimensions forming-in flows. Chronopolitics is grounded in the pace of exchange; how rapidly the flows can travel, expand, and unfold without meeting resistant barriers or closed borders becomes much more significant. Dominating the pace of process, setting the tempos of interaction, or managing the speed of exchange are the critical points of power in these informational systems of order. Here, barriers and borders are marked by user access or nonaccess, producer participation or nonparticipation, consumer linkage or nonlinkage, or symbolic complementarity or noncomplementarity.

Moving from place to flow, spaces to streams, introduces nonperspectival, antihierarchical, and disorganizational elements into traditional spatial/industrial/national notions of sovereignty. Without conceptually making these events a simple transition from one abstract state of political economy to another newer timeless state of abstract political economy, certain distinctions can be made. The ethnogeographic settings of self-rule defined by the classical Westphalian universe of borders, shorelines, and airspaces in spatially construed grids of/for sovereignty increasingly collide in the transnational multiverse of technoregions generated out of global monetary transactions, commodity exchanges, technical commerce, telecommunication links, and media markets. Having open and unconstrained access to the flows, not closed domination of places, perhaps becomes as crucial an attribute as sovereignty in informationalized societies. Likewise, stability spins through the codes by maintaining dynamic equilibria of access, linkage, turnover, connection, exchange, and service in accord with the diverse agendas of the various different encoders and decoders, whereas security slips alongside concerns over the assuring integrity of codes, openness of access, extent of service, scope of linkage, and increase of turnover. Caught in the currents of these hyperreal forces moving across the mediascapes and cyberspaces of informationalization, the nation-state – with more traditional geopolitical concerns for policing its territories, populations, and markets – often comes up short without total closure. When moving on these terrains, as Der Derian claims, one might supplement existing categories exclusively tied to geopolitics and the control of space by adopting alternative notions linked to chronopolitics and the control of pace.

The directions of flow are intentionally

guided, place-oriented, and socially sited at one, several, or many places. Nonetheless, when considering the flow,

> the organizational logic is placeless, being fundamentally dependent on the space of flows that characterizes information networks. But such flows are structured, not undetermined. They possess directionality, conferred both by the hierarchical logic of the organization as reflected in instructions given, and by the material characteristics of the information systems infrastructure.[13]

Given these larger structural trends, the reality of place, expressed in terms of a sociocultural context of spatial location, gradually is being resituated within the hyperreality of flow, understood in terms of iconic/symbolic access to or process through networks of informational circulation. The latter is not displacing or destroying the former, but rather they are coexisting together. Therefore, from these building contradictions within a dialectic of organizational centralization and informational decentralization "between places and flows" one might uncover in the workings of global change "the gradual transformation of the flows of power into the power of flows."[14]

As the hyperrealities of informational exchange unfold in the cyberspaces of informationalized processes, many new questions arise. Again, as Castells asserted,

> there is a shift, in fact, away from the centrality of the organizational unit to the network of information and decision. In other words, flows, rather than organizations, become the units of work, decision, and output accounting. Is the same trend developing in relation to the spatial dimension of organizations? Are flows substituting for localities in the information economy? Under the impact of information systems, are organizations not timeless but also placeless?[15]

In fact, flows are becoming the bases of defining new core, semiperipheral, peripheral, and external areas as they restructure the economic status and market niches of cities, regions, and countries. Without a tie into its currents, many once peripheral areas, particularly in Africa and Southeast Asia, seem to be slipping back into externalized zones of precapitalist existence, capturing only TV transmissions, food aid hand outs, or black market links with the outside world. Actually, the shadow trade in drugs, political influence, illegal weapons, environmentally banned contraband, or even human labor increasingly anchors many regions' tenuous cash nexus with global flows.

Most importantly, the cycles of informationalization seemingly entail the creation of entirely new sociospatial, semiopolitical, and sociochronic logics that simultaneously generate the (s)pace of contemporary power, ideology, and exchange dynamics. The flow is partly postspatial, partly postsovereign, and partly perhaps the beginning of a new kind of international community. Reich contrasted the "nominal nationality" held by many modern major corporations with their "actual transnationality" as global parts sourcing, foreign markets, expatriate management, multinational labor recruiting, and worldwide financial operations increasingly typify their operations.[16] This observation also can be extended to many scientific communities, cultural networks, technological innovations, telecommunication links, and media markets. Nominal nationality, or geopolitical spatiality, increasingly competes with actual transnationality, or chronopolitical flowality, in the processes of many international events and trends. It is a decentering, despatializing, and dematerializing force as it works alongside and against the geopolitical codes of spatial sovereignty.

Within the flows, there are new universals and new particulars being created by the networks of transnational exchange as fresh identities, unities, and values emerge from sharing access to the same symbols, markets, and commodities. On the one hand, the flow might be seen as essentially constituting entire "nations" or "countries," such as Panama, Hong Kong, Grand Cayman, or Singapore, as necessary flags of convenience. Or, on the other hand, single nations, such as West Germany, Japan, or the United States, have attempted to exclusively nationalize currents in the flow as "trading states" or "neomercantilists" to re-create their national power. Outside of the state, and inside of shared technological goals, common ecological challenges, similar symbolic systems, parallel coding orders, and comparable product meanings, the flow does create new transnational communities that are blurring the old geographics of "them" and "us," "other" and "I," or "friend" and "foe." Most of its component currents can still be traced back to ethnogeographic settings or the spaces of nominal nationality; however, their effects, taken together in the streams of the global flow, are being felt *postnationally* or locally and globally (perhaps "glocally"), as actual transnationality. In these contradictions alone, the battlelines over political community, psychosocial identity, and social justice can be sharply redrawn. Postmodern politics, as a global product of informationalization, is proving highly explosive as the flow implodes the geopolitics of nation-states.

If these narratives on informationalization can be trusted, the cyberspaces of interoperational coding in global flows now seem to define social purpose and performance as much as the more traditional geographics of organizational and national boundaries. Beyond the realities of territorial statics, fixed to structured processes inside of tightly enscribed borders, are the hyperrealities of flow dynamics fluctuating within coded links along loosely coupled networks. Many events over the past twenty-five years have been calling attention to these sorts of shifts; however, the invasion of Kuwait by Iraq in 1990 and the global response by thirty different nations to liberate Kuwait from Iraqi domination during 1991 provide several glimpses at these shifting logics in national security, particularly when seen closely juxtaposed to the ending of the Cold War in Eastern and Central Europe.

SOVEREIGNTY IN CYBERSPACE: NATION-NESS AS HYPERREAL ESTATE?

Kuwait is a uniquely suggestive case for discussing the nature of nation-states in contemporary informational world-systems. In one sense, a good measure of its suggestive qualities can be tied to the Kuwaitis' uniquely feudalistic ruling family. Once the emir and his extended family were bundled into their limousine motorcade and rushed out of the country, the legitimate state authorities – along with their patrimonial state's essential records – were free to operate in exile. Although much of Kuwait's oil wealth is tied to al-Sabah family accounts, there are also public funds established for the common good of those few hundred thousands of Kuwaiti citizens once fortunate to be its nationals. This combination of great wealth, fixed incomes from oil resources, large remittances from foreign investments, a relatively small number of national constituents, a patrimonial ruling elite, and an aristocratic state located on a tiny expanse of territory provides some additional evidence of how the geopolitical state and chronoeconomic flow interpenetrate.[17]

After gaining independence in 1962 as a constitutional monarchy, in which the emir of the al-Sabah family occasionally consulted with a parliament elected by only about 10 per cent of Kuwait's population, Kuwait developed into an oil-driven, patrimonial welfare state for its 600,000 resident citizens. These "native" inhabitants, in turn, provided jobs for an estimated 1.2 million expatriate workers, who ultimately did almost all of the work in the Kuwaiti economy and state. Nearly 400,000 of these expatriates were Palestinians; the remainder came from all over the world. Yet, the hostilities radically changed this demographic profile of Kuwait. At the war's end, most of these expatriates were gone, as were most of the native Kuwaitis. Estimates in early March 1991 indicated that virtually all non-Arab expatriates had left during the Iraqi occupation, probably half (or 200,000) of the Palestinians were no longer in the emirate, and perhaps two-thirds (or 400,000) Kuwaitis had left Kuwait during the crisis.

As an ethnographic place, then, Kuwait has had an ambiguous identity since its inception in the 1920s, when its territorial scope and native citizenry were basically regarded as all those lands and inhabitants within a fifty-mile radius of Kuwait City. At first, local merchant families worked in harmony with the al-Sabah family, which survived on the meager revenues produced by local commerce, pearl harvesting, and fisheries. The oil boom, of course, changed everything, as billions of dollars began flowing into the Kuwaiti economy from the Western developed oil fields after World War II. By 1990, the emirate had more than 900 major oil wells plus highly sophisticated pipelines, refineries, and other oil-shipping infrastructure in place and producing nearly $18 billion a year in oil revenues. Although the emirate, in a sense, "controlled" these activities within its borders, in fact, Kuwait was a small albeit major subassembly in the transnational machinery of global energy markets.[18]

Because of its tremendous oil revenues and truly limited abilities to absorb new investments after most "native" Kuwaitis had been served by the welfare state, in the early 1970s Kuwait began a diversified global investment strategy in foreign real estate, industrial firms, service businesses, and cash reserves that amounted to an estimated net worth of around $100 billion by 1990. One fund alone, the Fund for Future Generations, receives about 10 per cent of Kuwait's annual oil income; it stood at more than $40 billion when the Gulf War began.[19] Actually, even though Kuwait was the world's sixth largest oil producer, the emirate's financial earnings exceeded its oil income during the 1970s and 1980s. Having little territory, population, or industry to develop within the spaces of Kuwait, the Kuwait state acquired title to streams of wealth production flowing abroad in larger, more populous, and industrialized economies. These developments, at the same time, helped remake Kuwait, in some ways, into an unusual new series of cyberspatial portfolios, electronically variable capital masses, or hyperreal estates accumulated in a massive informational cross section from its commercial interoperations within various streams of the transnational flow. Although they are place oriented, and initially place generated, once entered into the flow, the benefits derived from these financial assets also become postspatial, and even potentially postnational, as the Iraqi invasion of August 2, 1990, illustrated.

Saddam Hussein, by invading and holding Kuwait's territories, also hoped to gain and keep its electronically accumulated wealth, but Kuwait's riches are, to a large extent, no more than its access codes to the flow. Saddam was foiled on this front, as the al-Sabah family and its bureaucratic retainers retreated spatially into Saudi Arabia, simultaneously barricading themselves and Kuwait's assets electronically inside the flow. Kuwait, as territorial real estate, was subjected to the preinformational logics of sovereign control by its neighbor Iraq, whose conquering hordes had to content themselves with looting Kuwait City of its gold faucets, hospital equipment, Rolls Royces, 747s, and consumer electronics. Yet, even as Kuwait as "the place in space" was being annexed, or deterritorialized, Kuwait as "a stream in flow" simply changed its passwords, recoded its access protocols to open at other nodes, and respecified its service-delivery points.

Already highly postspatial, extraterritorial, and dematerialized in its prewar activities, Kuwait as the coded streams of stockholdings, bank accounts, oil business, service obligations, and ethnonational symbols only had to "disk dump" everything to more secure nodes in flows of the computer and global media networks. Remaining free and at large in new corporate offices located in Washington, DC, and the Sheraton Al-Hada Hotel in Taif, Saudi Arabia, Kuwait the cyberspace was able to keep up a minimum level of "state services," paid for with its extensive assets, to its exiled populace in Saudi Arabia, North America, and Western Europe, while rearming itself within the ranks of the international coalition and hiring public relations consultants to articulate wounded rage over Iraq's violation of its territories.[20] During its exile, of course, the Kuwaiti state's fragility was totally exposed. As ministries were run out of double rooms with a bulletin board and a phone in the Taif Sheraton, the emirate could do very little but invalidate its old currency, try to plan for its return, and lobby on global talk shows for its territorial liberation.

By nullifying its old currency and printing a new one, the emirate readied to restart its management of the flow into its own Kuwaiti territory, but overcoming the disruption of the occupation without a major political restructuring will be very difficult.[21] Indeed, internal instability and mass political dissatisfaction will undoubtedly worsen in the emirate as it rebuilds, because for so long the availability of almost unlimited petrodollars covered many sins – basically by buying social peace among Kuwaitis and their expatriate helpers – prior to August 2, 1990. After the invasion, however, all of the essential contradictions and conflicts lying latent within Kuwaiti public life have been exposed, leaving Kuwait after its liberation a far more volatile and unstable place. Nonetheless, as a cyberspace existing largely on the TV screens of its sympathizers and the monitors of its foreign bankers, Kuwait then helped these allies retake its material-national space, which had, in turn, been reinscribed with the mark of Baghdad's authority as Iraq's "province 19." No longer able to show the flag or its force on its own real estate, Kuwait nonetheless continued flying the image of its flag and making tremendous shows of force from its hyperreal estate.

The essentially fictive nature of many contemporary nation-states, then, is more fully exposed by the Kuwaiti and Iraqi experiences in the Gulf War. As a classically styled authoritarian state, using modernist myths of military conquest, supreme leadership, national mission, and chiliastic global change to create a sense of nationhood out of its various ethnic, religious, and linguistic minorities, Iraq – like fascist Spain, Portugal, Argentina, Japan, Italy, or Germany before it – demonstrated the bankruptcy of spatial expansion, place domination, and territorial imperialism in the informational flows of contemporary world systems. Kuwait, on the other hand, as a bizarrely postmodern fusion of pre-modern feudalism with informational capitalism, is more of a place-oriented stream within the global flow of money, ideas, goods, symbols, and power. As a point of production and consumption in the flow, however, Kuwait far outclassed Iraq in global significance, even though it has fewer people, less territory, and a smaller military force.[22] Kuwait's informational cross section, or electronic signature, in the global flow by far exceeds that cut by Iraq, which has instead chosen a spatial path of nation-building by investing in its own advanced means of destruction rather than the advanced transnational means of information to drive its modes of production.

Confronting the flows of power coaligned in the thirty coalition partners, Iraq – with territory the size of California, population the size of the Netherlands, and GNP equal to that of Portugal – had little hope for success, particularly once its military ties and diplomatic friendship with the fixed, but brittle, geometry of alliance with the Soviet Union and former Warsaw Pact countries were broken by coalition maneuvering.[23] Iraq took Kuwait's real estate, but failed to capture its hyperreal estate. It conquered Kuwait's territorial space, but lost access to the genuine riches on deposit in its informational cyberspace. And, in turn, it provoked a rapidly deployed counterstrike by a flexibly configured variable geometry alliance of North American, South American, Australasian, African, European, and Asian nations, which was mostly put together on the phone by President Bush or on shuttle junkets by Secretary of State James Baker. Although many nations in the alliance provided only token contributions, its diplomatic writs did carry most of the major electronic signatures in the flow. Such a massing of global power, at the same time, so tightly compressed its impact in both space and time that only 100 hours were needed to erase Iraq's spatial annexation of Kuwait as well as most of Baghdad's carefully accumulated means of military destruction.

POST WAR AS RETRO WAR

In many respects, political ideologies in contemporary informational society unfold "as an immense accumulation of spectacles" in the mythological discourses of the mass media. As ideologies, various spectacular treatments of social relations can be continuously coded and recoded in streams of images, which constitute a peculiar political discourse about, but also within, the regime of transnational corporate society. One of the most fundamental and long-lasting scripts for these transnational discourses in the twentieth century is the mythology of World War II.[24] "The spectacle," as Guy Debord asserts, "is not a collection of images, but a social relation among people, mediated by images."[25] As a social relation among people,

World War II texts can be used to create social relations between nations as their mediagenic mythologies of collective purpose, common identity, and communal cooperation provide new ideas-in-form-and-action for those who recall these texts in times of crisis. As Roland Barthes suggested, the generation and reception of such meanings can be systematically studied: "It is a matter of studying human groups, of defining motives and attitudes, and of trying to link the behavior of these groups to the social totality of which they are a part."[26] Whenever and wherever a small, weak nation is threatened by a larger, stronger nation, or a militaristic authoritarian dictator challenges a relatively peaceful neighboring society, the discursive work-ups of World War II can be flexibly deployed to interpret, explain, and legitimate elite and mass responses in readily accessible and virtually uncontestable rhetorical terms. For example, Prime Minister Thatcher's and President Bush's rapid reinterpretation of Saddam Hussein's invasion of Kuwait used narrative tropes such as "naked aggression by the State of Iraq," in Bush's words.[27] Their meeting in Aspen, Colorado, during August 1990, as well as the international coalition's repeated rehearsal of these themes in its war of reconquest only underscore the irresistible power of this sort of "good war" rhetoric.

To fight a "good war," Bush rhetorically turned to one of World War II's greatest surviving international organizations: the United Nations. In October 1990, President Bush asked the United Nations for its assistance in his crusade against Saddam, because its resolute aid could help "bring about a new day.... A new world order and a long era of peace."[28] And, like World War II, with its alliance of capitalism and socialism against fascism, Bush quickly enlisted the aid of the Soviet Union. Seeing the changed terrain in Eastern Europe after 1989, he also signed up Poland and Czechoslovakia in the anti-Saddam coalition. When he visited Prague in November 1990, Bush announced the world had "a historic opportunity" in the Persian Gulf: "The opportunity to draw upon the great and growing strength of the commonwealth of freedom and forge for all nations a new world order far more stable and secure than any we all have known."[29] Everyone, even old Cold War enemies and quasi-feudal Islamic kingdoms, could join together with the West in this transnational commonwealth of freedom by mustering out to fight this "good war."

World War II remains deeply entrenched, symbolically and rhetorically, as the Western world's vision of a "just war." Drawing parallels in any present conflict to events, persons, or organizations in World War II can generate tremendous symbolic energies to direct against the opponent as well as to fuel domestic support. From the invasion's beginnings in August 1990, Saddam Hussein and Iraq provided a hermeneutic field day for US and global discourses to recharge these potent engines of World War II mythology. As the supreme leader of a secular one-party state, which rules in large part through terror and propaganda, Saddam Hussein immediately became Adolf Hitler. Baghdad, in turn, became Berlin, full of fascistic architecture, mindlessly loyal crowds, and imperialistic designs upon its neighbors' territories. Iraq, then, became today's expansionistic equivalent of Nazi Germany, and Kuwait assumed the role of the totalitarian empire's weak helpless victim, like Czechoslovakia, Poland, Norway, Denmark, Belgium, or the Netherlands. The emir and his people, like the hordes of refugees of World War II, fled their homeland into exile to await the liberation of their homeland by their own Free Kuwaiti Forces and a new international coalition of antitotalitarian nations.

Hitler, of course, was the cruel dictator who went unchallenged until it was almost too late. Moreover, he ruthlessly terrorized his own people, killed millions in his death camps, violated the legal structures weakly established by the League of Nations, developed new weapons of mass destruction, and sought *Lebensraum* all across Europe. The parallels, real and imagined, between Hitler and Saddam Hussein were continually hit upon in US rhetoric. Seeing the power of the Hitler mythololology, Bush recognized that Saddam had to be cast in the same role in Bush's rehearsal of the World War II script. In his November 1990 *Newsweek* essay, for example, Bush tied Saddam directly to the horrors of Iraq's invasion, which was, in turn, cast in the same light as the Nazi invasions of Western Europe in 1939–1940:

> Iraq's occupation of Kuwait has been a nightmare. Hundreds of thousands of Kuwaiti men, women and children have been driven from their country; Saddam has brought in tens of thousands of Iraqis and other foreigners to settle in their place. Homes, buildings and factories have been looted. Babies have been torn from incubators; children shot in front of their parents, disappearances and graphic accounts of torture are widespread.[30]

Thus, Hitler the barbarian provided a fitting geopolitical costume for Saddam, who, once placed in the role of savage dictator, could serve as the rhetorical linchpin of World War II-style scripts for the US response to his many crimes.

Because of these discursive frames, Kuwait became something worth fighting and dying for. Like World War II, the Gulf War had most of the symbolic attributes of being the right war fought at the right time and with the right strategies, leadership, and weapons. The battle lines were quite clear, the fronts were sharply defined, and the costs or benefits of victory were materially obvious. As these discourses framed the question, or as President Bush himself asked, "Can the world afford to allow Saddam Hussein a stranglehold around the world's economic lifeline? This is exactly what would happen if we failed.... Energy security is national security, and we must be prepared to act accordingly."[31] The war was about oil, because oil means, as Secretary Baker and President Bush claimed, jobs, growth, or even our "very way of life." Hence, Kuwait was the place to stand and fight to preserve these important values. Moreover, as the discourses of World War II show, and as Bush saw it, "innocent lives are at stake," and as long as Saddam was in Kuwait, Americans, Kuwaitis, and other victims of his aggression could not live "free from fear."[32]

Hence, most importantly, in the first serious threat to the ground rules of global cooperation since the Soviet–American Malta summit, the World War II discourses called to mind the Czechoslovakian crises of 1937–1938 or the invasion of Poland in 1939 – that is, as President Bush called it, "the world must not reward aggression," as the League of Nations and great powers had tolerated in Hitler, Mussolini, and Stalin. Therefore, Kuwait assumed world-historical importance for President Bush as he stuck to the scripts of World War II discourses:

> Iraq's invasion and occupation of Kuwait is not just a case of petty aggression. The civilized world is now in the process of fashioning the rules that will govern the new world order beginning to emerge in the aftermath of the Cold War. The history of this century shows clearly that rewarding aggression encourages more aggression. If the world looks the other way in this first crisis of the Post Cold War era, other would-be Saddams will conclude, correctly, that aggression pays. We must either be prepared to respond now or face a much greater set of challenges down the road.[33]

Fighting for, and then liberating Kuwait, therefore, becomes equivalent to invading and taking back Europe from fascism. From the victory, an entirely new world order will be born, based upon the notion of collective defense against aggression and flexible containment of current-day, or would-be, Saddams seeking to make aggression pay by invading and occupying their neighbors' territories. Yet, just as Bush sought to link his struggle to liberate Kuwait to the powerful World War II scripts, he also worked to sharply differentiate it from the narratives of Vietnam. As he maintained in early December 1990, "I know that there are fears of another Vietnam.... Let me assure you, should military action be required, this will not be another Vietnam. This will not be a protracted, drawn-out war," because in contrast to Vietnam all of the forces arrayed there "are different; the opposition is different; the resupply of Saddam's military would be very different; the countries united against him in the United Nations are different; the topography of Kuwait is different, and the motivation of our all-volunteer force is superb."[34] And, as a sympathetic reporter's sidebar essay affirmed, "a war against Iraq is winnable, and Vietnam never was," but, fortunately, "the United States will be going after victory with young troops free of doubts of the Vietnam era. An army trained to take on the Soviet superpower should be able to beat – and beat quickly – a Third World Force."[35]

Bush sounded these themes again in his address to the Reserve Officers Association a week after the air war began:

> This will not be another Vietnam. Never again will our armed forces be sent out to do a job with one hand tied behind their back. They will continue to have the support they need to get the job done – get it done quickly and with as little loss of life as possible.[36]

Similarly, in his masterful briefing of the press following his Desert Storm victory, General Norman Schwarzkopf was portrayed in *The Washington Post* as discharging "an institutional mission in describing how American power had won a Third World war that many Americans feared would turn into 'another Vietnam.' That is a phrase that Schwarzkopf may have helped lay to rest."[37] Having rhetorically caged the fearful monster of "another Vietnam," the World War II discourses kept the parallels between Nazi Germany and Baathist Iraq spinning on the screens of the global media markets, partly by rhetorical design and partly through bizarre coincidence. First, the US expeditionary forces introduced in early August, and augmented through the fall of 1990, faced an obvious enemy, "a new Hitler," along a well-defined front with a plainly apparent objective – the recapture of Kuwait. The sharply drawn battlefront, at the same time, helped to create, again like World War II, a vital, supportive home front in the United States. In penance for Vietnam, and in continuation of the spectacular style of patriotism sparked in the Reagan era, many US citizens threw themselves spiritually into the war – with displays of yellow ribbons, letter writing campaigns to the troops, or sending packages of gifts – as an opportunity to redeem the nation from its many perceived failures since Vietnam. By finding a "good war," to be fought, discursively and strategically, like the just war struggles of World War II, the "bad war" of Vietnam, and its allegedly lingering syndromes of defeatism, doubt, and cynicism might be exorcised for real and for good.

Second, the wartime behavior of Saddam Hussein and the coalition also were constantly reconstituted in the still-living imagery created by World War II mythologies. Like Hitler, the seemingly mad butcher of Baghdad cowered in his German-built Fuehrerbunker as he directed his almost Waffen-SS-like Republican Guards to fight to the last man. Whereas Hitler gassed millions of Jews in sealed death chambers and rocketed Allied cities with V-ls and V-2s, Saddam shot SCUDs against Israel, where Jews sat in sealed rooms wearing gas masks against chemical warheads made possible by West German-built factories in Iraq. Whereas the Nazis also blitzed the valiant British empire as the RAF fought off Luftwaffe attacks on London, the Baathists in Baghdad sent IRBMs against Saudi Arabia, where the residents of Riyadh and Dharan toughed it out against rocket attacks as Western newsmen gave blow-by-blow accounts of Patriot antimissile missiles rising against the SCUDs. Similarly, a multinational air force, including the emirate's own small, but frequently photographed, escadrille of A-4 Skyhawks, in turn, waged an intense 4-day air war against Iraqi forces in Iraq and Kuwait, like the US and British air forces bombing Hitler's Festung Europa, which proved its potency to evening news viewers in gun-camera or LANTRIN-sight videos of killed MiGs and exploding buildings. Meanwhile, like the Free French forces or Yugoslav partisans, the Kuwaiti underground fought doggedly against rapacious brutal Iraqi occupiers intent upon raping, pillaging, and ruining Kuwait's people, economy, and society. Indeed, a mini-Holocaust of sorts also has been discovered after Kuwait's liberation in the Iraqis' execution, torture, or abduction of thousands of Kuwaitis.

Third, the amazing 100-hour blitzkrieg of the coalition's ground forces resurrected buried emotions of global triumphalism first sparked by old newsreel films of Patton, Zhukov, and Montgomery, smashing over the Rhine and Elbe into Hitler's heartland, capturing thousands of prisoners, and obliterating entire Nazi armies as fighting formations. In less time that it takes for many major TV miniseries to reach their dramatic climax, the Gulf War revitalized, refought, and reaffirmed all of the old World War II articles of faith about air-land blitzkrieg in the deserts of Araby. Even though the aircraft now were supersonic jets, the field rations freeze-dried, the rifles lightweight alloys and plastics, the helmets Kevlar, the tanks turbine-driven, and the bombs smart, the coalition's script was one of triumphant World War II-style liberation as CNN 24 Hour Headline News replayed clips of quaking Iraqi prisoners begging for mercy before their coalition captors and ebullient Kuwaitis throwing kisses and flowers upon the victorious armored columns liberating Kuwait City. Meanwhile, during the commercial breaks on CNN, the retro-atmospherics were underscored by commercials for a slickly packaged compilation of old World War II-vintage big band hits and top-40 love songs, which was entitled *The White Cliffs of Dover* and wrapped in cover illustrations showing a valiant US airman hugging his sweetheart against a backdrop of the white cliffs of Dover while a B-52 winged its way toward Axis Europe.

This retrowar rhetoric blended contemporary images of triumph with a rebirth of World War II historical importance for the United States – all too often left on the sidelines during recent years as Prague, Bonn, Moscow, Pretoria, Beijing, or Teheran "made history" instead of Washington. The global war against Saddam, therefore, was everywhere – in school, at the grocery store, in fast-food outlets, on TV and radio – as a total environment, an electronic mantra to refocus the nation's citizens as truly loyal fans eager to "kick butt" and "be number one" after being humiliated by Ho Chi Minh, Pol Pot, and the Ayatollah Khomeini during the 1960s, 1970s, and 1980s. War under these conditions approached becoming its own enveloping virtual reality as US mass media audiences could sink into an almost seamless capsule of stimuli flashed from Washington, Riyadh, Dharan, Baghdad, Tel Aviv, London, and New York in real time as powerful currents of image, sound, voice, and icons.

With so few casualties and so little response from the Iraqis, the technological spectacle of the Gulf War was not so much combat between equals as a demolition derby or monster truck show staged by the coalition war machine romping roughshod over Iraqi military men and materiel. Indeed, upon entering Kuwait City, Captain Kevin Davis remarked, "I hate to say it, but once we got rolling it was like a training exercise with live people running around. Our training exercises are a lot harder."[38] Ironically, much of this success can be chalked up to the US military's new focus on simulation and war-gaming. After coincidentally simulating in hi-tech, computerized war games a US victory in Kuwait against "foreign aggressors" at Central Command headquarters in Tampa, Florida, during July 1990, General Schwarzkopf was called to Saudi Arabia in August 1990 to transform his simulated victories into a "real world scenario" as Operation Desert Storm.[39] Because press coverage did not reveal, or could not document, the 100,000 plus Iraqi soldiers being killed by these assaults, this televised war often seemed more like a technological spectacle on the order of a space shot, moon landing, or shuttle mission rather than a war. And, in assuming such video game formats, the Kuwaiti campaigns became that quick, clean kind of killing that militaristic Reagan-era films, like *Top Gun*, *Iron Eagle*, *Navy Seals*, or *Firebirds*, had prepared Western audiences for throughout the 1980s by showing the talents of the Pentagon's "hi-tech pack," – that is, its F-14 pilots, the F-16, electronic night vision snipers, or Apache attack helicopters. War, at its best, according to these military workout videos, is an exciting hi-tech job, performed on an F-16 or Apache sortie, while being dramatically documented by the videotaped testimony/celebration of a laser-guided glide bomb or Hellfire missile zooming down to its hapless target with pin-point accuracy. Afterwards, as the rock-and-roll soundtrack and diplomatic credits roll over images of the freeze-framed kill, the warrior returns to base for "Miller Time" with the babes sporting the yellow ribbons.

On watching the winning of such televisual victories in Kuwait, one US viewer argued,

> It's taken the monkey off our back that's been there since Korea and Vietnam and Beirut and a few places in between [furthermore, while] there's been talk of us losing our economic leadership, but this has reasserted our preeminence of a sort. VCRs may be made in Japan and Mercedes have their stamp of origin, but what's going on in the

> Middle East is undeniably made in the USA. I think that's a source of pride we have not had since World War II.[40]

World War II here, of course, is understood as the good vibes of Normandy, Iwo Jima, and Rosie the Riveter rather than Stalingrad, Auschwitz, or Hiroshima. Similarly, the Kuwait campaign will be remembered televisually as Patriots shredding SCUDs, Iraqi POWs kissing US soldiers' boots, and Kuwaitis celebrating the US Marines' arrival in Kuwait City, not the massive oil slicks, the nameless Iraqi victims of B-52 strikes, or broiled corpses in the burned-out buses on the "highway of death" on the road to Basra.

[...]

FROM GEOPOLITICAL REALISM TO CHRONOPOLITICAL HYPERREALISM?

Given all of these speculations, what can we learn from Kuwait? The Westphalian system of autonomous nation-states, organized around spatial logics of domination and development on the basis of geopolitical agendas, is not yet dead, although it does seem to be dying. Yet, a post-Westphalian system of global networks, transnational flows, and informational communities, tied into a fluid logic of influence and interaction set into the code of chronopolitical programs, is also not yet fully formed, even though it does appear to be rapidly developing. These distinctions, of course, are crude. The notions of geopolitics undoubtedly are themselves the products of increased technological velocities in the nineteenth century, as steamship travel, telegraphy, telephony, and railroadization immensely increased a nation-state's capabilities for inscribing its power on the globe by rapidly responding to each new opportunity to define and defend territorial space. Likewise, the frameworks of chronopolitics have their own geopolitical gloss inasmuch as extremely rapid telecommunications, jet travel and missile velocities, or computerized transactions project their own hyperreal spaces, which states must continuously manufacture and maintain as sources of authority, unity, or prosperity. In some sense, then, geopolitics might be chronopolitics at nineteenth-century paces, and chronopolitics could be geopolitics in twenty-first-century spaces.

Where might these interlaced, but still appositive, tendencies end up? Prior to the consolidation of modern nation-states during and after the Thirty Years' War, sovereignty, security, and stability were also deeply conflicted as power was exercised legitimately by extrastatal institutions such as the church, feudal manors, and urban guilds. On one level, the international politics today against the backdrop of the flow might presage an "info-medievalism" or "cyber-feudalism," in which quasistatal, nonstatal, poststatal, or semistatal forces engage in violent political struggles over land, resources, and population as the modern nation-state system collapses. Here, religious, ideological, economic, technological, and even lifestyle identities and interests might war over the control of markets, territory, and minds with low-intensity conflict and high-intensity persuasion.[41] Yet, on another level, the politics of the flow also might preview a global order of "hyperrealism," where many of the old signs and symbols of realist realpolitik are retained as a simulation of autonomous nation-states with their own territories, militaries, and currencies. In practice, however, the flow reduces their differences to basic equivalents, corrodes their borders as meaningful barriers, and eliminates older geopolitical divisions of "them" and "us," "inside" and "outside," "foreign" and "domestic." Whereas the shapes and sounds of international relations as geopolitics remain as nominal nationality, the stuff and substance of actual transnationality may now dominate intraglocal interoperations as chronopolitics.

In hyperrealism today, the simulation of reality, like President Bush's new world order as a spectacular resurrection of World War II's grand alliance, becomes "real," but real only within the imagery known to be readily accessible and comprehensible to the flow's mass clienteles in the ever-flexible World War II scripts. One "nation-state," the United States – which actually behaves like a truly transnational economic, military, technological, and financial empire of immense informational capacity – organized an international alliance of other nation-states, which mainly were informationalized core economies or post-World War I and World War II neocolonial constructs in Eastern Europe and the Third World, to aid another nation-state, namely Kuwait, the cyberspatial point of transnational oil dealing and hyperreal estate of a few hundred thousand access-holding Arabs. Where nominal nationality means something, like the White House or the Sheraton Al-Hada, it is played upon to revitalize national will or repossess national territories. Yet, it is the actual transnationality of the flow, and the variable geometry alliances of diverse sets of nation-states within its streams, that allows nominal nationality to serve these purposes. The nominal nationality of the United States clearly is critical, as it remains the semiurgic core of this new unipolar new world order, but the "Latin Americanization" of its society, the "Beirutification" of its major cities, and the "Japanization" of its economy in the interactive workings of the flow suggest that *Pax Americana* does not mean what it did in 1945–1946. Instead, fifty years later, as we might learn from Kuwait, the actual transnationality of the United States perhaps is becoming much more significant. The codes of containment and discourse of security, once pounded out in a metallurgic/technicurgic era of geopolitical realism, are being recoded to boot into the semiurgic hegemony of informational power, responding to the political possibilities intrinsic to chronopolitical hyperrealism.

NOTES

1 David Campbell, "Global Inscription: How Foreign Policy Constitutes the United States," *Alternatives* 15 (Summer 1990): 280.

2 See, for further discussion, Richard K. Ashley, "Untying the Sovereign State: A Double Reading of the Anarchy Problematique," *Millennium: Journal of International Studies* 17 (1988): 227–262; Timothy W. Luke, "On Post-War: The Significance of Symbolic Action in War and Deterrence," *Alternatives* 14 (July 1989): 343–362; Simon Dalby, "American Security Discourse: The Persistence of Geopolitics," *Political Geography Quarterly* 9 (April 1990): 171–188; and Bradley S. Klein, "How the West Was One: Representational Politics of NATO," *International Studies Quarterly* 34 (September 1990): 311–325.

3 Francis Fukuyama, "The End of History?" *The National Interest* 16 (Summer 1989): 3–18.

4 See Peter Drucker, *The New Society: The Anatomy of the Industrial Order* (New York: Harper, 1950); Ralf Dahrendorf, *Class and Class Conflict in Industrial Society* (Stanford, Calif.: Stanford University Press, 1959); Marshall McLuhan, *Understanding Media: The Extensions of Man* (New York: McGraw-Hill, 1964); Zbigniew Brzezinski, *Between Two Ages: America's Role in the Technetronic Era* (New York: Viking, 1970); Alain Touraine, *The Post-Industrial Society* (New York: Random House, 1971); and Daniel Bell, *The Coming of Post-Industrial Society: A Venture in Social Forecasting* (New York: Basic, 1973).

5 See Fredric Jameson, *Postmodernism or the Cultural Logic of Late Capitalism* (Durham, N.C.: Duke University Press, 1991); David Harvey, *The Condition of Postmodernity* (Oxford: Blackwell, 1984); and Jean François Lyotard, *The Post Modern Condition* (Minneapolis: University of Minnesota Press, 1984).

6 Timothy W. Luke, *Screens of Power: Ideology Domination and Resistance in Informational Society* (Urbana: University of Illinois Press, 1989), pp. 3–14.

7 Luke, note 6, pp. 46–54.

8 Campbell, note 1, p. 279.

9 R. B. J. Walker, "Security, Sovereignty, and the Challenge of World Politics," *Alternatives* 15 (Winter 1990): 11.

10 See James Der Derian, "The (S)pace of International Relations: Simulation, Surveillance, and Speed," *International Studies Quarterly* 34 (September 1990): 295–310.

11 Luke, note 6, pp. 46–54.

12 See Robert B. Reich, *The Work of Nations: Preparing Ourselves for 21st-Century Capitalism* (New York: Knopf, 1991); and Kevin P. Phillips, *The Politics of Rich and Poor* (New York: Random House, 1990).

13 Manuel Castells, *The Informational City: Information Technology Economic Restructuring and the Urban–Regional Process* (Oxford: Blackwell, 1989), pp. 170–171.

14 Castells, note 13, p. 171.

15 Castells, note 13, p. 142.

16 Reich, note 12, *Work of Nations*, p. 131.

17 See Timothy W. Luke, "Dependent Development and the Arab OPEC States," *Journal of Politics* 45 (September 1983): 979–1005; and Timothy W. Luke, "OPEC: The Basis of the Arab Developmental World – A Transnational Model," in Forest Grieves, ed., *Transnationalism in World Politics and Business* (New York: Pergamon Press, 1979), pp. 89–113.
18 See Steven Schneider, *The Oil Price Revlolution* (Baltimore: Johns Hopkins University Press, 1983); and John Blair, *The Control of Oil* (New York: Random House, 1976).
19 Judith Miller and Laurie Mylroie, *Saddam Hussein and the Crisis in the Gulf* (New York: Times Books, 1990), p. 202.
20 See, for example, *The Washington Post*, May 17, 1991, p. A15, for a full page, full-color appreciation of Desert Storm "prepared by the Lonergan-Dickerson Group" that fetes "the Thunder and Lightning of Desert Storm," or the American Armed Forces under President George Bush. Similar grandiose communiques ran in *The Washington Post* throughout the Gulf War and its aftermath.
21 *The Wall Street Journal*, March 14, 1991, pp. A1, A8
22 Luke, note 17, "Dependent Development and OPEC," pp. 979–1005.
23 See Adel Darwich and Gregory Alexander, *Unholy Alliance: The Secret History of Saddam's War* (New York: St. Martin's, 1991).
24 Luke, note 6, pp. 159–180.
25 Guy Debord, *Society of the Spectacle* (Detroit: Red and Black, 1967), No. 2.
26 Roland Barthes, *Image–Music–Text* (New York: Hill and Wang, 1978), p. 15.
27 *Vital Speeches*, September 1, 1990, p. 676.
28 Fred Barnes, "Brave New Gimmick," *New Republic* 204 (February 25, 1991): 15.
29 Barnes, note 28.
30 *Newsweek*, November 26, 1990, p. 28.
31 Ibid., p. 29.
32 Ibid.
33 Ibid.
34 *Newsweek*, December 10, 1990, pp. 24–25.
35 *Newsweek*, note 34, p. 30.
36 *The Washington Post*, January 24, 1991, p. A24.
37 *The Washington Post*, February 28, 1991, p. A34.
38 As quoted in *Newsweek*, March 23, 1991, p. 17.
39 James Der Derian, "War Games May Prove Deadly," *Newsday* 9 (December 1990): 19.
40 *The New York Times*, February 24, 1991, p. 43.
41 See James Der Derian, "The Terrorist Discourse: Signs, States, and Systems of Global Political Violence," in Michael T. Klare and Daniel C. Thomas, eds, *World Security: Trends and Challenges at Century's End* (New York: St. Martin's, 1991), pp. 237–265; and Robert Kuttner, *The End of Laissez-Faire: National Purpose and the Global Economy After the Cold War* (New York: Knopf, 1991).

Cartoon 10 Your Sheikness (by Wuerker)
While the rhetoric of the Gulf War replayed themes from World War II, the slippage in George Bush's homage before the King of Saudi Arabia reveals a more materialist motivation. Rhetorically represented as a retro-war, the Gulf War was always ultimately a petro-war.
Source: M. Wuerker

20 MICHAEL T. KLARE

"The New 'Rogue-State' Doctrine"

from *The Nation* (1995)

Washington is once again in the grip of "rogue mania." The recent alarms sounded about Iran as a potential nuclear power are the latest manifestations of US policy-makers' obsession with a handful of "rogue states," which are portrayed as major threats to US and Western security. These countries, particularly Iraq, Iran, Libya, Syria and North Korea, have become the dominant enemy image in Washington. All but unknown a few years ago, the rogue-state doctrine enjoys bipartisan support in Congress and is being pushed by a politically defensive White House. Unless it is tempered by domestic or international developments, it could embroil the United States in a Gulf War-like military clash with a rising Third World power. It also serves to sustain military spending at Cold War levels at a time when social programs are being severely cut.

The rogue-state concept is a product of a determined Pentagon effort to create a new foreign threat to justify military spending in the wake of the Cold War. To protect the mammoth defense establishment, US military officials began seeking a new kind of enemy within days of the Berlin wall's collapse. Under the direction of Gen. Colin Powell, then Chairman of the Joint Chiefs of Staff, the J-5 (Strategic Plans and Policy) Directorate of the Joint Staff worked throughout the winter and early spring of 1990 to devise a new military posture based on a non-Soviet threat.

In developing a new strategic concept, the J-5 staff was governed by some very significant considerations. In particular, they had to identify an enemy type that was powerful enough to justify retention of a large military establishment and sinister enough to arouse Congressional concern. By process of elimination, this led to a focus on rising Third World powers equipped with modern weapons and known for a history of antagonism to the United States – such as the above-mentioned nations. By May 1990, Pentagon leaders had reached a consensus on the adoption of a military posture aimed at states of this sort, and in June President Bush gave his approval to what then was termed the "New Regional Strategy."

Following the Gulf War, Pentagon officials unveiled a long range defense plan calling for a sustained preparation for continuing series of Desert Storm-like engagements. "The Gulf War presaged very much the type of conflict we are most likely to confront again in this new era," then Secretary of Defense Dick Cheney told Congress on March 19, 1991 – "major regional contingencies against foes well-armed with advanced conventional and unconventional munitions."

In accordance with this outlook, the Bush Administration proposed a permanent military establishment (or "Base Force") of about 1.6 million soldiers – enough, it was said, to fight two Desert Storms simultaneously. President Clinton pledged to take another look at US military requirements after assuming office in 1993, but he, too, endorsed a military posture based on the need to fight two "major regional conflicts" at the same time. Under the "bottom-up review" endorsed by Clinton in August 1993, US strength will drop to 1.4 million soldiers, but the basic design of the Bush strategy will remain intact. The cost to US taxpayers: approximately $260–270 billion per year, or about what the United States spent, in constant dollars, on defense during much of the Cold War era (Korean and Vietnam War years excepted).

To justify this vast expense, the Clinton Administration must be able to demonstrate that the United States is indeed threatened by potent foreign enemies. Hence the periodic alarms in Washington over the military power and aggressive designs of Iran, Iraq, Libya and North Korea. Only when Congress and the American people can be shown an authentic – and sufficiently menacing – threat on the horizon will they be prepared to subsidize indefinitely a Cold War-level military establishment.

As described by senior US policy-makers, rogue states possess large, modern military establishments; covet weapons of mass destruction; and violate various international "norms." "Our policy must face the reality of recalcitrant and outlaw states that not only choose to remain outside the family [of nations] but also assault its basic values," National Security Adviser Anthony Lake asserted in 1994. These states, he said, "exhibit a chronic inability to engage constructively with the outside world," as demonstrated most clearly by their support of terrorism and pursuit of nuclear and chemical weapons. Just as the United States once took the leadership in "containing" the Soviet Union, he argued, it now bears a "special responsibility" to "neutralize" and "contain" the band of "outlaw states."

When first introduced, in early 1990, the anti-rogue posture was aimed primarily at Iraq and North Korea. Beginning in March of that year – five months before Saddam Hussein ordered the invasion of Kuwait – US forces began planning for a possible war with Iraq. "Before a single Iraqi soldier entered Kuwait," the Pentagon acknowledged in its history of the Gulf War, "the basic concepts of Operations Desert Shield and Desert Storm were established." These included the deployment of a large US force in Saudi Arabia and a heavy reliance on air power and armored units.

Once the Gulf War was over, attention turned almost immediately to North Korea – the country widely viewed as the "next Iraq" by many in Washington. Throughout much of 1993 and 1994, US officials warned of the incipient nuclear threat posed by North Korea and accelerated planning for a "Second Korean War." But North Korea agreed last October to dismantle its nuclear weapons capabilities (in exchange for US aid in acquiring light-water nuclear power reactors), which has led to a gradual reduction in tension on the Korean Peninsula.

As tensions in Korea began to subside, the Clinton Administration next turned up the pressure on Iran. This March, the President used the threat of an executive order to force an American oil company, Conoco, to cancel a planned $1 billion offshore oil exploration project with the Iranian government. Although perfectly legal under current law, the Conoco project was said to aid Iran in its efforts to acquire funds and technology for the development of chemical and nuclear weapons. "We draw the line at countries with policies that are beyond the pale," a senior Administration official explained.

The Administration has also taken extraordinary steps – including the sharing of secret intelligence data – to persuade Russia to cancel its plans to build four nuclear power reactors in Iran. The Russian project, thought to be worth $1 billion or more, would not violate the nuclear Non-Proliferation Treaty or other restrictions on nuclear technology transfers. However, US officials argue that the Iranians would be able to use the project to gain expertise in nuclear matters, thus advancing their efforts to produce nuclear weapons. "Russia will rue the day it cooperated with the terrorist state of Iran if Iran builds nuclear weapons with Russian expertise and Russian equipment," Secretary of State Warren Christopher warned in March. (Russian leaders have so far refused to cancel the project, but have agreed to limit the transfer of technology that could be used for military purposes.)

In yet another expression of Washington's anti-Iran campaign, the Defense Department has hinted at vigorous military action to counter any threat by Iran to impede oil shipping in the Persian Gulf area. Charging that the Iranians have acquired two Russian-built "Kilo"-class submarines and reinforced their garrisons on several small islands in the gulf – islands that were seized from the United Arab Emirates by the Shah of Iran in the 1970s with implicit US approval – Pentagon officials have asserted their readiness to defend the oil lanes.

All of this is set against a backdrop of alarmist press stories on Iran's nuclear weapons program and expanding military capabilities. "Iran May Be Able to Build an Atomic Bomb in 5 Years, US and Israeli Officials Fear," a recent headline in *The New York Times* declared. The impression being given in these accounts by unnamed US officials – quite deliberately, it would appear – is that Iran is just about where Iraq was in 1990 in terms of its nuclear weapons research and all military capabilities. But a careful examination of the available data reveals a very different picture.

According to the 1994–95 edition of *The Military Balance*, the highly regarded annual publication of the International Institute of Strategic Studies in London, Iran today has an army of 345,000 – about a third of that fielded by Iraq in 1990. Its military possesses some 1,250 tanks (about one-fourth of the 1990 Iraqi force) and 900 armored personnel carriers (compared with 8,000 in the prewar Iraqi force). Much equipment in Iranian arsenals was acquired by the Shah in the 1960s and 1970s, and is now rusting away in storage depots due to lack of spare parts and maintenance.

Even more important in terms of real military capability, the Iranians have had to slash their military spending because of severe economic difficulties and growing popular dissatisfaction with government policies. From a high of about $5.8 billion in 1991, Iranian military spending has dropped to $2 billion today – less than 1 per cent of what the United States spends and less than one-sixth of what Saddam Hussein was spending in the late 1980s.

As for Tehran's nuclear weapons program, there is no evidence that the Iranians are anywhere close to Iraq's capabilities when the United States launched Desert Storm. Although Iran has sought to obtain some of the same technologies and equipment pursued by Iraqi technicians in the 1980s, their bomb program is still at a very early stage and lacks many of the components needed to produce a functioning weapon. As Defense Secretary William Perry said in January, Iran is "many years" away from developing a nuclear bomb, even under the best of circumstances.

At this point, it is unclear what moves the Administration will take next in its campaign against Iran. There is talk at the White House and the Pentagon of a complete US economic and trade embargo on the country, of the imposition of economic sanctions on any US or foreign companies that trade with Iran, and even of pre-emptive military strikes. Whether or not any of these actions materialize, it is likely that Washington will step up its diplomatic and propaganda campaign against the Iranians.

The Clinton Administration is also likely to step up its pressure on two other "rogues," Iraq and Libya. White House officials have recently indicated that the Iraqis might be trying to rebuild their capacity to manufacture weapons of mass destruction. While voting recently at the UN in favor of allowing Iraq to sell $2 billion worth of oil, Washington warned of vigorous action – including military action – if these reports proved accurate. The Administration has also explored the possibility of imposing additional economic and trade sanctions on Libya as punishment for its continuing refusal to extradite two Libyan intelligence officials accused of complicity in the 1988 Lockerbie aircraft explosion.

These moves, and others like them, are predictable manifestations of a security doctrine organized around the need to "neutralize" and "contain" rogue states. While Administration officials have made several highly publicized efforts to articulate a grand theme for US security policy in the post-Cold War era – most notably the 1993 effort by Anthony Lake to articulate a strategy of "enlargement" of the global community of market-oriented democracies – the White House has repeatedly fallen back on the rogue doctrine as its basic strategic design. President Clinton is also driven by political difficulties on the domestic front. With the Republicans in control of Congress and public support for his domestic policies crumbling, Clinton – like many Presidents before him – is tempted to gain media attention and favorable performance ratings by focusing on foreign policy concerns. And what better way to win support with the American people than to vilify familiar enemies like Saddam Hussein, Muammar el Qaddafi and the Iranian clerics?

The danger arising from this impulse to brand certain states as rogues and outlaws, and to threaten them with severe military punishment, is that the policy can take on a life of its own – spurring countermoves and counter-threats until the White House is forced to back up its words with a show of force. Where this will all lead is anyone's guess, but it would be a terrible mistake to assume that the next Desert Storm will produce as rapid and lopsided a victory as the last one did.

Even if the rogue doctrine does not lead to war, it will have other pernicious consequences for American society. Financing this strategy will require a constant drain on the Treasury, forcing avid budget-cutters from both parties to rip even deeper into domestic programs that support America's neediest populations. And the obsessive focus on a handful of secondary powers will divert attention and resources from the really important foreign policy problems facing America, such as the political and economic disintegration in Mexico, right wing nationalism in Russia and predatory trade practices in Japan. President Clinton has yet to articulate an effective strategy for dealing with these problems, so we can expect further digressions on the rogue-state threat. But no one should be fooled into thinking that the rogue doctrine provides a realistic framework for foreign policy decision-making in the post-Cold War era.

21

SAMUEL P. HUNTINGTON

"The Clash of Civilizations?"

from *Foreign Affairs* (1993)

THE NEXT PATTERN OF CONFLICT

World politics is entering a new phase, and intellectuals have not hesitated to proliferate visions of what it will be – the end of history, the return of traditional rivalries between nation states, and the decline of the nation state from the conflicting pulls of tribalism and globalism, among others. Each of these visions catches aspects of the emerging reality. Yet they all miss a crucial, indeed a central, aspect of what global politics is likely to be in the coming years.

It is my hypothesis that the fundamental source of conflict in this new world will not be primarily ideological or primarily economic. The great divisions among humankind and the dominating source of conflict will be cultural. Nation states will remain the most powerful actors in world affairs, but the principal conflicts of global politics will occur between nations and groups of different civilizations. The clash of civilizations will dominate global politics. The fault lines between civilizations will be the battle lines of the future.

Conflict between civilizations will be the latest phase in the evolution of conflict in the modern world. For a century and a half after the emergence of the modern international system with the Peace of Westphalia, the conflicts of the Western world were largely among princes – emperors, absolute monarchs and constitutional monarchs attempting to expand their bureaucracies, their armies, their mercantilist economic strength and, most important, the territory they ruled. In the process they created nation states, and beginning with the French Revolution the principal lines of conflict were between nations rather than princes. In 1793, as R. R. Palmer put it, "The wars of kings were over; the wars of peoples had begun." This nineteenth-century pattern lasted until the end of World War I. Then, as a result of the Russian Revolution and the reaction against it, the conflict of nations yielded to the conflict of ideologies, first among communism, fascism-Nazism and liberal democracy, and then between communism and liberal democracy. During the Cold War, this latter conflict became embodied in the struggle between the two superpowers, neither of which was a nation state in the classical European sense and each of which defined its identity in terms of its ideology.

These conflicts between princes, nation states and ideologies were primarily conflicts within Western civilization, "Western civil wars," as William Lind has labeled them. This was as true of the Cold War as it was of the world wars and the earlier wars of the seventeenth, eighteenth and nineteenth centuries. With the end of the Cold War, international politics moves out of its Western phase, and its center piece becomes the interaction between the West and non-Western civilizations and among non-Western civilizations. In the politics of civilizations, the peoples and governments of non-Western civilizations no longer remain the objects of history as targets of Western colonialism but join the West as movers and shapers of history.

THE NATURE OF CIVILIZATIONS

During the Cold War the world was divided into the First, Second and Third Worlds. Those divisions are no longer relevant. It is far more meaningful now to group countries not in terms of their political or economic systems or in terms of their level of economic development but rather in terms of their culture and civilization.

What do we mean when we talk of a civilization? A civilization is a cultural entity. Villages, regions, ethnic groups, nationalities, religious groups, all have distinct cultures at different levels of cultural heterogeneity. The culture of a village in southern Italy may be different from that of a village in northern Italy, but both will share in a common Italian culture that distinguishes them from German villages. European communities, in turn, will share cultural features that distinguish them from Arab or Chinese communities. Arabs, Chinese and Westerners, however, are not part of any broader cultural entity. They constitute civilizations. A civilization is thus the highest cultural grouping of people and the broadest level of cultural identity people have short of that which distinguishes humans from other species. It is defined both by common objective elements, such as language, history, religion, customs, institutions, and by the subjective self-identification of people. People have levels of identity: a resident of Rome may define himself with varying degrees of intensity as a Roman, an Italian, a Catholic, a Christian, a European, a Westerner. The civilization to which he belongs is the broadest level of identification with which he intensely identifies. People can and do redefine their identities and, as a result, the composition and boundaries of civilizations change.

Civilizations may involve a large number of people, as with China ("a civilization pretending to be a state," as Lucian Pye put it), or a very small number of people, such as the Anglophone Caribbean. A civilization may include several nation states, as is the case with Western, Latin American and Arab civilizations, or only one, as is the case with Japanese civilization. Civilizations obviously blend and overlap, and may include subcivilizations. Western civilization has two major variants, European and North American, and Islam has its Arab, Turkic and Malay subdivisions. Civilizations are nonetheless meaningful entities, and while the lines between them are seldom sharp, they are real. Civilizations are dynamic; they rise and fall; they divide and merge. And, as any student of history knows, civilizations disappear and are buried in the sands of time.

Westerners tend to think of nation states as the principal actors in global affairs. They have been that, however, for only a few centuries. The broader reaches of human history have been the history of civilizations. In *A Study of History*, Arnold Toynbee identified 21 major civilizations; only six of them exist in the contemporary world.

WHY CIVILIZATIONS WILL CLASH

Civilization identity will be increasingly important in the future, and the world will be shaped in large measure by the interactions among seven or eight major civilizations. These include Western, Confucian, Japanese, Islamic, Hindu, Slavic-Orthodox, Latin American and possibly African civilizations. The most important conflicts of the future will occur along the cultural fault lines separating these civilizations from one another.

Why will this be the case?

First, differences among civilizations are not only real; they are basic. Civilizations are differentiated from each other by history, language, culture, tradition and, most important, religion. The people of different civilization have different views on the relations between God and man, the individual and the group, the citizen and the state, parents and children, husband and wife, as well as differing views on the relative importance of rights and responsibilities, liberty and authority, equality and hierarchy. These differences are the product of centuries. They will not soon disappear. They are far more fundamental than differences among political ideologies and political regimes. Differences do not necessarily mean conflict, and conflict does not necessarily mean violence. Over the centuries, however, differences among civilizations have generated the most prolonged and the most violent conflicts.

Second, the world is becoming a smaller place. The interactions between peoples of different civilizations are increasing; these increasing interactions intensify civilization consciousness and awareness of differences between civilizations and commonalities within civilizations. North African immigration to France generates hostility among Frenchmen and at the same time increased receptivity to immigration by "good" European Catholic Poles. Americans react far more negatively to Japanese investment than to larger investments from Canada and European countries. Similarly, as Donald Horowitz has pointed out, "An Ibo may be . . . an Owerri Ibo or an Onitsha Ibo in what was the Eastern region of Nigeria. In Lagos, he is simply an Ibo. In London, he is a Nigerian. In New York, he is an African." The interactions among peoples of different civilizations enhance the civilization-consciousness of people that, in turn, invigorates differences and animosities stretching or thought to stretch back deep into history.

Third, the processes of economic modernization and social change throughout the world are separating people from longstanding local identities. They also weaken the nation state as a source of identity. In much of the world religion has moved in to fill this gap, often in the form of movements that are labeled "fundamentalist." Such movements are found in Western Christianity, Judaism, Buddhism and Hinduism, as well as in Islam. In most countries and most religions the people active in fundamentalist movements are young, college-educated, middle-class technicians, professionals and business persons. The "unsecularization of the world," George Weigel has remarked, "is one of the dominant social facts of life in the late twentieth century." The revival of religion, "la revanche de Dieu," as Gilles Kepel labeled it, provides a basis for identity and commitment that transcends national boundaries and unites civilizations.

Fourth, the growth of civilization-consciousness is enhanced by the dual role of the West. On the one hand, the West is at a peak of power. At the same time, however, and perhaps as a result, a return to the roots phenomenon is occurring among non-Western civilizations. Increasingly one hears references to trends toward a turning inward and "Asianization" in Japan, the end of the Nehru legacy and the "Hinduization" of India, the failure of Western ideas of socialism and nationalism and hence "re-Islamization" of the Middle East, and now a debate over Westernization versus Russianization in Boris Yeltsin's country. A West at the peak of its power confronts non-Wests that increasingly have the desire, the will and the resources to shape the world in non-Western ways.

In the past, the elites of non-Western societies were usually the people who were most involved with the West, had been educated at Oxford, the Sorbonne or Sandhurst, and had absorbed Western attitudes and values. At the same time, the populace in non-Western countries often remained deeply imbued with the indigenous culture. Now, however, these relationships are being reversed. A de-Westernization and indigenization of elites is occurring in many non-Western countries at the same time that Western, usually American, cultures, styles and habits become more popular among the mass of the people.

Fifth, cultural characteristics and differences are less mutable and hence less easily compromised and resolved than political and economic ones. In the former Soviet Union, communists can become democrats, the rich can become poor and the poor rich, but Russians cannot become Estonians and Azeris cannot become Armenians. In class and ideological conflicts, the key question was "Which side are you on?" and people could and did choose sides and change sides. In conflicts between civilizations, the question is "What are you?" That is a given that cannot be changed. And as we know, from Bosnia to the Caucasus to the Sudan, the wrong answer to that question can mean a bullet in the head. Even more than ethnicity, religion discriminates sharply and exclusively among people. A person can be half-French and half-Arab and simultaneously even a citizen of two countries. It is more difficult to be half-Catholic and half-Muslim.

Finally, economic regionalism is increasing. The proportions of total trade that were intra-regional rose between 1980 and 1989 from 51 per cent to 59 per cent in Europe, 33 per cent to 37 per cent in East Asia, and 32 per cent to 36 per cent in North America. The importance of regional economic blocs is likely to continue to increase in the future. On the one hand, successful economic regionalism will reinforce civilization-consciousness. On the other hand, economic regionalism may succeed only when it is rooted in a common civilization. The European Community rests on the shared foundation of European culture and Western Christianity. The success of the North American Free Trade Area depends on the convergence now underway of Mexican, Canadian and American cultures. Japan, in contrast, faces difficulties in creating a comparable economic entity in East Asia because Japan is a society and civilization unique to itself. However strong the trade and investment links Japan may develop with other East Asian countries, its cultural differences with those countries inhibit and perhaps preclude its promoting regional economic integration like that in Europe and North America.

Common culture, in contrast, is clearly facilitating the rapid expansion of the economic relations between the People's Republic of China and Hong Kong, Taiwan, Singapore and the overseas Chinese communities in other Asian countries. With the Cold War over, cultural commonalities increasingly overcome ideological differences, and mainland China and Taiwan move closer together. If cultural commonality is a prerequisite for economic integration, the principal East Asian economic bloc of the future is likely to be centered on China. This bloc is, in fact, already coming into existence....

[...]

Culture and religion also form the basis of the Economic Cooperation Organization, which brings together ten non-Arab Muslim countries: Iran, Pakistan, Turkey, Azerbaijan, Kazakhstan, Kyrgyzstan, Turkmenistan, Tadjikistan, Uzbekistan and Afghanistan. One impetus to the revival and expansion of this organization, founded originally in the 1960s by Turkey, Pakistan and Iran, is the realization by the leaders of several of these countries that they had no chance of admission to the European Community. Similarly, Caricom, the Central American Common Market and Mercosur rest on common cultural foundations. Efforts to build a broader Caribbean–Central American economic entity bridging the Anglo–Latin divide, however, have to date failed.

As people define their identity in ethnic and religious terms, they are likely to see an "us" versus "them" relation existing between themselves and people of different ethnicity or religion. The end of ideologically defined states in Eastern Europe and the former Soviet Union permits traditional ethnic identities and animosities to come to the fore. Differences in culture and religion create differences over policy issues, ranging from human rights to immigration to trade and commerce to the environment. Geographical propinquity gives rise to conflicting territorial claims from Bosnia to Mindanao. Most important, the efforts of the West to promote its values of democracy and liberalism as universal values, to maintain its military predominance and to advance its economic interests engender countering responses from other civilizations. Decreasingly able to mobilize support and form coalitions on the basis of ideology, governments and groups will increasingly attempt to mobilize support by appealing to common religion and civilization identity.

The clash of civilizations thus occurs at two levels. At the micro-level, adjacent groups along the fault lines between civilizations struggle, often violently, over the control of territory and each other. At the macro-level, states from different civilizations compete for relative military and economic power, struggle over the control of international institutions and third parties, and competitively promote their particular political and religious values....

THE FAULT LINES BETWEEN CIVILIZATIONS

The fault lines between civilizations are replacing the political and ideological boundaries of the Cold War as the flash points for crisis and bloodshed. The Cold War began when the Iron Curtain divided Europe politically and ideologically. The Cold War ended with the end of the Iron Curtain. As the ideological division of

Europe has disappeared, the cultural division of Europe between Western Christianity, on the one hand, and Orthodox Christianity and Islam, on the other, has reemerged. The most significant dividing line in Europe, as William Wallace has suggested, may well be the eastern boundary of Western Christianity in the year 1500. This line runs along what are now the boundaries between Finland and Russia and between the Baltic states and Russia, cuts through Belarus and Ukraine separating the more Catholic western Ukraine from Orthodox eastern Ukraine, swings westward separating Transylvania from the rest of Romania, and then goes through Yugoslavia almost exactly along the line now separating Croatia and Slovenia from the rest of Yugoslavia. In the Balkans this line, of course, coincides with the historic boundary between the Hapsburg and Ottoman empires. The peoples to the north and west of this line are Protestant or Catholic; they shared the common experiences of European history – feudalism, the Renaissance, the Reformation, the Enlightenment, the French Revolution, the Industrial Revolution; they are generally economically better off than the peoples to the east; and they may now look forward to increasing involvement in a common European economy and to the consolidation of democratic political systems. The peoples to the east and south of this line are Orthodox or Muslim; they historically belonged to the Ottoman or Tsarist empires and were only lightly touched by the shaping events in the rest of Europe; they are generally less advanced economically; they seem much less likely to develop stable democratic political systems. The Velvet Curtain of culture has replaced the Iron Curtain of ideology as the most significant dividing line in Europe. As the events in Yugoslavia show, it is not only a line of difference; it is also at times a line of bloody conflict.

Conflict along the fault line between Western and Islamic civilizations has been going on for 1300 years. After the founding of Islam, the Arab and Moorish surge west and north only ended at Tours in 732. From the eleventh to the thirteenth century the Crusaders attempted with temporary success to bring Christianity and Christian rule to the Holy Land. From the fourteenth to the seventeenth century, the Ottoman Turks reversed the balance, extended their sway over the Middle East and the Balkans, captured Constantinople, and twice laid siege to Vienna. In the nineteenth and early twentieth centuries as Ottoman power declined, Britain, France and Italy established Western control over most of North Africa and the Middle East.

After World War II, the West, in turn, began to retreat; the colonial empires disappeared; first Arab nationalism and then Islamic fundamentalism manifested themselves; the West became heavily dependent on the Persian Gulf countries for its energy; the oil-rich Muslim countries became money-rich and, when they wished to, weapons-rich. Several wars occurred between Arabs and Israel (created by the West). France fought a bloody and ruthless war in Algeria for most of the 1950s; British and French forces invaded Egypt in 1956; American forces went into Lebanon in 1958; subsequently American forces returned to Lebanon, attacked Libya, and engaged in various military encounters with Iran; Arab and Islamic terrorists, supported by at least three Middle Eastern governments, employed the weapon of the weak and bombed Western planes and installations and seized Western hostages. This warfare between Arabs and the West culminated in 1990, when the United States sent a massive army to the Persian Gulf to defend some Arab countries against aggression by another. In its aftermath NATO planning is increasingly directed to potential threats and instability along its "southern tier."

This centuries-old military interaction between the West and Islam is unlikely to decline. It could become more virulent. The Gulf War left some Arabs feeling proud that Saddam Hussein had attacked Israel and stood up to the West. It also left many feeling humiliated and resentful of the West's military presence in the Persian Gulf, the West's overwhelming military dominance, and their apparent inability to shape their own destiny. . . .

CIVILIZATION RALLYING: THE KIN-COUNTRY SYNDROME

Groups or states belonging to one civilization that become involved in war with people from a different civilization naturally try to rally support from other members of their own civilization. As the post-Cold War world evolves, civilization commonality, what H. D. S. Greenway has termed the "kin-country" syndrome, is replacing political ideology and traditional balance of power considerations as the principal basis for cooperation and coalitions. It can be seen gradually emerging in the post-Cold War conflicts in the Persian Gulf, the Caucasus and Bosnia. None of these was a full-scale war between civilizations, but each involved some elements of civilizational rallying, which seemed to become more important as the conflict continued and which may provide a foretaste of the future....

Civilization rallying to date has been limited, but it has been growing, and it clearly has the potential to spread much further. As the conflicts in the Persian Gulf, the Caucasus and Bosnia continued, the positions of nations and the cleavages between them increasingly were along civilizational lines. Populist politicians, religious leaders and the media have found it a potent means of arousing mass support and of pressuring hesitant governments. In the coming years, the local conflicts most likely to escalate into major wars will be those, as in Bosnia and the Caucasus, along the fault lines between civilizations. The next world war, if there is one, will be a war between civilizations.

THE WEST VERSUS THE REST

The West is now at an extraordinary peak of power in relation to other civilizations. Its superpower opponent has disappeared from the map. Military conflict among Western states is unthinkable, and Western military power is unrivaled. Apart from Japan, the West faces no economic challenge. It dominates international political and security institutions and with Japan international economic institutions. Global political and security issues are effectively settled by a directorate of the United States, Britain and France, world economic issues by a directorate of the United States, Germany and Japan, all of which maintain extraordinarily close relations with each other to the exclusion of lesser and largely non-Western countries. Decisions made at the UN Security Council or in the International Monetary Fund that reflect the interests of the West are presented to the world as reflecting the desires of the world community. The very phrase "the world community" has become the euphemistic collective noun (replacing "the Free World") to give global legitimacy to actions reflecting the interests of the United States and other Western powers.[1] Through the IMF and other international economic institutions, the West promotes its economic interests and imposes on other nations the economic policies it thinks appropriate. In any poll of non-Western peoples, the IMF undoubtedly would win the support of finance ministers and a few others, but get an overwhelmingly unfavorable rating from just about everyone else, who would agree with Georgy Arbatov's characterization of IMF officials as "neo-Bolsheviks who love expropriating other people's money, imposing undemocratic and alien rules of economic and political conduct and stifling economic freedom."

Western domination of the UN Security Council and its decisions, tempered only by occasional abstention by China, produced UN legitimation of the West's use of force to drive Iraq out of Kuwait and its elimination of Iraq's sophisticated weapons and capacity to produce such weapons. It also produced the quite unprecedented action by the United States, Britain and France in getting the Security Council to demand that Libya hand over the Pan Am 103 bombing suspects and then to impose sanctions when Libya refused. After defeating the largest Arab army, the West did not hesitate to throw its weight around in the Arab world. The West in effect is using international institutions, military power and economic resources to run the world in ways that will maintain Western predominance, protect Western interests and promote Western political and economic values.

That at least is the way in which non-Westerners see the new world, and there is a significant element of truth in their view. Differences in power and struggles for military, economic and institutional power are thus one source of conflict between the West and other civilizations. Differences in culture, that is basic values and beliefs, are a second source of conflict. V. S. Naipaul has argued that Western civilization is the "universal civilization" that "fits all men." At a superficial level much of Western culture has indeed permeated the rest of the world. At a more basic level, however, Western concepts differ fundamentally from those prevalent in other civilizations. Western ideas of individualism, liberalism, constitutionalism, human rights, equality, liberty, the rule of law, democracy, free markets, the separation of church and state, often have little resonance in Islamic, Confucian, Japanese, Hindu, Buddhist or Orthodox cultures. Western efforts to propagate such ideas produce instead a reaction against "human rights imperialism" and a reaffirmation of indigenous values, as can be seen in the support for religious fundamentalism by the younger generation in non-Western cultures. The very notion that there could be a "universal civilization" is a Western idea, directly at odds with the particularism of most Asian societies and their emphasis on what distinguishes one people from another. Indeed, the author of a review of 100 comparative studies of values in different societies concluded that "the values that are most important in the West are least important worldwide."[2] In the political realm, of course, these differences are most manifest in the efforts of the United States and other Western powers to induce other peoples to adopt Western ideas concerning democracy and human rights. Modern democratic government originated in the West. When it has developed in non-Western societies it has usually been the product of Western colonialism or imposition.

The central axis of world politics in the future is likely to be, in Kishore Mahbubani's phrase, the conflict between "the West and the Rest" and the responses of non-Western civilizations to Western power and values.[3] Those responses generally take one or a combination of three forms. At one extreme, non-Western states can, like Burma and North Korea, attempt to pursue a course of isolation, to insulate their societies from penetration or "corruption" by the West, and, in effect, to opt out of participation in the Western-dominated global community. The costs of this course, however, are high, and few states have pursued it exclusively. A second alternative, the equivalent of "bandwagoning" in international relations theory, is to attempt to join the West and accept its values and institutions. The third alternative is to attempt to "balance" the West by developing economic and military power and cooperating with other non-Western societies against the West, while preserving indigenous values and institutions; in short, to modernize but not to Westernize.

THE TORN COUNTRIES

In the future, as people differentiate themselves by civilization, countries with large numbers of peoples of different civilizations, such as the Soviet Union and Yugoslavia, are candidates for dismemberment. Some other countries have a fair degree of cultural homogeneity but are divided over whether their society belongs to one civilization or another. These are torn countries. Their leaders typically wish to pursue a bandwagoning strategy and to make their countries members of the West, but the history, culture and traditions of their countries are non-Western. The most obvious and prototypical torn country is Turkey. The late twentieth-century leaders of Turkey have followed in the Attaturk tradition and defined Turkey as a modern, secular, Western nation state. They allied Turkey with the West in NATO and in the Gulf War; they applied for membership in the European Community. At the same time, however, elements in Turkish society have supported an Islamic revival and have argued that Turkey is basically a Middle Eastern Muslim society. In addition, while the elite of Turkey has defined Turkey as a Western society, the elite of the West refuses to accept Turkey as such. Turkey will not become a member of the European Community, and the real reason, as President Ozal said, "is that we are Muslim and they are Christian and they don't say that." Having rejected Mecca, and then being rejected by Brussels, where does Turkey look? Tashkent may be the answer. The end of the Soviet Union gives Turkey the opportunity to become the leader of a revived Turkic civilization involving seven countries from the borders of Greece to those of China. Encouraged by the West, Turkey is making strenuous efforts to carve out this new identity for itself.

During the past decade Mexico has assumed a position somewhat similar to that of Turkey. Just as Turkey abandoned its historic opposition to Europe and attempted to join Europe, Mexico has stopped defining itself by its opposition to the United States and is instead attempting to imitate the United States and to join it in the North American Free Trade Area. Mexican leaders are engaged in the great task of redefining Mexican identity and have introduced fundamental economic reforms that eventually will lead to fundamental political change. In 1991 a top adviser to President Carlos Salinas de Gortari described at length to me all the changes the Salinas government was making. When he finished, I remarked: "That's most impressive. It seems to me that basically you want to change Mexico from a Latin American country into a North American country." He looked at me with surprise and exclaimed: "Exactly! That's precisely what we are trying to do, but of course we could never say so publicly." As his remark indicates, in Mexico as in Turkey, significant elements in society resist the redefinition of their country's identity. In Turkey, European-oriented leaders have to make gestures to Islam (Ozal's pilgrimage to Mecca); so also Mexico's North American-oriented leaders have to make gestures to those who hold Mexico to be a Latin American country (Salinas' Ibero–American Guadalajara summit).

Historically Turkey has been the most profoundly torn country. For the United States, Mexico is the most immediate torn country. Globally the most important torn country is Russia. The question of whether Russia is part of the West or the leader of a distinct Slavic Orthodox civilization has been a recurring one in Russian history. That issue was obscured by the communist victory in Russia, which imported a Western ideology, adapted it to Russian conditions and then challenged the West in the name of that ideology. The dominance of communism shut off the historic debate over Westernization versus Russification. With communism discredited, Russians once again face that question. President Yeltsin is adopting Western principles and goals and seeking to make Russia a "normal" country and a part of the West. Yet both the Russian elite and the Russian public are divided on this issue. Among the more moderate dissenters, Sergei Stankevich argues that Russia should reject the "Atlanticist" course, which would lead it

> to become European, to become a part of the world economy in rapid and organized fashion, to become the eighth member of the Seven, and to put particular emphasis on Germany and the United States as the two dominant members of the Atlantic alliance.

While also rejecting an exclusively Eurasian policy, Stankevich nonetheless argues that Russia should give priority to the protection of Russians in other countries, emphasize its Turkic and Muslim connections, and promote "an appreciable redistribution of our resources, our options, our ties, and our interests in favor of Asia, of the eastern direction." People of this persuasion criticize Yeltsin for subordinating Russia's interests to those of the West, for reducing Russian military strength, for failing to support traditional friends such as Serbia, and for pushing economic and political reform in ways injurious to the Russian people. Indicative of this trend is the new popularity of the ideas of Petr Savitsky, who in the 1920s argued that Russia was a unique Eurasian civilization.[4] More extreme dissidents voice much more blatantly nationalist, anti-Western and anti-Semitic views, and urge Russia to redevelop its military strength and to establish closer ties with China and Muslim countries. The people of Russia are as divided as the elite. An opinion survey in European Russia in the spring of 1992 revealed that 40 per cent of the public had positive attitudes toward the West and 36 per cent had negative attitudes. As it has been for much of its history, Russia in the early 1990s is truly a torn country.

To redefine its civilization identity, a torn

country must meet three requirements. First, its political and economic elite has to be generally supportive of and enthusiastic about this move. Second, its public has to be willing to acquiesce in the redefinition. Third, the dominant groups in the recipient civilization have to be willing to embrace the convert. All three requirements in large part exist with respect to Mexico. The first two in large part exist with respect to Turkey. It is not clear that any of them exist with respect to Russia's joining the West. The conflict between liberal democracy and Marxism–Leninism was between ideologies which, despite their major differences, ostensibly shared ultimate goals of freedom, equality and prosperity. A traditional, authoritarian, nationalist Russia could have quite different goals. A Western democrat could carry on an intellectual debate with a Soviet Marxist. It would be virtually impossible for him to do that with a Russian traditionalist. If, as the Russians stop behaving like Marxists, they reject liberal democracy and begin behaving like Russians but not like Westerners, the relations between Russia and the West could again become distant and conflictual.[5]

THE CONFUCIAN–ISLAMIC CONNECTION

The obstacles to non-Western countries joining the West vary considerably. They are least for Latin American and East European countries. They are greater for the Orthodox countries of the former Soviet Union. They are still greater for Muslim, Confucian, Hindu and Buddhist societies. Japan has established a unique position for itself as an associate member of the West: it is in the West in some respects but clearly not of the West in important dimensions. Those countries that for reason of culture and power do not wish to, or can not, join the West compete with the West by developing their own economic, military and political power. They do this by promoting their internal development and by cooperating with other non-Western countries. The most prominent form of this cooperation is the Confucian–Islamic connection that has emerged to challenge Western interests, values and power.

Almost without exception, Western countries are reducing their military power; under Yeltsin's leadership so also is Russia. China, North Korea and several Middle Eastern states, however, are significantly expanding their military capabilities. They are doing this by the import of arms from Western and non-Western sources and by the development of indigenous arms industries. One result is the emergence of what Charles Krauthammer has called "Weapon States," and the Weapon States are not Western states. Another result is the redefinition of arms control, which is a Western concept and a Western goal. During the Cold War the primary purpose of arms control was to establish a stable military balance between the United States and its allies and the Soviet Union and its allies. In the post-Cold War world the primary objective of arms control is to prevent the development by non-Western societies of military capabilities that could threaten Western interests. The West attempts to do this through international agreements, economic pressure and controls on the transfer of arms and weapons technologies.

The conflict between the West and the Confucian–Islamic states focuses largely, although not exclusively, on nuclear, chemical and biological weapons, ballistic missiles and other sophisticated means for delivering them, and the guidance, intelligence and other electronic capabilities for achieving that goal. The West promotes nonproliferation as a universal norm and nonproliferation treaties and inspections as means of realizing that norm. It also threatens a variety of sanctions against those who promote the spread of sophisticated weapons and proposes some benefits for those who do not. The attention of the West focuses, naturally, on nations that are actually or potentially hostile to the West.

The non-Western nations, on the other hand, assert their right to acquire and to deploy whatever weapons they think necessary for their security. They also have absorbed, to the full, the truth of the response of the Indian defense minister when asked what lesson he learned from the Gulf War: "Don't fight the United States unless you have nuclear weapons."

Nuclear weapons, chemical weapons and missiles are viewed, probably erroneously, as the potential equalizer of superior Western conventional power. China, of course, already has nuclear weapons; Pakistan and India have the capability to deploy them. North Korea, Iran, Iraq, Libya and Algeria appear to be attempting to acquire them. A top Iranian official has declared that all Muslim states should acquire nuclear weapons, and in 1988 the president of Iran reportedly issued a directive calling for development of "offensive and defensive chemical, biological and radiological weapons."

Centrally important to the development of counter-West military capabilities is the sustained expansion of China's military power and its means to create military power. Buoyed by spectacular economic development, China is rapidly increasing its military spending and vigorously moving forward with the modernization of its armed forces. It is purchasing weapons from the former Soviet states; it is developing long-range missiles; in 1992 it tested a one-megaton nuclear device. It is developing power-projection capabilities, acquiring aerial refueling technology, and trying to purchase an aircraft carrier. Its military build-up and assertion of sovereignty over the South China Sea are provoking a multilateral regional arms race in East Asia. China is also a major exporter of arms and weapons technology. It has exported materials to Libya and Iraq that could be used to manufacture nuclear weapons and nerve gas. It has helped Algeria build a reactor suitable for nuclear weapons research and production. China has sold to Iran nuclear technology that American officials believe could only be used to create weapons and apparently has shipped components of 300-mile-range missiles to Pakistan. North Korea has had a nuclear weapons program under way for some while and has sold advanced missiles and missile technology to Syria and Iran. The flow of weapons and weapons technology is generally from East Asia to the Middle East. There is, however, some movement in the reverse direction; China has received Stinger missiles from Pakistan.

A Confucian–Islamic military connection has thus come into being, designed to promote acquisition by its members of the weapons and weapons technologies needed to counter the military power of the West. It may or may not last. At present, however, it is, as Dave McCurdy has said, "a renegades' mutual support pact, run by the proliferators and their backers." A new form of arms competition is thus occurring between Islamic–Confucian states and the West. In an old-fashioned arms race, each side developed its own arms to balance or to achieve superiority against the other side. In this new form of arms competition, one side is developing its arms and the other side is attempting not to balance but to limit and prevent that arms build-up while at the same time reducing its own military capability.

IMPLICATIONS FOR THE WEST

This article does not argue that civilization identities will replace all other identities, that nation states will disappear, that each civilization will become a single coherent political entity, that groups within a civilization will not conflict with and even fight each other. This paper does set forth the hypotheses that differences between civilizations are real and important; civilization-consciousness is increasing; conflict between civilizations will supplant ideological and other forms of conflict as the dominant global form of conflict; international relations, historically a game played out within Western civilization, will increasingly be de-Westernized and become a game in which non-Western civilizations are actors and not simply objects; successful political, security and economic international institutions are more likely to develop within civilizations than across civilizations; conflicts between groups in different civilizations will be more frequent, more sustained and more violent than conflicts between groups in the same civilization; violent conflicts between groups in different civilizations are the most likely and most dangerous source of escalation that could lead to global wars; the paramount axis of world politics will be the relations between the "West and the Rest"; the elites in some torn non-Western countries will try to make their countries part of the West, but in most cases face major obstacles to accomplishing this; a central focus of conflict for the immediate future will be between the West and several Islamic–Confucian states.

This is not to advocate the desirability of conflicts between civilizations. It is to set forth descriptive hypotheses as to what the future may be like. If these are plausible hypotheses, however, it is necessary to consider their implications for Western policy. These implications should be divided between short-term advantage and long-term accommodation. In the short term it is clearly in the interest of the West to promote greater cooperation and unity within its own civilization, particularly between its European and North American components; to incorporate into the West societies in Eastern Europe and Latin America whose cultures are close to those of the West; to promote and maintain cooperative relations with Russia and Japan; to prevent escalation of local inter-civilization conflicts into major inter-civilization wars; to limit the expansion of the military strength of Confucian and Islamic states; to moderate the reduction of Western military capabilities and maintain military superiority in East and Southwest Asia; to exploit differences and conflicts among Confucian and Islamic states; to support in other civilizations groups sympathetic to Western values and interests; to strengthen international institutions that reflect and legitimate Western interests and values and to promote the involvement of non-Western states in those institutions.

In the longer term other measures would be called for. Western civilization is both Western and modern. Non-Western civilizations have attempted to become modern without becoming Western. To date only Japan has fully succeeded in this quest. Non-Western civilizations will continue to attempt to acquire the wealth, technology, skills, machines and weapons that are part of being modern. They will also attempt to reconcile this modernity with their traditional culture and values. Their economic and military strength relative to the West will increase. Hence the West will increasingly have to accommodate these non-Western modern civilizations whose power approaches that of the West but whose values and interests differ significantly from those of the West. This will require the West to maintain the economic and military power necessary to protect its interests in relation to these civilizations. It will also, however, require the West to develop a more profound understanding of the basic religious and philosophical assumptions underlying other civilizations and the ways in which people in those civilizations see their interests. It will require an effort to identify elements of commonality between Western and other civilizations. For the relevant future, there will be no universal civilization, but instead a world of different civilizations, each of which will have to learn to coexist with the others.

NOTES

1 Almost invariably Western leaders claim that they are acting on behalf of "the world community." One minor lapse occurred during the run-up to the Gulf War. In an interview on "Good Morning America," Dec. 21, 1990, British Prime Minister John Major referred to the actions "the West" was taking against Saddam Hussein. He quickly corrected himself and subsequently referred to "the world community." He was, however, right when he erred.

2 Harry C. Triandis, *The New York Times*, Dec. 25, 1990, p. 41, and "Cross-Cultural Studies of Individualism and Collectivism," *Nebraska Symposium on Motivation*, vol. 37, 1989, pp. 41–133.

3 Kishore Mahbubani, "The West and the Rest," *The National Interest*, Summer 1992, pp. 3–13.

4 Sergei Stankevich, "Russia in Search of Itself," *The National Interest*, Summer 1992, pp. 47–51; Daniel Schneider, "A Russian Movement Rejects Western Tilt," *Christian Science Monitor*, Feb. 5, 1993, pp. 5–7.

5 Owen Harries has pointed out that Australia is trying (unwisely in his view) to become a torn country in reverse. Although it has been a full member not only of the West but also of the ABCA military intelligence core of the West, its current leaders are in effect proposing that it defect from the West, redefine itself as an Asian country and cultivate close ties with its neighbours. Australia's future, they argue, is with the dynamic economies of East Asia. But, as I have suggested, close economic cooperation requires a common cultural base. In addition, none of the three conditions necessary for a torn country to join another civilization is likely to exist in Australia's case.

22

GEARÓID Ó TUATHAIL

"Samuel Huntington and the 'Civilizing' of Global Space"

from *Critical Geopolitics* (1996)

In contrast to Edward Luttwak, Samuel Huntington is an intellectual of statecraft who specializes in questions of governance, particularly the problems of hegemonic governance. Educated at Yale (BA, 1946), the University of Chicago (MA, 1948) and Harvard (PHD, 1951), Huntington has spent his life within the elite circles of the Ivy League academic and foreign policy establishment. A neoconservative advisor to Democrats like Hubert Humphrey and Jimmy Carter, Huntington was retained by Harvard to work in the Department of Government. Co-founder of the journal *Foreign Policy*, associated at various times with Columbia, Oxford and Stanford universities, a member of the Council on Foreign Relations and the Trilateral Commission, Huntington has served as a consultant to various agencies of the US government and worked as coordinator of security planning at the National Security Council (1977–78) when that agency was headed by his one-time colleague and co-author Zbigniew Brzezinski. Huntington has a long history of association with the Center for International Affairs (CFIA) at Harvard, a research institute established in 1957 by Robert Bowie (chief policy planner in the State Department of John Foster Dulles) together with McGeorge Bundy and Henry Kissinger.[1] Through its research fellowships, the CFIA was designed to attract and cultivate an array of influential intellectuals from around the globe who subsequently would serve as a worldwide network of informal influence for Harvard and its "imperial" ideas of good government. Huntington served as associate director (1973–78) and subsequently as director (1978–89) of CFIA. He is currently the director of the John Olin Institute of Strategic Studies at Harvard.

Huntington's work is explicitly concerned with questions of governmentality in developing and developed states. In *Political Order in Changing Societies* (1968), he addresses the question of how viable political regimes can be established and made to last in developing countries, a problem that was of particular concern to the United States government in South Vietnam at the time. Huntington's stress on the establishment of democratic organizations and institutions of authority (particularly the military) was a direct response to the challenge Communism posed to "modernizing countries."[2] Organization, he concluded, was the road to political power. "In the modernizing world he controls the future who organizes its politics" (461). In *The Crisis of Democracy* (1975), a report on the governability of developed democracies to the Trilateral Commission (at the time under the directorship of Zbigniew Brzezinski) in the wake of the 1960s upheavals and Watergate, Huntington analyzes the social and political ferment of the period in terms of a growing disrespect for authority and a general "excess of democracy." The solution to the adversarial culture of the youth, the mass media and dissident intellectuals was "a greater degree of moderation in democracy" and restoration of a "democratic balance," code phrases for a neoconservative re-disciplining of society around elitist and hierarchial notions of "expertise, seniority and experience."[3] Like other neoconservatives, Huntington was critical of his former colleague Henry Kissinger's policy of *detente* with the Soviet Union. He supported the US military buildup begun under President Carter and continued under President Reagan as well as the US war against the Sandanistas and its tilt towards the governments of El Salvador and South Africa.[4]

The sudden end of the Cold War threw the imperial visions of the intellectuals and apologists of Cold War militarism into confusion in 1990. "The world changed in 1990," Huntington remarked, "and so did strategic discourse."[5] But, in Huntington's case at least, strategic discourse did not change that much. The Cold War with the Soviets may have ended but there were suddenly new (and modified old) Cold Wars to be fought in the murky and dangerous post-Cold War unknown. This emerging world, Huntington noted, "is likely to lack the clarity and stability of the Cold War and to be a more jungle-like world of multiple dangers, hidden traps, unpleasant surprises and moral ambiguities" (7). Huntington outlines three principal American strategic interests: (i) maintaining the United States as the premier global power, which "means countering the Japanese economic challenge;" (ii) preventing the "emergence of a political-military hegemonic power in Eurasia;" and (iii) protesting "concrete American interests in the Persian Gulf and Middle America" (8).

Like Luttwak, Huntington securitizes the US–Japan relationship and highlights the US's "economic performance gaps" with Japan, evoking the specter of *The Japan That Can Say No* as evidence that Japan is an emergent threat to US primacy in world affairs. This focus on Japan as a threat and preoccupation with maintaining the primacy of the United States in world affairs led Huntington to support Bill Clinton for the presidency in 1992. For Huntington, the economic renewal of America was an overriding priority but it was an economic renewal that was to be achieved not by downsizing the US military or breaking up the society of security. Rather, the United States had to intensify its concern with security by improving its "competitiveness" and confronting Japan. For "the first time in two hundred years," Huntington wrote in a symposium on advice for a Democratic President, "the United States faces a major economic threat." "In terms of economic power . . . Japan is rapidly overtaking the United States. And economic power is not only central to the relations among the major states, it is also the underpinning of virtually every other form of power."[6] Japanese strategy is a strategy of economic warfare. Buttressed by appropriate citations by Japanese figures declaring themselves the new economic superpower and the United States as a premier agrarian power ("a giant version of Denmark"), Huntington reasoned that Chamberlain and Daladier did not take Hitler seriously in the 1930s nor did Truman and his successors take it seriously when Stalin and Khrushchev said "We will bury you," but Americans "would do well to take equally seriously both Japanese declarations of their goal of achieving economic dominance and the strategy they are pursuing to achieve that goal."[7]

Huntington's re-charting of the threats faced by the United States from Japan, Eurasia and the Third World crystallized into a geopolitical world-picture which was prominently unveiled and publicized in the Council on Foreign Relations' journal *Foreign Affairs* in 1993. Entitled "The Clash of Civilizations," Huntington's essay was promoted – first as a leading article with solicited comments and reply in *Foreign Affairs*, second as a *New York Times* opinion editorial piece and subsequent syndicated columnist debate, third as a special Council on Foreign Relations reader, and fourth in what promises to be a much hyped "best seller" book on the thesis due out in 1996 – as a comprehensive vision of the "next pattern of conflict" in world politics.[8] His goal, in his own words, was to produce "the best simple map of the post-Cold War world."[9] The overarching ambition, conciseness and sloganistic simplicity of Huntington's mediagenic thesis accounts for its appeal to opinion makers, news journalists and professional politicians casting about for a new interpretative system by which to order global affairs given the waning interpretative power of the more optimistic visions of Fukuyama and Bush's new world order. In 1993, at least, Huntington's thesis was good copy with the media's foreign policy scribes.

Though it is provoked by the vertigo of postmodernity, Huntington's vision is projected with modernist assurance. The heteroglossia of global politics – what Huntington refers to as "bloomin' buzzin' confusion"[10] – is reduced to a total(izing) world-picture. The gaze Huntington employs is that of a natural scientist qua geologist observing a world of solid forms whose meaning can be declaratively and unambiguously stated. He writes as a self-certain subject who reveals the "basic" plate tectonics of civilizational blocs (the "product of centuries") that "stretch back into history" and are now clashing once more along ancient "fault lines." Like a natural scientist, Huntington employs a series of definitional, periodizing, classificatory and spatialization strategies to enframe human history and global space into an ordered geological exhibit. Civilizations, for Huntington, are foundational cultural totalities. A civilization is "the highest cultural grouping of people and the broadest level of cultural identity people have short of that which distinguishes humans from other species." Though they are supposedly centuries-old and primordial, the clash of civilizations is the very latest stage in the evolution of conflict in the modern world, following monarchial, popular and ideological stages of conflict. This fourth stage features a clash between seven or eight major civilizations classified as Western, Confucian, Japanese, Islamic, Hindu, Slavic–Orthodox, Latin American and possibly African. The clash of these civilizations occurs at a micro and macro level, the micro-level being the struggle of adjacent groups for territory (e.g. Bosnia), the macro-level being the struggle of states from different civilizations for power, international institutions and influence over third parties (e.g. the Gulf War or the US-Japan economic struggle for world markets). Huntington sloganizes a new axis of global politics ("the West versus the Rest") and proclaims that the Iron Curtain of ideology in Europe has been replaced by a "Velvet Curtain of culture" which is cartographically displayed in a map of Middle Europe with a line dividing "Western Christianity circa 1500" from "Orthodox Christianity and Islam."[11]

Huntington's concept of a "civilization" is a curious one that is crucial to his writing of global political space. A deterministic totality that is not reducible to either religion, ethnicity, geography or attitude, his classification gestures to all these factors. The Japanese, Latin American and African civilizations appear to be specified geographically whereas the Confucian, Islamic, Hindu and Slavic–Orthodox civilizations are specified in religious and ethnic terms. Western civilization is somewhat unique in that it has universalistic ambitions and is apparently secular. However, to specify Huntington's thesis in purely representational terms is to miss how the notion of "civilization" functions as a flexible free floating sign for Huntington, a sign which refers to other signs and not to any stable referent. To evoke a "civilization" is to call up a foundational identity, a mystical and mythical transcendental presence that is vague yet absolutely fundamental. To designate a conflict a civilizational one is to determine its character in a definitive and totalizing manner. It is to impose a closure upon events, situations and peoples. The geographical specificity and place-based particularity of conflicts are reduced to the terms of a civilizational script. In much the same way as Cold War discourse depluralized and homogenized global space, Huntington's civilizational discourse reduces the geographical specificity of conflicts to reified identities and attributes, transforming their ambiguities and indeterminacies into graspable certainties and solid truths. The multiplicity of identities that traverse the world's peoples are diminished to a set of essential differences and distinctions. States are stamped with civilizational labels: Western states, Confucian states, Islamic states, Hindu states, Latin American states, Orthodox–Slavic states (and sometimes combinations like Islamic–Confucian states). Global space is "civilized."

This civilizing of global space produces, as we might expect, highly problematic interpretations of the world's various conflicts. For example, the conflict in Yugoslavia is located on the fault line between Western Christianity, Orthodox Christianity and Islam. For Huntington, this location becomes an explanation. However, to reduce the war in Bosnia to an ancient fault line civilizational struggle is to read it in the same terms as those who wish to produce it as an essential civilizational war of the Orthodox Slavic Serbs against Islam. The possibilities of "Bosnia" as a multicultural state and "Bosnian" as a multicultural identity are precluded. Huntington, in other words, accepts the Bosnian Serb leadership's interpretation of the war and then cites this as illustrative of his thesis. The multiculturalism of the region's past is ignored and the complexity of its present struggle is reduced to a sight/site/cite of an ahistorical essential antagonism. Huntington does the same for the Gulf War accepting, as Fouad Ajami noted, Saddam Hussein's interpretation of the conflict as one "between the Arabs and the West."[12] The Gulf War is written as a case of civilization rallying, a kin-and-country syndrome whereby groups or states belonging to one civilization rally to the support of other members of their civilization when they are involved in wars and conflicts. Such a claim, however, simplifies the multiplicity of different ways in which the Gulf War was understood and interpreted by different groups in different places. In yet another example, Huntington arbitrarily reads US–China relations in civilizational terms with the end of the Cold War seeing the "reassertion" of "underlying differences" between the United States and China (a vague claim that is dubious given the Bush and Clinton administrations' reassurances to China). A reported statement by Deng Xaioping – China's reformist Marxist leader who supposedly represents alien non-Western "Confucian" values – that a new Cold War is under way between both countries is cited as evidence of a civilizational clash.

Huntington's civilizing of the deterritorializing space of the post-Cold War unknown is also extended as an explanation of development struggles in certain states. The efforts by governing elites in countries like Turkey, Mexico, Russia and other states to open up their territorial economies to global markets mark these countries as "torn countries" within Huntington's civilizational tableaux. These are countries where two civilizations – Western and non-Western – are at war for the identity of the society and state. Such reasoning is problematic in a number of ways. First, it assumes that most states are not torn but isolated entities with stable and fully formed identities. This has never been the case in the history of the modern world system where the identity of states is a product of ongoing histories of struggle and mutual interaction not isolationism. Second, it reduces economic transformations to cultural wars. Postmodernity and globalization become *Kulturkampf* which then itself becomes explanation. The materiality of the cultural and ideological battles Huntington identifies is ignored. The problematic of countries like Turkey, Mexico and Russia is much more complex than one of elites creating "torn countries" by trying to make their non-Western countries part of the West.

As many commentators have noted, Huntington's thesis is remarkably simplistic and comprehensively flawed. It is significant, nevertheless, as an example of how neoconservative intellectuals of statecraft are endeavoring to chart global space after the Cold War. What is most interesting about this act of geopower is how it uses the assumptions, goals and methods of Cold War strategic culture to re-territorialize the global scene in a way which perpetuates the society of security and politics as *Kulturkampf*. We find this in Huntington in three significant ways. First, like Luttwak, Huntington's thesis re-territorializes global space by triangulating from the same ahistorical realism that produced Cold War strategic discourse. This charts international politics as a perpetual struggle for power between coherent and isolated units each seeking to advance their interests in a condition of anarchy. As Rubenstein and Crocker note, "Huntington has replaced the nation-state, the primary playing piece in the old game of realist politics, with a larger counter: the civilization. But in crucial respects, the game itself goes on as always."[13] This notion of an even more intense playing of a now redundant game is important. In producing a civilizationalized global scene, Huntington is creatively seeking to save political realism from the condition of postmodernity by generating a hyper-real civilizational order in the face of the spatial vertigo of the new world disorder. In using terms that are without definitive referents (like "the West" or "the non-West"),[14] Huntington simulates civilizations and an ordered global political scene. In Baudrillard's terms, he produces truth-effects that hide the truth's non-existence.

Second, the purpose of Huntington's post-

Cold War strategic discourse is the same as Cold War strategic discourse: to perpetuate the primacy of the United States in world affairs. This is to be achieved by the United States renewing its Western civilization from within and actively containing, dividing and playing off other civilizations against each other. What is different from the Cold War is how the "West" is re-written by Huntington to partially exclude Japan, the US's traditional Cold War ally. Japan has an ambivalent status for Huntington. On the one hand, he notes, the West faces no economic challenge "apart from Japan" yet, on the other hand, he describes world economic issues as settled by "a directorate of the United States, Germany and Japan, all of which maintain extraordinary close relations with each other to the exclusion of lesser and largely non-Western countries."[15] The reason for Huntington's ambivalence about the territorial extent of "the West" is that it refers more to an imaginative and idealized cultural order than it does to any territorial entity. Rooted in the political mythology of Western Europe and North America, the "West" is not simply a geographical community but a universalistic creed of individualism, liberalism, constitutionalism, human rights, democracy and free markets.[16] It is simultaneously an imaginary cultural order before it is a real place. "The West versus the Rest" is not simply a spatial struggle between a distinct "here" (the West) and an identifiable "there" (the Rest) but a cultural and spatial struggle that occurs *everywhere* (just like the Cold War, for it too, for Huntington, was a clash of civilizations). Huntington's neoconservative anxiety is a product of the emergent disjuncture between the real and the imaginary West brought about by globalization and postmodernity. That the American economy is now dependent upon Japanese industrial and finance capital and that significant proportions of the US population belong to other civilizations (Latin American, African, Confucian, Japanese and Islamic) is a cause for alarm, because an idealized "West" is being weakened and undermined from within the real. The struggle between "the West and the Rest," therefore, begins on the home front in the fight for domestic economic renewal (reducing America's dependence on foreign capital) and against "multiculturalism" which Huntington associates with "the de-Westernization of the United States." The internal *Kulturkampf* Huntington describes is viewed in apocalyptic terms. "If . . . Americans cease to adhere to their liberal democratic and European-rooted political ideology, the United States as we have known it will cease to exist and will follow the other ideologically defined superpower onto the ash heap of history."[17] Huntington's reasoning envisions a geopolitics of exclusion. De-territorializing geographical space is to be hardened against foreign civilizations and re-territorialized along the lines of an imaginary Euro-America cultural and political order. The real is to be re-disciplined to fit the imaginary.

Third, Huntington's thesis is a writing of a world of threats to the United States, a world of potential and actual Cold Wars that require a renewal of the society of security within the "West." Interestingly, Huntington's earlier preoccupation with the "economic Cold War" with Japan is not as prominent as before. The new danger is a "Confucian–Islamic connection" which features a militaristic Chinese economy exporting arms to Islamic states who are determined to seek nuclear, chemical and biological weapons capabilities. "A Confucian–Islamic military connection has . . . come into being, designed to promote acquisition by its members of the weapons and weapons technologies needed to counter the military power of the West. . . . A new form of arms competition is thus occurring between Islamic–Confucian states and the West" (47). This is occurring at a time when Western states are reducing their military power. Huntington's response, amongst other things, is to call for a moderation in this reduction of Western military capabilities and for the West to "maintain military superiority in East and Southeast Asia" (49).

Like Luttwak's [argument], Huntington's thesis is interesting not for its explanatory value but as a map of a certain structure of feeling within the US foreign policy community. This structure of feeling is a reactionary one in the sense that it reacts to the vertigo and complexity of postmodernity with a longstanding conservative pessimism and fundamentalism. As

Kurth points out, the term "Western civilization" was only invented at the beginning of the twentieth century and was itself a sign of a pessimistic feeling of decline within Europe (most pointedly expressed by Oswald Spengler).[18] This pessimism remains at the end of the twentieth century within many elite academic and governmental circles in the United States as it faces an increasingly disorderly world over which it has less and less control. Huntington's partly resigned but also partly defiant response to this unhappy condition is to return to the imaginary fundamentals of earlier history and re-cycle them in the hope of re-territorializing global space in such a way that his neoconservative agenda of cultural and ideological war against those who would challenge Western fundamentalism (its national security state and society of security) becomes the only option. Huntington's thesis is not about the clash of civilizations. It is about making global politics a clash of civilizations.

NOTES

1 For an account of Kissinger's role in the CFIA see Isaacson, *Kissinger: A Biography* (New York: Simon and Schuster, 1992), 94–107.

2 "The real challenge which communists pose to modernizing countries is not that they are so good at overthrowing governments (which is easy), but that they are so good at making governments (which is far more difficult). They may not provide liberty, but they do provide authority; they do create governments that can govern." Samuel Huntington, *Political Order in Changing Societies* (New Haven: Yale University Press, 1968), 8. On the place of Huntington's ideas within the ideology of liberal modernization in US foreign policy see Bradley Klein, *Strategic Studies and World Order: The Global Politics of Deterrence* (Cambridge: Cambridge University Press, 1994), 99–101.

3 Samuel Huntington, "The United States," in *The Crisis of Democracy*, Michael Crozier, Samuel Huntington and Joji Watanuki (New York: New York University Press, 1975), 113.

4 Huntington's views on these issues are expressed in the interview, "A Better America: Ideas for the President From Experts; Arms Control Must Not Be 'Oversold,'" *US News and World Report*, 7 January 1985, 75.

5 Samuel Huntington, "America's Changing Strategic Interests," *Survival* 33 (January/February 1991): 3.

6 Samuel Huntington, "The Economic Renewal of America," *The National Interest* 28 (Spring 1992): 14–18.

7 Samuel Huntington, "Why International Primacy Matters," *International Security* 17 (4, Spring 1993): 76.

8 Samuel Huntington, "The Clash of Civilizations?" *Foreign Affairs* (Summer 1993): 22–49; "The Coming Clash of Civilizations or, the West Against the Rest," *New York Times*, 6 June 1993, E19; "If Not Civilizations, What?: Paradigms of the Post-Cold War World," *Foreign Affairs* 72 (November–December, 1993); *The Clash of Civilizations: The Debate* (New York: Council on Foreign Relations, 1993).

9 Huntington, "If Not Civilizations, What?" 187.

10 Huntington, "If Not Civilizations, What?" 186, 191. Huntington appropriates the term from William James who used it to describe thought without concepts, theories, models and paradigms.

11 Huntington, "The Clash of Civilizations?" 31. Huntington's map is strange in that it arbitrarily chooses 1500 as a dividing time for a Velvet Curtain that is only emerging in 1990 and it lumps both Orthodox Christianity and Islam together on the same side of the divide.

12 Fouad Ajami, "The Summoning," *Foreign Affairs* 72 (September–October 1993): 7.

13 Richard Rubenstein and Jarle Crocker, "Challenging Huntington," *Foreign Policy* 96 (Fall 1994): 115. They sum up Huntington's argument thus: "The old Cold War is dead, he loudly declares. Then – *sotto voce* – Long live the new Cold War!" (117).

14 Interestingly this is now conceded by many conservatives. See Owen Harries, "The Collapse of 'The West,'" *Foreign Affairs* 72 (July–August 1993): 41–53, and James Kurth, "The *Real* Clash." A sycophantic essay on Huntington's thesis, Kurth writes: "The tale of the decline of 'Western civilization' as a term is part of the longer tale of the decline of Western civilization itself" (10). He connects this to the transformation from an industrial to a post-industrial economy and from an international economy to a global one which meant that "the most advanced countries are becoming less modern (i.e. post-modern), while the less advanced countries are becoming more modern" (12). The

West, he suggests, is now being "deconstructed" by a multicultural coalition and a feminist movement that is marginalizing Western civilization in the universities and media of America. At the very moment of its greatest triumph, Western civilization is becoming non-Western because it has become global and because it has become post-modern. The real clash, for Kurth, is the clash between the West and the post-West, or the modern and the post-modern within the West itself.

15 Huntington, "The Clash of Civilizations?" 39. At another point he writes: "Japan has established a unique position for itself as an associate member of the West: it is in the West in some respects but clearly not of the West in important dimensions" (45). This locational uniqueness of Japan makes it potentially even more of a threat in that it is not of the West yet it operates within its territorial economies and within its community of confidence. The potential of Japan to divide the West and undermine its security from within is consequently all the greater. Tom Clancy plays upon this important imaginative fear in *Debt of Honor* by having American lobbyists and Japanese corporations attempt to subvert US democracy. See also Pat Choate, *Agents of Influence* (New York: Alfred Knopf, 1990).

16 That Huntington supports any of these ideals is, of course, contestable given his historical support for brutal and repressive Third World military governments.

17 Huntington, "If Not Civilizations, What?" 190.

18 Kurth, "The Real Clash" 10.

PART 4
Environmental Geopolitics

PART 4

INTRODUCTION

Simon Dalby

Like environmental determinism and imperialist geopolitics, the concept of ecology and related ideas that humanity could collectively do large scale damage to natural systems, dates back to the nineteenth century (Kuehls, 1996). But such ideas have become widespread matters of concern only since World War II and a serious matter for public geopolitical discussion in the last few decades. In the process the important concepts of "global" problems and "global" security have become part of the geopolitical lexicon. Additional environmental and scientific expertise in natural systems has now been added to the discussions of national and international security in Western states.

During the 1950s and 1960s episodes such as mercury poisoning at Minimata in Japan, fears about widespread use of pesticides in the United States, killer smogs in London and oil spills from a number of high profile tanker accidents introduced environmental themes onto the political agenda in many states (Sandbach, 1980). In the 1960s the issue of nuclear fallout from Cold War weapon "test" explosions which affected people worldwide connected the fate of all inhabitants to the consequences of geopolitical rivalry. The agreement to ban atmospheric tests by the US, the Soviet Union and the UK in 1963 is an example of an early international environmental agreement dealing with a problem that had global ramifications because radioactive fallout from weapons tests travelled round the world in the atmosphere. Alarmist predictions of looming natural, resource, population and pollution "limits to growth" also published in the industrialized states in the late 1960s and early 1970s drew considerable international attention (Ehrlich, 1968; Meadows *et al.*, 1974).

New forms of expertise in pollution monitoring and especially in environmental impact assessment emerged and became part of the political and administrative processes of modern industrial states. But this expertise was often challenged by activists and citizens who were unconvinced that state programs were either reliable or doing enough to curb industrial damage. High profile actions by Greenpeace and other environmental organizations raised the profile of ecosystem destruction and resource destruction in headline grabbing ways that often bypassed the more technical debates (Dale, 1996; Wapner, 1996). These actions generated political pressure for institutional and policy change on specific items and often made the international dimensions of issues such as whaling obvious. Since the early 1970s many environmental questions have been understood to be matters of "global" concern. Research on environmental issues at a local level often led to research at larger scales which has subsequently produced a series of new environmental "threats" such as ozone depletion, biodiversity loss and global climate change.

What is especially important for the discussion here is the emergence of the "global environment" as an object for analysis and policy prescription (Porter and Brown, 1995; Vogler and Imber, 1996). These new modes of knowledge, in which the globe is now the topic for discussion and analysis, and crucially of "management" by international agreements and agencies set up for the purpose, suggest that a new form of power/knowledge is now part of twentieth-century geopolitics (Litfin, 1994). These new modes of "knowing" the world and their political specification of peoples and societies as "threats" or in need of "management" are not merely technical issues requiring research, analysis and coordination by appropriately qualified experts. These may also be understood as a new form of global politics in which interests are engaged in a variety of forums and in which, like other geopolitical arenas, knowledge is not neutral but appears in various forms of power/knowledge used by protagonists in the politics of environment at local as well as state and international scales (Redclift and Benton, 1994).

This is not to suggest that environmental problems are only "political" and do not in some sense "exist" in the real world. Forests are being cut down and people displaced. The potential for disruptions as a result of climate change needs to be taken seriously. Ozone holes are a real danger to ecosystems and both directly and indirectly to human health. But, as is clarified below, and in the readings that follow this introduction, how these issues are described and who is designated as either the source of the problem, or provider of the potential solution to the problem, is an important matter in how environmental themes are argued about and in who gets to make decisions about what should be done by whom (Seager, 1993). At the large scale this is very much a matter of geopolitics.

For example, if climate change is understood as being a problem caused mainly by car exhaust in industrial cities and their suburbs, or by international oil company policies in search of huge profits, or by peasants cutting down tropical rainforests to grow food, very different solutions are likely to be suggested. If energy conservation and environmental city planning are widely introduced and oil companies taxed heavily to provide an international "green tax" for environmental projects, results will be very different than if "Northern" states attempt to use economic sanctions to try to get "Southern" governments to stop tropical deforestation. Once again we can see that geopolitics is about supposedly factual arguments and descriptions of the world that at first glance do not appear to be at all political. But careful analysis of the geographical assumptions in these arguments suggests that knowledge is not neutral, but a political resource used in political arguments and in policy decisions.

In the 1970s many people from underdeveloped states objected to the logic of the arguments about the "limits to growth" and the concerns that finite planetary supplies of key resources required a reduction in industrial production. This they saw as a direct challenge to their need for economic growth to provide wealth for the millions of impoverished people in their states. These objections were loudly voiced at the first United Nations Conference on Environment and Development (UNCED) held in Stockholm in 1972. Environment and development were often portrayed as opposites; pollution and environmental degradation were accepted by many as the price of progress. Attempts by rich "Northern" states to argue for population curbs were also sometimes dismissed as racist. There was a very obvious geography to this whole debate, with people from "the North" having very different priorities to those from "the South" (Miller, 1995).

Some of these themes became connected to traditional geopolitical matters in the mid-1970s. "Limits to growth" seemed to be imminent to many people in Western states following disruptions of oil supplies as a result of the oil embargo and price rises introduced by the Organization of Petroleum Exporting Countries (OPEC) and manipulations of oil prices by the trans-national oil companies during and after the Yom Kippur/October war between Syria, Egypt and Israel in 1973. Increased attention was paid, by American foreign policy makers in particular, to the possibilities of using military intervention around the world to ensure that supplies of crucial resources, especially oil from the Middle East, would not be interrupted by either local political instability, or by Soviet political and military action. The geopolitical assumptions present here were that the flow of oil from outside the West had to be maintained come what may and despite what people in the states that had the supplies might have to say about the matter (Yergin, 1991).

In the mid-1970s concerns were also raised in the US about supplies of minerals from African states and elsewhere needed for military equipment production. These scenarios of resource "strangulation" fit well with the Cold War geopolitical understanding of the world as one of geopolitical rivalry between the Cold War blocs, with the "Third World" as the arena in which the contest for global domination was played out. But some prominent environmentalists in this period argued that the best method of ensuring resource security, at least in the sense of oil supplies, was to work hard at improving conservation measures and introducing such things as efficient building heating systems and automobile engines (Lovins, 1977). They argued that doing so would reduce the need for military interventions in OPEC states or elsewhere as part of the geopolitical rivalry of the Cold War, and simultanously clean up pollution in the industrialized states while costing much less than military preparations for intervention.

By the early 1980s many people were arguing that development and environmental concerns would have to be understood as complementary. Development in many cases had to mean alleviating some of the worst environmental problems that were caused by poverty and the inability of poor people to use efficient technologies precisely because of their poverty. The concept of sustainable development encapsulated these concerns in a convenient phrase that was subsequently popularized by the report by the World Commission on Environment and Development (WCED) on *Our Common Future* (1987). Under the chairpersonship of the Prime Minister of Norway, Gro Harlem Brundtland, this report which is often simply known as "the Brundtland report", drew leading public figures from around the world to endorse a program that was a compromise between those who wished to emphasize environment and those who argued that development needed priority. The assumption on all sides of the debate was that development could be sustained, albeit with some modifications to take into account environmental difficulties and limitiations. But to accomplish such tasks would, it was assumed, need widespread government action and global agreements on a package of programs and their funding. International negotiations took the form of a series of preparatory conferences leading up to the second UNCED held in Rio de Janerio in 1992.

Many heads of government went to Brazil in June of that year to sign a number of international agreements and set in motion a series of follow up meetings and programs of action on forestry, atmospheric change and other matters (Grubb *et al.*, 1993). A new "Global Environmental Facility" under control of the World Bank was also established to fund some projects on climate change, biodiversity, ozone depletion and ocean pollution in international waters. Environmental organizations and numerous representatives of social movements and indigenous peoples also attended a parallel "Global Forum" where discussion of the possibilities of non-governmental actions drew attention from some international media, although apparently had only limited effect on the governmental negotiations on the other side of the city (Thomas, 1994).

But understanding the environmental dimensions of contemporary geopolitics requires understanding more than the emergence of "sustainable development" and the UNCED conferences. Many environmental issues now seem to require political attention because they are understood to be the source of threats to health and wealth which have international dimensions. As the Cold War geopolitical order unravelled in the late 1980s and early 1990s, numerous political phenomena were understood to be newly threatening to the international political order. Extending the concept of security to encompass new threats in need of management and control by states and international political organization was one obvious response to the new circumstances (Myers, 1993).

Numerous facets of the contemporary scene can fairly easily be interpreted as threats requiring control and "management" by the dominant powers. Most alarming to those who usually think of the priority in matters of geopolitics being to maintain the political stability of modern states is the potential for environmental and demographic changes to lead to destabilizing population movements and possible military confrontations (Kennedy, 1993). Refugees are rapidly increasing in number, and while only some of the causes of their flight can be directly connected to environmental factors, there is widespread concern that environmental degradation may trigger many millions more environmental refugees.

An especially alarming article on these themes was written by Robert D. Kaplan and published by the *Atlantic Monthly* in February 1994 under the title "The Coming Anarchy." Widely read in Washington, this article crystallized concern about environmental causes of chaos and state breakdown. As the excerpts reprinted here show (Reading 23), this is powerful prose that is compelling reading. But, as my critique of it suggests (Reading 24), it is inadequate as a rigorous analysis because of its many omissions and its failure to provide clear links between many of the things that it discusses and their apparent causes. Nonetheless it has been an influential article in policy-making circles, not least because it explicitly argues that the West faces new security threats in the form of crime, drugs, economic instabilities, diseases and "failed states." This feeds into the discussions by academics and policy makers about re-thinking the key concepts of national security and international security after the Cold War period (Klare and Thomas, 1994; Renner, 1996). Here are apparently obvious new threats to the political order requiring a geopolitical view of the global scene and management strategies in some cases backed by military preparedness. Many of these newly defined "threats" are interpreted as global phenomena; population growth, ozone holes, biodiversity loss and climate change are only the most obvious matters of what is now often called "environmental security."

Concern has been widely expressed about the possibilities of international armed conflict over water resources which are being ever more heavily used by growing urbanized and increasingly industrialized populations using water directly and relying on irrigated crops for food. Climate change as a result of human activities in changing the global atmosphere may lead to weather changes that upset global agricultural productivity and induce political strife. The Middle East (or South West Asia) is often considered a high risk area for disputes over water to lead to warfare (Gleick, 1993). It has been argued that the 1967 war between Israel and its neighbors was partly caused by disputes over the use of the water in the Jordan river. Clearly water issues are an important component of peace discussions in the region in the 1990s. Some alarmist scenarios of the future draw from arguments about environmental degradation to suggest that resource shortages may lead to major conflict in places like China, where regional disparities and consequent political tensions are likely to be aggravated by pressures on the environment due to rapid industrialization and urbanization (Brown, 1995; Goldstone, 1995).

Scenarios of declining resource bases leading to heightened awareness of communal identity and resulting group conflicts have been proposed by many writers in the last few years. Migrants in a number of places have come into conflict with host populations. Sorting out how environmental factors are influencing these processes is not easy, as migrants usually move for complicated combinations of reasons (Wood, 1994). In addition, doing detailed research in the middle of these conflicts is often very difficult. Researchers have undertaken a number of case studies in specific places in the last few years, but the precise role of the environment as a specific cause of conflict and refugee migration is not easy to figure out in general terms. Nonetheless, Thomas Homer-Dixon (Reading 25) is now prepared to offer some clear arguments about what can be concluded from this research.

But many of the assumptions in the arguments about environment as a security threat or a cause of conflict are, according to critics like Vaclav Smil (Reading 26), highly doubtful. General arguments about global environmental change are often so imprecise when applied to specific places as to be practically useless. Local economic situations, or the disruptions caused by development projects, often generate poverty that is then blamed on environmental degradation. Specific environmental degradations are undoubtedly important in particular places, but generalizations as to how to respond are often not helpful. The potential for increased agricultural production in many parts of the world is still considerable, and statistics about arable land are often misleading because of large inaccuracies.

There are also, as Gareth Porter (Reading 27) notes, arguments that make the case that understanding all these things in terms of traditional Cold War themes of security are not helpful to either discovering the causes of contemporary problems or suggesting solutions. While most writers argue that any security crisis resulting from environmental degradation will need to be handled by cooperative measures rather than traditional "security" responses by the armed forces, nonetheless, as Matthias Finger warns in his article (Reading 28), if the military is seen as an essential institution in dealing with these problems, they are much more likely to be perpetuated than alleviated. This is the case in part because of the appalling records of environmental destruction by many militaries through the period of the Cold War. But, in addition, security is often defined in terms of state security requiring a modern military armed with expensive industrial weapons. This leads to the perpetuation of industrial state policies as the "solution" to "security problems" when these are the very cause of much widespread environmental degradation.

But viewed from many places in "the South," the "discourses of danger" that structure the environmental security literature can be seen as little more than attempts to reassert Northern corporations' and political institutions' colonial domination of Southern societies, albeit now sometimes in the name of protecting the planet (Faber, 1993; Rich, 1994). These specifications of the new "green" dimensions of geopolitics are not innocent constructions or "true" statements about how the world is organized. They are understandings of the world that relate to the traditional institutions of global politics but with new terms and language. They are related to political power and enmeshed within the global political economy. This form of what Vandana Shiva (Reading 29) calls "green imperialism" is often based on a simple but powerful geographical "sleight of hand" where the particular interests of the rich in the North are portrayed as the common interests of all humanity. Shiva's point is that the expansion of global economic activity has negative consequences for many peoples and places, while it enriches the beneficiaries of the global economy elsewhere.

The arguments about global dangers are, not surprisingly, understood in very different terms by the mainly poor people in "the South" who are the source of these new "threats." They can easily argue that the geopolitical specifications of the "dangers" are part of the problem because the relationships that have long maintained the inequities between rich and poor in the global economy, which cause much suffering and "insecurity" for so many people, are simply reinforced by the new discourses of sustainable development, environmental security and the "global" environment (Chatterjee and Finger, 1994). Based on the assumption that the planet is a "resource" that can be administered, as Visvanathan argues (Reading 30), the discourse of sustainable development can become a dangerous formulation that allows injustice and environmental degradation to continue as part of the ideologically refurbished processes of "development."

Development as it has been practiced for the last half century assumes a separation of humanity and "nature." As Visvanathan explains nature has been transformed into "environment" by a sophisticated series of forms of knowledge, and the endless writing of "reports" that empower its users to divide and control "nature" in order to "develop" and modernize it. Using the environment as a resource adds commercial value to a national economy and increases national wealth and economic measures such as the Gross National Product that are so widely still taken as indications of progress and development.

Traditional practices of food gathering and agriculture often produce food for subsistence but not for sale on the commercial market. Such activities are not understood to be "productive" by most conventional economic measures because they do not show up as money transactions or as sources of state revenue in the form of taxes. Peoples who survive in traditional economies often also do not have formal legal titles to land and operate on assumptions that environments are not owned but are there to be lived in, and cared for, in a communal manner. They rely on intact natural ecosystems for food, medicines and clean sources of water (Gadgil and Guha, 1995).

Such modes of existence, when taken seriously at all by conventional thinking about development, are often dismissed as primitive and irrelevant. The task of modernization is usually understood as being to convert these environments and their inhabitants into "productive" commercial enterprises. The commodization of traditional foods and the appropriation of traditional knowledge of seeds by multinational agricultural corporations, the clearance of forests which displace indigenous peoples, and the enclosure of traditional peasant common lands all threaten the security of indigenous peoples and traditional peasant cultures (Johnston, 1994).

The processes of enclosure and displacement are also a form of modern geopolitics where geographical space is divided up and controlled. Although working on a smaller scale than the divisions of political space into sovereign territorial states that traditional discussions of geopolitics usually deal with, these spatial divisions of the globe are part of the same global political economy. Property relationships imposed on indigenous peoples or peasants using common resources divide up land, water, shelter and forests and use them for private commercial gain rather than collective survival in ways that are loosely similar to states dividing up resources and spaces on larger scales. In many cases the "owners" of the land are not the same people as those who traditionally used it before development and modernization arrived and imposed a very different social understanding of the environment and the appropriate ways of using it (*The Ecologist*, 1993).

Sometimes traditional peoples are displaced to make way for large resource developments such as dams, mines or forestry plantations leading to what are now sometimes called the new "resource wars" or sometimes "environmental conflicts." Where these conflicts challenge the control by states over sections of their territory or disrupt supplies of resources for global markets they can become traditional armed conflicts, understood in traditional geopolitical terms of access to and control over resources at the large scale, and as matters of national security for the particular state concerned. What one considers the appropriate way of responding to these issues depends to a substantial degree on how the question is phrased in a geopolitical framework.

If the global market for commodities and the "right" of transnational corporations to access resources and markets is taken for granted then these issues become one of ensuring that states have the power to quell rebels who might disrupt the production and export of minerals or timber. If phrased in terms of the "rights" of local indigenous peoples to use the forest or environments that they have traditionally lived and survived in without interference from states or international corporations, then matters look very different. When links are made between the "global" environmental concerns to keep substantial parts of the remaining forests of the world intact and the rights of indigenous peoples who have lived in relative harmony with these forests for many generations, interesting international political alliances form. Political elites in states interested in "developing" "their" resources often portray such alliances as interference in the sovereign realm of the state, and environmentalists are then consequently seen as a threat to the national security of the state.

Because the most powerful social agencies involved in the UNCED process, and subsequent treaty making on environmental issues, are usually states, many critics are concerned that the latter view, with states understood as the sole managers of "sustainable develoment" and "environmental security," has come to dominate contemporary politics. If this is the case then states now have additional powerful "global environment" arguments to use to justify their policies when dressed up in the language of "sustainable development." The assumption that only states, and in current circumstances the behind the scenes operations of global corporations, matter in determining geopolitical priorities is once again re-affirmed and opposition rendered more difficult. But, given the importance of the issues and the practical matters of daily survival that peasant and indigenous peoples face, it is highly unlikely that these interpretations of global politics are going to avoid repeated challenges from grass roots organizations and environmental movements.

REFERENCES AND FURTHER READINGS

On the recent history of environment and politics

Athanasiou, T. (1996) *Divided Planet: The Ecology of Rich and Poor*, Boston: Little Brown.

Conca, K., Alberty, M. and Dabelko, G.D. (eds) (1995) *Green Planet Blues: Environmental Politics from Stockholm to Rio*, Boulder: Westview.

Dale, S. (1996) *McLuhan's Children: The Greenpeace Message and the Media*, Toronto: Between the Lines.

Ehrlich, P.R. (1968) *The Population Bomb*, New York: Ballantine.

Lovins, A.B. (1977) *Soft Energy Paths: Toward a Durable Peace*, London: Penguin.

Meadows, D.H., Meadows, D.L,. Randers, J. and Behrens III, W.W. (1974) *The Limits to Growth*, London: Pan.

Sandbach, F. (1980) *Environment, Ideology and Policy*, Oxford: Basil Blackwell.

Wapner, P. (1996) *Environmental Activism and World Politics*, Albany: State University of New York Press.

Yergin, D. (1991) *The Prize: The Epic Quest for Oil, Money and Power*, New York: Simon and Schuster.

On contemporary environmental politics

Kuehls, T. (1996) *Beyond Sovereign Territory: The Space of Ecopolitics*, Minneapolis: University of Minnesota Press.

Litfin, K. (1994) *Ozone Discourses: Science and Politics in Global Environmental Cooperation*, New York: Columbia University Press.

Miller, M.A.L. (1995) *The Third World in Global Environmental Politics*, Boulder: Lynne Rienner.

Porter, G. and Brown, J.W. (1995) *Global Environmental Politics*, Boulder: Westview.

Redclift, M. and Benton, T. (eds) (1994) *Social Theory and Global Environment*, London: Routledge.

Rifkin, J. (1991) *Biosphere Politics: A New Consciousness for a New Century*, New York: Crown.

Seager, J. (1993) *Earth Follies: Making Feminist Sense of Environmental Politics*, New York: Routledge.

Vogler, J. and Imber, M.E. (eds) (1996) *The Environment and International Relations*, London: Routledge.

On population and environment as "new threats"

Brown, L. (1995) *Who Will Feed China? Wake Up Call for a Small Planet*, New York: Norton.

Connelly, M. and Kennedy, P. (1994) "Must it be the West Against the Rest?" *The Atlantic Monthly*, 274(6): 61–83.

Gleick, P. (1993) "Water and Conflict: Fresh Water Resources and International Security," *International Security*, 18(1): 79–112.

Goldstone, J. (1995) "The Coming Chinese Collapse," *Foreign Policy*, 99: 35–53.

Homer-Dixon, T. (1994) "Environmental Scarcities and Violent Conflict: Evidence from Cases," *International Security*, 19(1): 5–40.

Kennedy, P. (1993) *Preparing for the Twenty First Century*, New York: Harper Collins.

Klare, M.T. and Thomas, D.C. (eds) (1994) *World Security: Challenges for a New Century*, New York: St Martin's Press.

Renner, M. (1996) *Fighting for Survival: Environmental Decline, Social Conflict, and the New Age of Insecurity*, New York: Norton.

Romm, J.J. (1993) *Defining National Security: The Non-Military Aspects*, New York: Council on Foreign Relations.

Wood, W. (1994) "Forced Migration: Local Conflicts and International Dilemmas," *Annals of the Association of American Geographers*, 84(4): 607–634.

On the military, environment and security

Dalby, S. (1992) "Security, Modernity, Ecology: The Dilemmas of Post-Cold War Security Discourse," *Alternatives: Social Transformation and Humane Governance*, 17(1): 95–134.

Deudney, D. (1992) "The Mirage of Eco-War: The Weak Relationship Among Global Environmental Change, National Security and Interstate Violence," in Rowlands, I.H. and Greene, M. (eds) *Global Environmental Change and International Relations*, London: Macmillan.

Deudney, D. and Matthew, R. (eds) (1997) *Contested Grounds: Security and Conflict in the New Environmental Politics*, Albany: State University of New York Press.

Ehrlich, A.H. and Birks, J.W. (eds) (1990) *Hidden Dangers: Environmental Consequences of Preparing for War*, San Francisco: Sierra Club Books.

Feshbach, M. and Friendly, A. (1992) *Ecocide in the USSR: Health and Nature under Siege*, New York: Basic.

Kakonen, J. (ed.) (1994) *Green Security or Militarized Environment*, Aldershot: Dartmouth.

Myers, N. (1993) *Ultimate Security: The Environment as the Basis of Political Stability*, New York: Norton.

Prins, G. (ed.) (1993) *Threats Without Enemies: Facing Environmental Insecurity*, London: Earthscan.

On development and environmental politics

The Ecologist (ed.) (1993) *Whose Common Future? Reclaiming the Commons*, London: Earthscan.

Faber, D. (1993) *Environment Under Fire: Imperialism and the Ecological Crisis in Central America*, New York: Monthly Review Press.

Gadgil, M. and Guha, R. (1995) *Ecology and Equity: The Use and Abuse of Nature in Contemporary India*, London: Routledge.

Hampson, F.O. and Reppy, J. (eds) (1996) *Earthly Goods: Environmental Change and Social Justice*, Ithaca: Cornell University Press.

Johnston, B.R. (ed.) (1994) *Who Pays the Price? The Sociocultural Context of Environmental Crisis*, Washington: Island.

Peet, R. and Watts, M. (eds) (1996) *Liberation Ecologies: Environment, Development and Social Movements*, London: Routledge.

Rich, B. (1994) *Mortgaging the Earth: The World Bank, Environmental Impoverishment and the Crisis of Development*, London: Earthscan.

On the UNCED process

Chatterjee, P. and Finger, M. (1994) *The Earth Brokers: Power, Politics and World Development*, London: Routledge.

Dalby, S. (1996) "Reading Rio, Writing the World: *The New York Times* and the 'Earth Summit'," *Political Geography*, 15(6&7): 593–614.

Grubb, M., Koch, M., Munson, A., Sullivan, F. and Thomson, K. (1993) *The Earth Summit Agreements: A Guide and Assessment*, London: Earthscan.

Middleton, N., O'Keefe, P. and Moyo, S. (1993) *Tears of the Crocodile: From Rio to Reality in the Developing World*, London: Pluto.

Sachs, W. (ed.) (1993) *Global Ecology: A New Arena of Political Conflict*, London: Zed.

Thomas, C. (ed.) (1994) *Rio: Unravelling the Consequences*, London: Frank Cass.

World Commission on Environment and Development (1987) *Our Common Future*, New York: Oxford University Press.

Cartoon 11 Rio: Save energy, do nothing (by TOM)
US President George Bush was very reluctant to even agree to go to the Earth Summit conference in Rio de Janeiro in 1992. Once there he was also reluctant to endorse substantial international agreements on energy conservation to reduce waste gases being released into the atmosphere.
Source: Tom, Cartoon & Writers' Syndicate

23 ROBERT D. KAPLAN

"The Coming Anarchy"

from *The Atlantic Monthly* (1994)

The minister's eyes were like egg yolks, an after effect of some of the many illnesses, malaria especially, endemic in his country. There was also an irrefutable sadness in his eyes. He spoke in a slow and creaking voice, the voice of hope about to expire. "In 45 years I have never seen things so bad. We did not manage ourselves well after the British departed. But what we have now is something worse – the revenge of the poor, of the social failures, of the people least able to bring up children in a modern society." Then he referred to the recent coup in the West African country Sierra Leone. "The boys who took power in Sierra Leone come from houses like this." The minister jabbed his finger at a corrugated metal shack teeming with children. "In three months these boys confiscated all the official Mercedeses, Volvos and BMWs and wilfully wrecked them on the road." The minister mentioned one of the coup's leaders, Solomon Anthony Joseph Musa, who shot the people who had paid for his schooling, "in order to erase the humiliation and mitigate the power his middle class sponsors held over him." . . .

[. . .]

The cities of West Africa at night are some of the unsafest places in the world. Streets are unlit; the police often lack gasoline for their vehicles; armed burglars, carjackers and muggers proliferate. Direct flights between the United States and the Murtala Muhammed Airport, in neighboring Nigeria's largest city, Lagos, have been suspended by order of the US Secretary of Transportation because of ineffective security at the terminal and its environs. A State Department report cited the airport for "extortion by law enforcement and immigration officials." This is one of the few times the US government has embargoed a foreign airport for reasons that are linked purely to crime. In Abidjan, effectively the capital of the Cote d'Ivoire, or Ivory Coast, restaurants have stick and gun wielding guards who walk you the 15 feet or so between your car and the entrance, giving you an eerie taste of what American cities might be like in the future. An Italian ambassador was killed by gunfire when robbers invaded an Abidjan restaurant. The family of the Nigerian ambassador was tied up and robbed at gunpoint in the ambassador's residence. . . .

"In the poor quarters of Arab North Africa," the minister continued, "there is much less crime, because Islam provides a social anchor of education and indoctrination. Here in West Africa we have a lot of superficial Islam and superficial Christianity. Western religion is undermined by animist beliefs not suitable to a moral society, because they are based on irrational spirit power. Here spirits are used to wreak vengeance by one person against another, or one group against another."

Finally the minister mentioned polygamy. Designed for a pastoral way of life, polygamy continues to thrive in sub-Saharan Africa even though it is increasingly uncommon in Arab North Africa. Most youths I met on the road in West Africa told me that they were from "extended" families, with a mother in one place and a father in another. Translated to an urban environment, loose family structures are largely responsible for the world's highest birth rates and the explosion of the HIV virus on the continent. Like the communalism and animism, they provide a weak shield against the corrosive social effects of life in cities.

A PREMONITION OF THE FUTURE

West Africa is becoming the symbol of worldwide demographic, environmental and societal stress, in which criminal anarchy emerges as the real "strategic" danger. Disease, overpopulation, unprovoked crime, scarcity of resources, refugee migrations, the increasing erosion of nation states and international borders, and the empowerment of private armies, security firms and international drug cartels are now most tellingly demonstrated through a West African prism. West Africa provides an appropriate introduction to the issues, often extremely unpleasant to discuss, that will soon confront our civilization.

There is no other place on the planet where political maps are so deceptive – where, in fact, they tell such lies – as in West Africa. Start with Sierra Leone. According to the map, it is a nation state of defined borders, with a government in control of its territory. In truth the Sierra Leonian government, run by a 27-year-old army captain, Valentine Strasser, controls Freetown by day and by day also controls part of the rural interior. In the government's territory the national army is an unruly rabble threatening drivers and passengers at most checkpoints. In the other part of the country, units of two separate armies from the war in Liberia have taken up residence, as has an army of Sierra Leonian rebels. The government force fighting the rebels is full of renegade commanders who have aligned themselves with disaffected village chiefs. A premodern formlessness governs the battlefield, evoking the wars in medieval Europe prior to the 1648 Peace of Westphalia, which ushered in the era of organized nation states.

As a consequence, roughly 400,000 Sierra Leonians are internally displaced, 280,000 more have fled to neighboring Guinea, and another 100,000 have fled to Liberia, even as 400,000 Liberians have fled to Sierra Leone. The third largest city in Sierra Leone, Gondama, is a displaced-persons camp. With an additional 600,000 Liberians in Guinea and 250,000 in the Ivory Coast, the borders dividing these four countries have become largely meaningless. Even in quiet zones none of the governments except the Ivory Coast's maintains the schools, bridges, roads and police forces in a manner necessary for functional sovereignty.

In Sierra Leone, as in Guinea, as in the Ivory Coast, as in Ghana, most of the primary rain forest and the secondary bush is being destroyed at an alarming rate. When Sierra Leone achieved its independence, in 1961, as much as 60 per cent of the country was primary rain forest. Now 6 per cent is. In the Ivory Coast the proportion has fallen from 38 per cent to 8 per cent. The deforestation has led to soil erosion, which has led to more flooding and more mosquitoes. Virtually everyone in the West African interior has some form of malaria.

Sierra Leone is a microcosm of what is occurring, albeit in a more tempered and gradual manner, throughout West Africa and much of the underdeveloped world: the withering away of central governments, the rise of tribal and regional domains, the unchecked spread of disease, and the growing pervasiveness of war. West Africa is reverting to the Africa of the Victorian atlas. It consists now of a series of coastal trading posts, such as Freetown and Conakry, and an interior that, owing to violence, volatility, and disease, is again becoming, as Graham Greene once observed, "blank" and "unexplored." However, whereas Greene's vision implies a certain romance, as in the somnolent and charmingly seedy Freetown of his celebrated novel *The Heart of the Matter*, it is Thomas Malthus, the philosopher of demographic doomsday, who is now the prophet of West Africa's future. And West Africa's future, eventually, will also be that of most of the rest of the world.

Consider "Chicago." I refer not to Chicago, Illinois, but to a slum district of Abidjan, which the young toughs in the area have named after the American city. ("Washington" is another poor section of Abidjan.) Chicago, like more and more of Abidjan, is a slum in the bush: a checkerwork of corrugated zinc roofs and walls made of cardboard and black plastic wrap. It is located in a gully teeming with coconut palms and oil palms, and is ravaged by flooding. Few residents have easy access to electricity, a sewage system or a clean water supply....

Fifty-five per cent of the Ivory Coast's population is urban, and the proportion is expected to reach 62 per cent by 2000. The yearly net population growth is 3.6 percent. This means that the Ivory Coast's 13.5 million people will become 39 million by 2025, when much of the population will consist of urbanized peasants like those of Chicago. But don't count on the Ivory Coast's still existing then. Chicago, which is more indicative of Africa's and the Third World's demographic present – and even more in the future – than any idyllic junglescape of women balancing earthen jugs on their heads, illustrates why the Ivory Coast, once a model of Third World success, is becoming a case study in Third World catastrophe....

Because the military is small and the non-Ivorian population large, there is neither an obvious force to maintain order nor a sense of nationhood that would lessen the need for such enforcement. The economy has been shrinking since the mid-1980s. Though the French are working assiduously to preserve stability, the Ivory Coast faces a possibility worse than a coup: an anarchic implosion of criminal violence – an urbanized version of what has already happened in Somalia....

As many internal African borders begin to crumble, a more impenetrable boundary is being erected that threatens to isolate the continent as a whole: the wall of disease.... Africa may today be more dangerous in this regard than it was in 1862, before antibiotics.... Of the approximately 12 million people worldwide whose blood is HIV positive, 8 million are in Africa. In the capital of the Ivory Coast, whose modern road system only helps to spread the disease, 10 per cent of the population is HIV positive. And war and refugee movements help the virus break through to more remote areas of Africa. It is malaria that is most responsible for the disease wall that threatens to separate Africa and other parts of the Third World from more developed regions of the planet in the twenty-first century. Carried by mosquitoes, malaria, unlike AIDS, is easy to catch. Most people in sub-Saharan Africa have recurring bouts of the disease throughout their entire lives, and it is mutating into increasingly deadly forms.

Africa may be as relevant to the future character of world politics as the Balkans were a hundred years ago, prior to the two Balkan wars and the First World War. Then the threat was the collapse of empires and the birth of nations based solely on tribe. Now the threat is more elemental: *nature unchecked*. Africa's immediate future could be very bad. The coming upheaval, in which foreign embassies are shut down, states collapse, and contact with the outside world takes place through dangerous, disease ridden coastal trading posts, looms large in the century we are entering. Precisely because much of Africa is set to go over the edge at a time when the Cold War has ended, when environmental and demographic stress in other parts of the globe is becoming critical, and when the post-First World War system of nation states – not just in the Balkans but perhaps also in the Middle East – is about to be toppled, Africa suggests what war, borders, and ethnic politics will be like a few decades hence....

THE ENVIRONMENT AS A HOSTILE POWER

For a while the media will continue to ascribe riots and other violent upheavals abroad mainly to ethnic and religious conflict. But as these conflicts multiply, it will become apparent that something else is afoot, making more and more places like Nigeria, India and Brazil ungovernable....

It is time to understand "the environment" for what it is: the national security issue of the early twenty-first century. The political and strategic impact of surging populations, spreading disease, deforestation and soil erosion, water depletion, air pollution and, possibly, rising sea levels in critical, overcrowded regions such as the Nile Delta and Bangladesh – developments that will prompt mass migrations and, in turn, incite group conflicts – will be the core foreign policy challenge from which most others will ultimately emanate, arousing the public and uniting assorted interests left over from the Cold War. In the twenty-first century, water will be in dangerously short supply in such diverse locales as Saudi Arabia, Central Asia and the south-western United States. A war could erupt between Egypt and Ethiopia over Nile River water. Even in Europe tensions have arisen between Hungary and Slovakia over the damming of the Danube, a classic case of how environmental disputes fuse with ethnic and historical ones.

Our Cold War foreign policy truly began with George F. Kennan's famous article, signed "X," published in *Foreign Affairs* in July of 1947, in which Kennan argued for a "firm and vigilant containment" of a Soviet Union that was imperially, rather than ideologically, motivated. It may be that our post-Cold War foreign policy will one day be seen to have had its beginnings in an even bolder and more detailed piece of written analysis: one that appeared in the journal *International Security.* The article, published in the fall of 1991 by Thomas Fraser Homer-Dixon, who is the head of the Peace and Conflict Studies Program at the University of Toronto, was titled "On the Threshold: Environmental Changes as Causes of Acute Conflict." Homer-Dixon has, more successfully than other analysts, integrated two hitherto separate fields – military conflict studies and the study of the physical environment.

In Homer-Dixon's view, future wars and civil violence will often arise from scarcities of resources such as water, cropland, forests and fish. Just as there will be environmentally driven wars and refugee flows, there will be environmentally induced praetorian regimes – or, as he puts it, "hard regimes." Countries with the highest probability of acquiring hard regimes, according to Homer-Dixon, are those that are threatened by a declining resource base yet also have "a history of state (read 'military') strength." Candidates include Indonesia, Brazil and, of course, Nigeria. Though each of these nations has exhibited democratizing tendencies of late, Homer-Dixon argues that such tendencies are likely to be superficial "epiphenomena" having nothing to do with long term processes that include soaring populations and shrinking raw materials. Democracy is problematic; scarcity is more certain.

Indeed, the Saddam Husseins of the future will have more, not fewer, opportunities. In addition to engendering tribal strife, scarcer resources will place a great strain on many peoples who never had much of a democratic or institutional tradition to begin with. Over the next 50 years the Earth's population will soar from 5.5 billion to more than 9 billion. Though optimists have hopes for new resource technologies and free market development in the global village, they fail to note that, as the National Academy of Sciences has pointed out, 95 per cent of the population increase will be in the poorest regions of the world, where governments now – just look at Africa –show little ability to function, let alone to implement even marginal improvements. Homer-Dixon writes ominously, "neo-Malthusians may underestimate human adaptability in today's environmental social system, but as time passes their analysis may become ever more compelling."

While a minority of the human population will be, as Francis Fukuyama would put it, sufficiently sheltered so as to enter a "post-historical" realm, living in cities and suburbs in which the environment has been mastered and ethnic animosities have been quelled by bourgeois prosperity, an increasingly large number of people will be stuck in history, living in shantytowns where attempts to rise above poverty, cultural dysfunction and ethnic strife will be doomed by a lack of water to drink, soil to till and space to survive in. In the developing world, environmental stress will present people with a choice that is increasingly among totalitarianism (as in Iraq), fascist tending mini states (as in Serb-held Bosnia) and road warrior cultures (as in Somalia). Homer-Dixon concludes that "as environmental degradation proceeds, the size of the potential social disruption will increase."

Quoting Daniel Deudney, another pioneering expert on the security aspects of the environment, Homer-Dixon says that

> for too long we've been prisoners of "social–social" theory, which assumes there are only social causes for social and political changes, rather than natural causes, too. This social–social mentality emerged with the Industrial Revolution, which separated us from nature. But nature is coming

back with a vengeance, tied to population growth. It will have incredible security implications.

Think of a stretch limo in the potholed streets of New York City, where homeless beggars live. Inside the limo are the air conditioned post-industrial regions of North America, Europe, the emerging Pacific Rim and a few other isolated places, with their trade summitry and computer information highways. Outside is the rest of mankind, going in a completely different direction.

SKINHEAD COSSACKS, JUJU WARRIORS

In the summer 1993 issue of *Foreign Affairs*, Samuel P. Huntington, of Harvard's Olin Institute for Strategic Studies, published a thought-provoking article called "The Clash of Civilizations?" The world, he argues, has been moving during the course of this century from nation state conflict to ideological conflict to, finally, cultural conflict. I would add that as refugee flows increase and as peasants continue migrating to cities around the world – turning them into sprawling villages – national borders are the most tangible and intractable ones: those of culture and tribe. Huntingdon writes, "First, differences among civilizations are not only real; they are basic, " involving, among other things, history, language, and religion. "Second, ... interactions between peoples of different civilizations are increasing; these interactions intensify civilization consciousness." Economic modernization is not necessarily a panacea, since it fuels individual and group ambitions while weakening traditional loyalties to the state. It is worth noting, for example, that it is precisely the wealthiest and fastest developing city in India, Bombay, that has seen the worst intercommunal violence between Hindus and Muslims. Consider that Indian cities, like African and Chinese ones, are ecological timebombs – Delhi and Calcutta, and also Beijing, suffer the worst air quality of any cities in the world – and it is apparent how surging populations, environmental degradation and ethnic conflict are deeply related.

Huntington points to interlocking conflicts among Hindu, Muslim, Slavic Orthodox, Western, Japanese, Confucian, Latin American and possibly African civilizations: for instance, Hindus clashing with Muslims in India, Turkic Muslims clashing with Slavic Orthodox Russians in Central Asian cities, the West clashing with Asia. (Even in the United States, African–Americans find themselves besieged by an influx of competing Latinos.) Whatever the laws, refugees find a way to crash official borders, bringing their passions with them, meaning that Europe and the United States will be weakened by cultural disputes. . . .

Most people believe that the political Earth since 1989 has undergone immense change. But it is minor compared with what is yet to come. The breaking apart and remaking of the atlas is only now beginning. The crack up of the Soviet empire and the coming end of Arab–Israeli military confrontation are merely prologues to the really big changes that lie ahead. Michael Vlahos, a long range thinker for the US Navy, warns, "We are not in charge of the environment, and the world is not following us. It is going in many directions. Do not assume that democratic capitalism is the last word in human social evolution." ...

THE PAST IS DEAD

Built on steep, muddy hills, the shantytowns of Ankara, the Turkish capital, exude visual drama. Altindag, or "Golden Mountain," is a pyramid of dreams, fashioned from cinder blocks and corrugated iron, rising as though each shack were built on top of another, all reaching awkwardly and painfully toward heaven – the heaven of wealthier Turks who live elsewhere in the city. For reasons that I will explain, the Turkish shack town is a psychological universe away from the African one.

Slum quarters in the Ivory Coast's Abidjan terrify and repel the outsider. In Turkey it is the opposite. Golden Mountain was a real neighborhood. The inside of one house told the story: The architectural bedlam of cinder block and sheet metal and cardboard walls was deceiving. Inside was a *home* – order, that is, bespeaking dignity. I saw a working refrigerator, a television, a wall cabinet with a few books and lots of family pictures, a few plants by a window, and a stove. Though the streets become rivers of mud when it rains, the floors inside this house were spotless.

My point in bringing up a rather wholesome, crime free slum is this: Its existence demonstrates how formidable is the fabric of which Turkish Muslim culture is made. A culture this strong has the potential to dominate the Middle East once again. Slums are litmus tests for innate cultural strengths and weaknesses. Those peoples whose cultures can harbor extensive slum life without decomposing will be, relatively speaking, the future's winners. Those whose cultures cannot will be the future's victims. . . .

In Turkey, . . . Islam is painfully and awkwardly forging a consensus with modernization, a trend that is less apparent in the Arab and Persian worlds (and virtually invisible in Africa). In Iran the oil boom –because it put development and urbanization on a fast track, making the culture shock more intense – fuelled the 1978 Islamic revolution. But Turkey, unlike Iran and the Arab world, has little oil. Therefore, its development and urbanization have been more gradual. Islamists have been integrated into the parliamentary system for decades.

Resource distribution is strengthening Turks in another way vis-à-vis Arabs and Persians. Turks may have little oil, but their Anatolian heartland has lots of water – the most important fluid of the twenty-first century. Turkey's Southeast Anatolia Project, involving 22 major dams and irrigation systems, is impounding the waters of the Tigris and Euphrates rivers. Much of the water that Arabs and perhaps Israelis will need to drink in the future is controlled by Turks. The project's centerpiece is the mile wide, 16 story Ataturk Dam, upon which are emblazoned the words of modern Turkey's founder: "Ne Mutlu Turkum Diyene" ("Lucky is the one who is a Turk"). . . . Power is certainly moving north in the Middle East, from the oil fields of Dhahran, on the Persian Gulf, to the water plain of Harran, in southern Anatolia – near the site of the Ataturk Dam. But will the nation state of Turkey, as presently constituted, be the inheritor of this wealth? I very much doubt it.

THE LIES OF MAPMAKERS

According to the map, the great hydropower complex emblemized by the Ataturk Dam is situated in Turkey. Forget the map. This southeastern region of Turkey is populated almost completely by Kurds. About half of the world's 20 million Kurds live in "Turkey." The Kurds are predominant in an ellipse of territory that overlaps not only with Turkey but also with Iraq, Iran, Syria and the former Soviet Union. The Western enforced Kurdish enclave in northern Iraq, a consequence of the 1991 Persian Gulf War, has already exposed the fictitious nature of that supposed nation state.

On a recent visit to the Turkish–Iranian border, it occurred to me what a risky idea the nation state is. Here I was on the legal fault line between two clashing civilizations, Turkic and Iranian. Yet the reality was more subtle: As in West Africa, the border was porous and smuggling abounded, but here the people doing the smuggling, on both sides of the border, were Kurds. In such a moonscape, over which peoples have migrated and settled in patterns that obliterate borders, the end of the Cold War will bring on a cruel process of natural selection among existing states. No longer will these states be so firmly propped up by the West or the Soviet Union. Because the Kurds overlap with nearly everybody in the Middle East, on account of their being cheated out of a state in the post-First World War peace treaties, they are emerging, in effect, as the natural selector – the ultimate reality check. They have destabilized Iraq and may continue to disrupt states that do not offer them adequate breathing space, while strengthening states that do.

Because the Turks, owing to their water resources, their growing economy and the social cohesion evinced by the most crime free slums I have encountered, are on the verge of big power status, and because the 10 million Kurds within Turkey threaten that status, the outcome of the Turkish–Kurdish dispute will be more critical to the future of the Middle East than the eventual outcome of the recent Israeli–Palestinian agreement.

A NEW KIND OF WAR

To appreciate fully the political and cartographic implications of postmodernism – an epoch of themeless juxtapositions, in which the classificatory grid of nation states is going to be replaced by a jagged glass pattern of city states, shanty states, nebulous and anarchic regionalisms – it is necessary to consider, finally, the whole question of war.

The intense savagery of the fighting in such diverse cultural settings as Liberia, Bosnia, the Caucasus and Sri Lanka – to say nothing of what obtains in American inner cities – indicates something very troubling that those of us concerned with issues such as middle-class entitlements and the future of interactive cable television lack the stomach to contemplate. It is this: A large number of people on this planet, to whom the comfort and stability of a middle-class life are utterly unknown, find war and a barracks existence a step up rather than a step down.

"Just as it makes no sense to ask 'why people eat' or 'what they sleep for,'" writes Martin van Creveld, a military historian at the Hebrew University in Jerusalem, in "The Transformation of War," "so fighting in many ways is not a means but an end. Throughout history, for every person who has expressed his horror of war there is another who found in it the most marvellous of all the experiences that are vouchsafed to man, even to the point that he later spent a lifetime boring his descendants by recounting his exploits." ...

Van Creveld's book begins by demolishing the notion that men don't like to fight. "By compelling the senses to focus themselves on the here and now," van Creveld writes, war "can cause a man to take his leave of them." As anybody who has had experience with Chetniks in Serbia, "technicals" in Somalia, Tontons Macoutes in Haiti or soldiers in Sierra Leone can tell you, in places where the Western Enlightenment has not penetrated and where there has always been mass poverty, people find liberation in violence. Physical aggression is a part of being human. Only when people attain a certain economic, educational and cultural standard is this trait tranquillized. In light of the fact that 95 per cent of the Earth's population growth will be in the poorest areas of the globe, the question is not whether there will be war (there will be a lot of it) but what kind of war. And who will fight whom?

Debunking the great military strategist Carl von Clausewitz, van Creveld, who may be the most original thinker on war since that early nineteenth-century Prussian, writes, "Clausewitz's ideas ... were wholly rooted in the fact that, ever since 1648, war had been waged overwhelmingly by states." But, as van Creveld explains, the period of nation states and, therefore, of state conflict is now ending, and with it the clear "threefold division into government, army and people" which state directed wars enforce. Thus, to see the future, the first step is to look back to the past immediately prior to the birth of modernism – the wars in medieval Europe that began during the Reformation and reached their culmination in the Thirty Years' War.

Van Creveld writes:

> In all these struggles political, social, economic and religious motives were hopelessly entangled. Since this was an age when armies consisted of mercenaries, all were also attended by swarms of military entrepreneurs.... Many of them paid little but lip service to the organizations for whom they had contracted to fight. Instead, they robbed the countryside on their own behalf.... Given such conditions, any fine distinctions ... between armies on the one hand and peoples on the other were bound to break down. Engulfed by war, civilians suffered terrible atrocities.

Back then, in other words, there was no "politics" as we have come to understand the term, just as there is less and less "politics" today in Liberia, Sierra Leone, Somalia, Sri Lanka, the Balkans and the Caucasus, among other places.

Because, as van Creveld notes, the radius of trust within tribal societies is narrowed to one's immediate family and guerrilla comrades, truces arranged with one Bosnian commander, say, may be broken immediately by another Bosnian commander. The plethora of short lived cease fires in the Balkans and the Caucasus constitute proof that we are no longer in a world where the old rules of state warfare apply. . . .

Also, war making entities will no longer be restricted to a specific territory. Loose and shadowy organisms such as Islamic terrorist organizations suggest why borders will mean increasingly little and sedimentary layers of tribalistic identity and control will mean more. "From the vantage point of the present, there appears every prospect that religious . . . fanaticisms will play a larger role in the motivation of armed conflict" in the West than at any time "for the last 300 years," van Creveld writes. . . .

Future wars will be those of communal survival, aggravated or, in many cases, caused by environmental scarcity. These wars will be sub-national, meaning that it will be hard for states and local governments to protect their own citizens physically. This is how many states will ultimately die. . . .

THE LAST MAP

In "Geography and the Human Spirit," Anne Buttimer, a professor at University College, Dublin, recalls the work of an early nineteenth-century German geographer, Carl Ritter, whose work implied "a divine plan for humanity" based on regionalism and a constant, living flow of forms. The map of the future, to the extent that a map is even possible, will represent a perverse twisting of Ritter's vision. Imagine cartography in three dimensions, as if in a hologram. In this hologram would be the overlapping sediments of group and other identities atop the merely two dimensional color markings of city states and the remaining nations, themselves confused in places by shadowy tentacles, hovering overhead, indicating the power of drug cartels, mafias and private security agencies. Instead of borders, there would be moving "centers" of power, as in the Middle Ages. Many of these layers would be in motion. Replacing fixed and abrupt lines on a flat space would be a shifting pattern of buffer entities, like the Kurdish and Azeri buffer entities between Turkey and Iran, the Turkic Uighur buffer entity between Central Asia and Inner China (itself distinct from coastal China), and the Latino buffer entity replacing a precise US Mexican border. To this protean cartographic hologram one must add other factors, such as migrations of populations, explosions of birth rates, vectors of disease. Henceforward the map of the world will never be static. This future map – in a sense, the "Last Map" – will be an ever mutating representation of chaos.

Indeed, it is not clear that the United States will survive the next century in exactly its present form. Because America is a multiethnic society, the nation state has always been more fragile here than it is in more homogeneous societies such as Germany and Japan. James Kurth, in an article published in *The National Interest* in 1992, explains that whereas nation state societies tend to be built around a mass conscription army and a standardized public school system, "multicultural regimes" feature a high tech, all volunteer army (and, I would add, private schools that teach competing values), operating in a culture in which the international media and entertainment industry have more influence than the "national political class." In other words, a nation state is a place where everyone has been educated along similar lines, where people take their cue from national leaders, and where everyone (every male, at least) has gone through the crucible of military service, making patriotism a simpler issue. Writing about his immigrant family in turn of the century Chicago, Saul Bellow states, "The country took us over. It was a country then, not a collection of 'cultures.'"

During the Second World War and the decade following it, the United States reached its apogee as a classic nation state. During the 1960s, as is now clear, America began a slow but unmistakable process of transformation. The signs hardly need belaboring: racial polarity, educational dysfunction, social fragmentation of many and various kinds. "Patriotism" will become increasingly regional as people in Alberta and Montana discover that they have far more in common with each other than they do with Ottawa or Washington, and Spanish speakers in the Southwest discover a greater commonality with Mexico City. As Washington's influence wanes, and with it the traditional symbols of American patriotism,

North Americans will take psychological refuge in their insulated communities and cultures.

Returning from West Africa last fall was an illuminating ordeal. After leaving Abidjan, my Air Afrique flight landed in Dakar, Senegal, where all passengers had to disembark in order to go through another security check, this one demanded by US authorities before they would permit the flight to set out for New York. Once we were in New York, despite the midnight hour, immigration officials at Kennedy Airport held up disembarkation by conducting quick interrogations of the aircraft's passengers – this was in addition to all the normal immigration and customs procedures. It was apparent that drug smuggling, disease and other factors had contributed to the toughest security procedures I have ever encountered when returning from overseas.

Then, for the first time in over a month, I spotted businesspeople with attaché cases and laptop computers. When I had left New York for Abidjan, all the business people were boarding planes for Seoul and Tokyo, which departed from gates near Air Afrique's. The only non-Africans off to West Africa had been relief workers in T-shirts and khakis. Although the borders within West Africa are increasingly unreal, those separating West Africa from the outside world are in various ways becoming more impenetrable.

But Afrocentrists are right in one respect: We ignore this dying region at our own risk. When the Berlin Wall was falling, in November of 1989, I happened to be in Kosovo, covering a riot between Serbs and Albanians. The future was in Kosovo, I told myself that night, not in Berlin. The same day that Yitzhak Rabin and Yasser Arafat clasped hands on the White House lawn, my Air Afrique plane was approaching Bamako, Mali, revealing corrugated zinc shacks at the edge of an expanding desert. The real news wasn't at the White House, I realized. It was right below.

24 SIMON DALBY

"Reading Robert Kaplan's 'Coming Anarchy'"

from *Ecumene* (1996)

ROBERT KAPLAN'S GEOPOLITICAL IMAGINATION

The world is not quite so conveniently simple as Kaplan's popularization of environmental degradation as the key national security issue for the future suggests. His article for all its dramatic prose and empirical observation is vulnerable to numerous critiques. Read as a cultural production of considerable political importance it is fairly easy to see how the logic of the analysis, premised on "eye witness" empirical observation, and drawing on an eclectic mixture of intellectual sources, leaves so much of significance unsaid. But the impression, as has traditionally been the case in geopolitical writing, generated from the juxtaposition of expert sources and empirical observation is that this is an "objective" detached geopolitical treatise. The focus in what follows is on the political implications of the widely shared geopolitical assumptions that structure this text and ultimately render the environment as a threat.

The most important geopolitical premise in the argument posits a "bifurcated world", one in which the rich in the prosperous "post-historical" cities and suburbs have mastered nature through the use of technology, while the rest of the population is stuck in poverty and ethnic strife in the shanty towns of the underdeveloped world. The presentation of the article in the magazine supports this basic formulation of the world into the rich, who read magazines like *Atlantic*, and the rest who don't. The closing image in the text of New York airport with its business people flying to Asia, but not to Africa, is very strongly reinforced through the article by the juxtaposition of the advertisements in the original magazine version of the article with the violent imagery of the photographs, and the themes in the text. The affluence of New York airport contrasts sharply with the poverty and dangers elsewhere.

But these phenomena are treated as completely separate in terms of economics. Poverty and affluence are only connected where poverty is seen as a threat to the affluence of the *Atlantic*'s North American readers. In all of Kaplan's article matters of international trade are barely mentioned. The wall of disease he writes about may bar many foreigners from all except some coastal "trading posts" of Africa in the future, but the significance of what is being traded and with what implications for the local environment is not investigated. "Hot cash", presumably laundered drug moneys from African states, apparently does flow to Europe we are told, but this has significance only because of the criminal dimension of the activity, not as part of a larger pattern of political economy. While the lack of business people flying to Africa is noted, comments about the high rate of logging are never connected to the export markets for such goods, or to the economic circumstances of indebted African states that distort local economies to pay international loans and meet the requirements for structural adjustment programs. Logging continues apace, but it is apparently driven only by some indigenous local desire to strip the environment of trees, not by any exogenous cause. A focus on the larger political economy driving forest destruction would lead the analysis in a very different direction, but it is a direction that is not taken by the focus on West Africa as a quasi-autonomous geopolitical entity driven by internal developments.

The political violence and environmental degradation are not related to larger economic processes anywhere in this text. This is not to suggest that the legacy of colonialism, or the subsequent neo-colonial economic arrangements, are solely to "blame" for current crises, although the history cannot be ignored as Kaplan is wont to do. It is to argue that these sections of Kaplan's text show a very limited geopolitical imagination, one that focuses solely on local phenomena in a determinist fashion that ignores the larger trans-boundary flows and the related social and economic causes of resource depletion. Kaplan ignores the legacy of the international food economy which has long played a large role in shaping the agricultural infrastructures, and the nutritional levels of many populations of different parts of the world in specific ways. He also ignores the impact of the economic crisis of the 1980s and the often deleterious impact of the debt crisis and structural adjustment policies. He completely misses their important impact on social patterns and the impact on rural women upon whom many of the worst impacts fell (Mackenzie, 1993).

Ironically, given his repeated comments about the inadequacies of cartographic designations of state boundaries in revealing cross-border ethnic and criminal flows, Kaplan effectively establishes economic boundaries precisely by not investigating economic phenomena that supposedly ought to be crucial to his specification of various regions in Malthusian terms. While Kaplan emphasizes the inadequacies of maps for understanding ethnic and cultural clashes, he never investigates their similar inadequacies for understanding economic interconnections as an important part of either the international relations or the foreign policies of these states. The crucial failure to do this allows for the attribution of the "failure" of societies to purely internal factors. Once again the local environment can be constructed as the cause of disaster without any reference to the historical patterns of development that may be partly responsible for the social processes of degradation (Crush, 1995; Slater, 1993).

Given the focus of most Malthusians on the shortage of "subsistence" and resources in general, there is remarkably little investigation of how the burgeoning populations of various parts of the world actually are provided for either in terms of food production or other daily necessities. Despite accounts of trips across Africa by "bush-taxi", agricultural production remains invisible to Kaplan's "eye witness". While cities are dismissed as "dysfunctional" the very fact that they continue to grow despite all their difficulties suggests that they do "function" in many ways. Informal arrangements and various patterns of "civil society" are ignored. People move to the cities, but quite why is never discussed in this article. There is no analysis here of traditional patterns of subsistence production and how they and access to land may be changing in the rural areas, particularly under the continuing influence of modernization. While it is made clear that traditional rural social patterns fray when people move to the very different circumstances of the city, the reasons for migration are assumed but never investigated. In Homer-Dixon's language, absolute scarcity is assumed and the possibilities of relative scarcity, with the negative consequences for poor populations due to unequal distribution or the marginalization of subsistence farmers as a result of expanded commercial farming, is never investigated. Why Malthus, in particular, should be the prophet of West Africa, given the complete failure to investigate the changing patterns of these rural economies, is far from clear. Disease and crowding there may be in the shanty towns of many cities, a phenomenon that is not exactly new, but not all the new urban population are dispossessed forest dwellers or refugees from criminal activities.

The focus on environment as the key factor in triggering violent changes is not entirely consistent with Kaplan's arguments elsewhere about the cohesive force of Islam, identified ironically in a few places, given the usual orientalizations in practice when discussing Islam, as a Western religion. His discussion of Turkey suggests that while urbanization is occurring rapidly, social cohesion and resistance to crime are being maintained by Islam, even as new geopolitical identities are being forged in the slums. While he suggests that these identities may transcend the force of Islam in the ongoing conflict between Turks and Kurds, his emphasis on non-environmental factors of social cohesion suggests that his argument is perhaps more concerned with traditional matters of ethnic identity and "civilizational clashes", than with environmental degradation.

Here resurgent cultural fears of "the Other" and assumptions about the persistence of cultural patterns of animosity and social cleavage are substituted for analysis of resources and rural political ecology. Precisely where the crucial connections between environmental change, migration and conflict should be investigated, the analysis turns away to look at ethnic rivalries and the collapse of social order. The connections are asserted, not demonstrated, and in so far as this is done the opportunity for detailed analysis is missed and the powerful rhetoric of the argument retraces familiar political territory instead of looking in detail at the environment as a factor in social change. In this failure to document the crucial causal connections in his cases Kaplan ironically follows Malthus who relied on his unproven key assumption that subsistence increases only at an arithmetic rate in contrast to geometric population growth.

Political angst about the collapse of order is substituted for an investigation of the specific reasons for rapid urbanization, a process that is by default rendered as a "natural" product of demographic pressures. This unstated "naturalization" then operates to support the Malthusian fear of poverty stricken mobs, or in Kaplan's terms, young homeless and rootless men forming criminal gangs, as a threat to political order. Economics becomes nature, nature in the form of political chaos becomes a threat, the provision of security from such threats thus becomes a policy priority. In this way "nature unchecked" can thus be read directly as a security threat to the political order of post-modernity.

GEOPOLITICS, MALTHUS AND KAPLAN

Kaplan explicitly links the Malthusian theme in his discussion of Africa to matters of national security, where a clear "external" threatening dimension of crime and terrorism is linked to the policy practices of security and strategic thinking. The logic of a simple Malthusian formulation is complicated by the geographical assumptions built into Kaplan's argument, while he has simultaneously avoided any explicit attempt to deal at all with the political economy of rural subsistence or contemporary population growth. Thus, in his formulation, the debate is shifted from matters of humanitarian concern, starvation, famine relief and aid projects and refocused on matters of military threat and concern for political order within Northern states.

What ultimately seems to matter in this new designation is whether political disorder and crime will spill over into the affluent North. The affluent world of the *Atlantic* advertisements with their high-technology consumer items (Saabs, Mazdas and Bose stereos etc.) is implicitly threatened by the spreading of "anarchy". The article implies that it has done so already in so far as American inner cities are plagued with violent crime. The reformulation once again posits a specific geopolitical framework for security thinking. Kaplan himself suggests that by his own logic the US may become more fragmented. What cannot be found in this article is any suggestion that the affluence of those in the limousine might in some way be part of the same political economy that produces the conditions of those outside. This connection is simply not present in the text of the article because of the spatial distinctions Kaplan makes between "here" and "there". He notes the dangers of the criminals from "there" compromising the safety of "here" but never countenances the possibility that the economic affluence of "here" is related to the poverty of "there". The spatial construction of his discourse precludes such consideration, only some factors violate the integrity of cartographic boundaries.

Although Kaplan is particularly short on policy prescription in his *Atlantic* article, some of the implications of his reworked Malthusianism do have clear policy implications. Instead of repression and the use of political methods to maintain inequalities in the face of demands for reform, Kaplan's implicit geopolitics suggest abandoning Africa to its fate. If more Northern states withdraw diplomatic and aid connections and, as he notes, stop direct flights to airports such as Lagos, the potential to isolate this troubled region may be considerable. If contact is restricted to coastal trading posts then the "wall of disease" will become a wall of separation keeping non-Africans out and restricting the possibilities for Africans to migrate. Once again security is understood in the geopolitical term of containment and exclusion.

In a subsequent article in the *Washington Post* (17 April 1994) Kaplan explicitly argues against US military interventions in Africa. He suggests that intervention in Bosnia would do some good, because the developed nature of the societies in conflict there allows some optimism that a political settlement is workable. The chances of intervention having much effect in Africa are dismissed because of the illiterate poverty stricken populations there. However, the pessimism of the *Atlantic* article is muted here by a contradictory suggestion that all available foreign policy money for Africa be devoted to population control, resource management and women's literacy. These programs will, Kaplan hopes, in the very long term resolve some of the worst problems allowing development to occur and "democracy" eventually to emerge.

The ethnocentrism of the suggestion that Africa's problems are solvable in terms of modernization, is coupled to the implication that West Africa is of no great importance to the larger global scheme of power and economy, and therefore can be ignored, at least as long as the cultural affinities between Africans and African Americans don't cause political spillovers into the United States. In this geopolitical argument Kaplan parallels Saul Cohen's geopolitical designation of Sub-Saharan Africa as part of a "quartersphere of marginality" consigning it to irrelevance in the post-Cold War order (Cohen, 1994). Precisely this marginalization is of concern to many African leaders and academics. But in stark contrast to Kaplan, many Africans emphasize the need to stop the export of wealth from the Continent, and the need to draw on indigenous traditions to rebuild shattered societies and economies (Adadeji, 1993; Amin, 1990; Taylor and Mackenzie, 1992).

Spatial strategies of containment are a long standing component of security thinking. Cutting anarchy ridden regions loose in the hopes that their political turmoil will remain internal makes sense in an argument that constructs these places as clearly external to the political arrangements that one wishes to render secure from threats. Given the specification of the political turmoil as caused internally within these areas, this argument makes logical sense. Also given the startling failure in this analysis to consider matters of international economics as a possible cause for some of the phenomena that are involved in the dissolution of political order, no sense of external responsibility applies. Kaplan deals with deterritorialized phenomena when they suit his argument, but conveniently ignores trans-boundary flows when they don't fit his cartographic scheme. They suit it here because they emphasize political violence and threats across frontiers that are in some cases disappearing.

Large scale geopolitical isolation as a *cordon sanitaire* might work as a Western security strategy in these circumstances; it seems less likely to help Africans, but that point is not high on Kaplan's scheme of priorities. But to advocate these "solutions" is once again to specify complex political phenomena in territorial terms, a strategy that is, as John Agnew argues, falling into the familiar "territorial trap" in international relations thinking where boundaries are confused with barriers and flows and linkages are obscured by the widespread assumption of autonomous states as the only actors of real importance in considering global politics (Agnew, 1994).

There is an ironic twist in Kaplan's geopolitical specifications of "wild zones". He argues that they are threats to political stability and in the case of Africa probably worth cutting loose from conventional political involvement. In the subsequent *Washington Post* article he argues against military interventions in Africa on the basis of their uselessness in the political situation of gangs, crime and the absence of centralized political authority. His suggestions imply that interventions are only considered in terms of political attempts to resolve conflicts and provide humanitarian aid. In this assumption Kaplan is at odds with Cold War geopolitical thinking. While ignoring the political economy of underdevelopment as a factor in the African situation, he also ignores the traditional justifications for US political and military involvement in Africa and much of the Third World. Through the Cold War these focused on questions of ensuring Western access to strategic minerals in the continent. This theme continues to appear in many other discussions of post-Cold War foreign policy and in US strategic planning. But Kaplan ignores both these economic interconnections and their strategic implications, preferring an oversimplified geopolitical specification of Malthusian-induced social collapse as the sole focus of concern.

But the specification of danger as an external "natural" phenomena works in an analogous way to the traditional political use of Neo-Malthusian logic. Once again threats are outside human regulation, inevitable and natural in some senses – if not anarchic in the neo-realist sense of state system structure then natural in a more fundamental sense of "nature unchecked". By the specific spatial assumptions built into his reasoning Kaplan accomplishes geopolitically what Malthusian thinking did earlier in economic terms. Coupled to prevalent American political concerns with security as "internal" vulnerability to violent crime, and "external" fears of various foreign military, terrorist, economic, racial, and immigration "threats", Kaplan re-articulates his modified Malthusianism in the powerful discursive currency of geopolitics. His themes fit neatly with media coverage of Rwanda and Somalia where his diagnosis of the future appeared in many media accounts to be occurring nearly immediately.

Understood as problems of "tribal" warfare such formulations reproduce the earlier tropes of "primitive savagery". As other commentators on contemporary conflict have noted, detailed historical analysis suggests that the formation of "tribes" , and many of the "tribal wars" that European colonists deplored, were often caused by the sociological disruptions triggered by earlier European intrusions. Denial or failure to understand the causal interconnections of this process allowed for the attribution of "savagery" to "Others" inaccurately specified as geographically separate. Kaplan notes that the disintegration of order is not a matter of a "primitive" situation, but following van Creveld, a matter of "reprimitivized" circumstances in which high-technology tools are used for gang and "tribal" rivalries. But the economic connections that allow such "tools" to become available are not mentioned. Thus re-primitivization is specified as the indirect result of environmental degradation, a process that is asserted frequently but not argued, demonstrated or investigated in any detail. Once again geopolitical shorthand is substituted for detailed geographical analysis. In Ó Tuathail and Agnew's (1992) terms, the irony of the policy discourse of geopolitics, as the antithesis to detailed geographical understanding, is in play once again in this text, although this time with environment as a reified concept.

BEYOND MALTHUS AND MACKINDER?

The continued possibilities of using Malthusian themes as ideological weapons by the powerful in justifying repression, or at the least, justifying inaction in the face of gross inequities, now have to be complimented by a recognition that these themes can be mobilized in foreign policy discourse to suggest the appropriateness of military solutions to demographic and "environmental" problems. At least in the earlier version of his famous essay, Malthus argued that population growth is inevitable, natural, and largely beyond human regulation. Politics is thus rendered as just a reaction to the consequences of the unchangeable patterns of fecundity. If the political consequences of population growth are disruptive to the Northern geopolitical order that is judged to be the only acceptable one, then Neo-Malthusianism acts as a powerful intellectual weapon in formulating policies to repress and politically control reformist demands for greater equality or economic redistribution. It can do so on the grounds that such policies only aggravate adverse demographic trends. When coupled to Kaplan's assertions that population growth is related to environmental degradation, the argument is strengthened.

If the more alarmist versions of some of Kaplan's arguments gain credence in Washington, or if the formulation of politics in terms of the Rest and the West becomes prominent, then the dangers of a new Cold War against the poor are considerable. The discussions of illegal immigration in the US in the early 1990s, and suggestions that the solution is increased border guards, denial of services to immigrants incapable of proving legal residence, and deportations, suggest that the geopolitical imagination of spatial exclusion is dominating the policy discourse once again. In particular this may be because of the propensity among American politicians to formulate American identity in antithesis to external perceived dangers. Through the history of the last two centuries this has been a powerful theme in the formulation of American foreign policy which has drawn on the related discourses of American exceptionalism (Agnew, 1983).

This geopolitical imagination has been frequently coupled to assertions of cultural superiority and ideological rectitude in the form of various articulations of moral certainty. The dangers of ethnocentrism, when coupled to geopolitical reasoning, are greatest precisely where they assert strategic certainty in ways that prevent analysis of the complex social, political and economic interactions that might lead to assessments that in at least some ways "the problem is us" (Hentsch, 1992). Through the course of the Cold War and subsequently in the 1991 Gulf War, these formulations have fueled arms races, the global politics of deterrence and "security" understood in terms of violent containment and military superiority (Campbell, 1992, 1993; Dalby, 1990). This is done by privileging territorial sovereignty over other modes of human organization.

But it is the focus on the failures of these strategies in many places that makes Kaplan's vision so troubling to conventional analysis. In Shapiro's (1991) terms he focuses on some flows or "exchanges" that transgress the frontiers of sovereignty unsettling the possibilities of political order constrained in the spatial imaginations of modern sovereignty. While the fear that traditional military protection of borders is no longer efficacious, and that social disorder will spread despite the spatial demarcations of boundaries, induces fear, it can also ironically draw on the traditional thinking to suggest that if current efforts are inadequate then what is needed is redoubled actions in the military sphere to reassert control. Such a policy of militarization suggests escalating violence rather than attempts to tackle large scale problems in more cooperative ways.

Kaplan's posing of these problems in terms of national security suggests such a strategy. Once again the sovereignty problematic can lead to specifications of dangers and violent solutions, rather than to any consideration of an ethics of post-sovereignty (Shapiro, 1994; Walker, 1993). The construction of the threat as "nature unchecked" simply adds to the specification of danger as beyond the possibilities of simple interventions and amelioration, hence a long lasting security threat that is particularly intractable. Kaplan's analysis doesn't escape classical geopolitical thinking. While his analysis of the collapse of geopolitical boundaries suggests a new departure in understanding politics, one that looks at the necessity of rethinking warfare and that gets beyond themes of geopolitical boundaries, his focus on organic communities and on Malthusian environmental causes of turmoil, phrased as security threats, leads the analysis back to the need to keep the feared threats at bay by strategies of spatial exclusion. A form of geographical determinism is once again linked to threats of geopolitical violence.

REFERENCES

Adadeji, A. (ed.) (1993) *Africa Within the World: Beyond Dispossession and Dependance*, London: Zed.

Agnew, J. (1983) "An Excess of 'National Exceptionalism': Towards a New Political Geography of American Foreign Policy", *Political Geography Quarterly*, 2(2). 151–66.

Agnew, J. (1994) "The Territorial Trap: The Geographical Assumptions of International Relations Theory", *Review of International Political Economy*, 1(1). 53–80.

Amin, S. (1990) *Maldevelopment: Anatomy of a Global Failure*, London: Zed.

Bennet, O. (ed.) (1991) *Greenwar: Environment and Conflict*, London: Panos.

Campbell, D. (1992) *Writing Security: American Foreign Policy and the Politics of Identity*, Minneapolis: University of Minnesota Press.

Campbell, D. (1993) *Politics Without Principle: Sovereignty, Ethics, and the Narratives of the Gulf War*, Boulder: Lynne Rienner.

Cohen, S. (1994) "Geopolitics in the New World Era: A New Perspective to an Old Discipline" in G.J. Demko and W.B. Wood (eds) *Reordering the World: Geopolitical Perspectives on the Twenty-first Century*, Boulder, Colorado: Westview, pp. 15–48.

Crush, J. (ed.) (1995) *Power of Development*, London: Routledge.

Dalby, S. (1990) *Creating the Second Cold War: The Discourse of Politics*, London: Pinter and New York: Guilford.

Hentsch, T. (1992) *Imagining the Middle East*, Trans. F.A. Reed, Montreal: Black Rose.

Mackenzie, F. (1993) "Exploring the Connections: Structural Adjustment, Gender and the Environment", *Geoforum*, 24(1). 71–87.

Ó Tuathail, G. and J. Agnew (1992) "Geopolitics and Discourse: Practical Geopolitical Reasoning in American Foreign Policy", *Political Geography*, 11(2). 190–204.

Shapiro, M.J. (1991) "Sovereignty and Exchange in the Orders of Modernity", *Alternatives*, 16(4). 447–477.

Shapiro, M.J. (1994) "Moral Geographies and the Ethics of Post-Sovereignty", *Public Culture*, 6. 479–502.

Slater, D. (1993) "The Geopolitical Imagination and the Enframing of Development Theory", *Transactions of the Institute of British Geographers New Series* 18. 419–437.

Taylor, D.R.F. and F. Mackenzie (eds) (1992) *Development From Within: Survival in Rural Africa*, London: Routledge.

Walker, R.B.J. (1993) *Inside/Outside: International Relations as Political Theory*, Cambridge: Cambridge University Press.

25 THOMAS F. HOMER-DIXON

"Environmental Scarcity and Mass Violence"

from *Current History* (1996)

Scarcities of critical environmental resources – especially cropland, fresh water, forests, and fish stocks – are powerfully contributing to mass violence in key areas of the world. While these "environmental scarcities" do not cause wars among countries, they do sometimes sharply aggravate stresses within countries, helping to stimulate ethnic clashes, urban unrest, and insurgencies. This violence affects Western national interests by destabilizing trade and economic relations, provoking distress migrations, and generating complex humanitarian disasters that distract our militaries and absorb huge amounts of aid.

Policy makers and citizens in the West ignore these pressures at their peril. In Chiapas, Mexico, Zapatista insurgents rose against land scarcity and insecure land tenure produced by ancient inequalities in land distribution, by rapid population growth among groups with the least land, and by changes in laws governing land access. The insurgency rocked Mexico to the core, helped trigger the peso crisis and reminded the world that Mexico remains – despite NAFTA and the pretenses of the country's economic elites – a poor and profoundly unstable developing country.

In Pakistan, shortages and maldistribution of good land, water, and forests in the countryside have encouraged migration of huge numbers of rural poor into major cities, such as Karachi and Hyderabad. The conjunction of this in-migration with high fertility rates is causing urban populations to grow at a staggering 4 to 5 per cent a year, producing fierce competition – and often violence – among ethnic groups over land, basic services, and political and economic power. This turmoil exacts a huge cost on the national economy. It also probably encourages the Pakistani regime to buttress its internal legitimacy by adopting a more belligerent foreign policy on issues such as Kashmir and nuclear proliferation.

In South Africa severe land, water, and fuelwood scarcity in the former black homelands has helped drive millions of poor blacks into teaming squatter settlements in the major cities. The settlements are often constructed on the worst urban land, in depressions prone to flooding, on hillsides vulnerable to slides, or near heavily polluting industries. Scarcities of land, water, and fuelwood in these settlements provoke interethnic rivalry and extraordinarily violent feuds among settlement warlords and their followers. This strife jeopardizes the country's transition to democratic stability and prosperity.

Over the last six years a diverse group of one hundred experts from fifteen countries has closely studied cases such as these. Organized by the Peace and Conflict Studies Program at the University of Toronto and the American Academy of Arts and Sciences in Cambridge, Massachusetts, this group has examined in detail fourteen cases, including those mentioned above, as well as the cases of Mauritania-Senegal, Rwanda, Bangladesh, India, Indonesia, Philippines, China, Haiti, Peru, Gaza, and the West Bank. Taken in conjunction with research by other groups, especially in Switzerland and Norway, a clear picture has emerged of how and where environmental scarcity produces social breakdown and violence. In this article, I survey these findings.

It is easy for the billion-odd people living in rich countries to forget that the wellbeing of about half of the world's population of 5.8 billion remains directly tied to local natural resources. Nearly 3 billion people rely on agriculture for their main income; perhaps 1 billion are subsistence farmers, which means they survive by eating what they grow. Over 40 per cent of people on the planet – some 2.2 billion – use fuelwood, charcoal, straw, or cow dung as their main source of energy; 50 to 60 per cent rely on these biomass fuels for at least some of their primary energy needs. Over 1.2 billion people lack access to clean drinking water; many are forced to walk miles to get what water they can find.

The cropland, forests, and water supplies that underpin the livelihoods and wellbeing of these billions are renewable. Unlike non-renewable resources such as oil and iron ore, renewables are replenished over time by natural processes. In most cases, if used prudently, they should sustain an adequate standard of living indefinitely. Unfortunately, in the majority of regions where people are highly dependent on renewable resources, they are being depleted or degraded faster than they are being renewed. From Gaza to the Philippines to Honduras, the evidence is stark: aquifers are being overdrawn and salinized, coastal fisheries are disappearing, and steep uplands have been stripped of their forests leaving their thin soils to erode into the sea.

These environmental scarcities usually have complex causes. Resource depletion and degradation are a function of the physical vulnerability of the resource, the size of the resource-consuming population, and the technologies and practices this population uses in its consumption behavior. The size of the population and its technologies and practices are, in turn, a result of a wide array of other variables, from women's status to the availability of human and financial capital.

Moreover, resource depletion and degradation are together only one of three sources of environmental scarcity. Depletion and degradation produce a decrease in total resource supply or, in other words, a decrease in the size of the total resource "pie." But population growth and changes in consumption behavior can also cause greater scarcity by boosting the demand for a resource. Thus if a rapidly growing population depends on a fixed amount of cropland, the amount of cropland per person – the size of each person's slice of the resource pie – falls inexorably. In many countries, resource availability is being squeezed by both these supply and demand pressures.

Finally, scarcity is often caused by a severe imbalance in the distribution of wealth and power that results in some groups in a society getting disproportionately large slices of the resource pie, while others get slices that are too small to sustain their livelihoods. Such unequal distribution – or what we call structural scarcity – is a key factor in every case our research team has examined. Often the imbalance is deeply rooted in institutions and class and ethnic relations inherited from the colonial period. Often it is sustained and reinforced by international economic relations that trap developing countries into dependence on a few raw material exports. It can also be reinforced by heavy external debts that encourage countries to use their most productive environmental resources – such as their best croplands and forests – to generate hard currency rather than to support the most impoverished segments of their populations.

In the past, scholars and policy makers have usually addressed these three sources of scarcity independently. But research shows that supply, demand, and structural scarcities interact and reinforce each other in extraordinarily pernicious ways.

One type of interaction is resource capture. It occurs when powerful groups within a society recognize that a key resource is becoming more scarce (due to both supply and demand pressures) and use their power to shift in their favor the regime governing resource access. This shift imposes severe structural scarcities on weaker groups. Thus in Chiapas worsening land scarcities, in part caused by rapid population growth, encouraged powerful land owners and ranchers to exploit weaknesses in the state's land laws in order to seize lands from campesinos and indigenous farmers. Gradually these peasants were forced deeper into the state's lowland rain forest, further away from the state's economic heartland and further into poverty.

In the Jordan River basin, Israel's critical dependence on groundwater flowing out of the West Bank – a dependence made acute by a rising Israeli population and salinizing aquifers along the Mediterranean coast – encouraged Israel to restrict groundwater withdrawals on the West Bank during the occupation. These restrictions were far more severe for Palestinians than for Israeli settlers. They contributed to the rapid decline in Palestinian agriculture in the region, to the increasing dependence of young Palestinians on day-labor within Israel and, ultimately, to rising frustrations in the Palestinian community.

Another kind of interaction, ecological marginalization, occurs when a structural imbalance in resource distribution joins with rapid population growth to drive resource-poor people into ecologically marginal areas, such as upland hillsides, areas at risk of desertification, and tropical rainforests. Higher population densities in these vulnerable areas – along with a lack of the capital and knowledge needed to protect local resources – causes local resource depletion, poverty, and eventually further migration, often to cities.

Ecological marginalization affects hundreds of millions of people around the world, across an extraordinary range of geographies and economic and political systems. We see the same process in the Himalayas, the Sahel, Central America, Brazil, Rajasthan, and Indonesia. For example, in the Philippines an extreme imbalance in cropland distribution between land owners and peasants has interacted with high population growth rates to force large numbers of the landless poor into interior upland regions of the archipelago. There, the migrants use slash and burn agriculture to clear land for crops. As more millions arrive from the lowlands, new land becomes hard to find, and as population densities on the steep slopes increase, erosion, landslides, and flash floods become critical. During the 1970s and 1980s, the resulting poverty helped drive many peasants into the arms of the communist New People's Army insurgency that had a stranglehold on upland regions. Poverty also drove countless others into wretched squatter settlements in cities like Manila.

Of course, numerous contextual factors – factors unique to the Filipino situation – have combined with environmental and demographic stress to produce these outcomes. Environmental scarcity is never a determining or sole cause of large migrations, poverty, or violence; it always joins with other economic, political, and cultural factors to produce its effects. In the Filipino case, for example, the lack of clear property rights in upland areas encouraged migration into these regions and discouraged migrants from conserving the land once they arrived. And President Marcos's corrupt and authoritarian leadership reduced regime legitimacy and closed off options for democratic action by aggrieved groups.

Analysts often overlook the importance of such contextual factors and, as a result, jump from evidence of simple correlation to unwarranted conclusions about causation. Thus some commentators have asserted that rapid population growth, severe land scarcity, and the resulting food shortfalls caused the Rwandan genocide. In an editorial in August 1994, the *Washington Post* argued that while the Rwandan civil war was "military, political, and personal in its execution," a key underlying cause was "a merciless struggle for land in a peasant society whose birthrates have put an unsustainable pressure on it." Yet, while environmental scarcities in Rwanda were serious, close analysis shows that the genocide arose mainly from a conventional struggle among elites for control of the Rwandan state. Land scarcity played at most a peripheral role by reducing regime legitimacy in the countryside and restricting alternatives for elite enrichment outside of the state.

Despite these caveats, in many cases environmental scarcity does powerfully contribute to mass violence. Moreover, it is not possible entirely to subordinate its role to a society's particular institutions and policies. Some skeptics claim that a society can fix its environmental problems by fixing its institutional and policy mistakes; thus, they assert, environmental scarcity's contribution to conflict does not merit independent attention. But our research shows that such arguments are incomplete at best.

Environmental scarcity is not only a consequence of institutions and policy, it also can reciprocally influence these institutions and policies in harmful ways. For example, during the 1970s and 1980s the prospect of chronic food shortages and a serious drought encouraged governments along the Senegal River to build a series of irrigation and flood-control dams. Due to critical land scarcities elsewhere in the region, land values in the basin shot up. The Mauritanian government, controlled by Moors of Arab origin, then captured this resource by changing the laws governing land ownership and abrogating the traditional rights of black Mauritanians to farm, herd, and fish along the Mauritanian side of the river.

Moreover, environmental scarcity should not be subordinated to institutions and policies because it is partly a function of the physical context in which a society is embedded. The original depth of soils in the Filipino uplands and the physical characteristics that make Israel's aquifers vulnerable to salt intrusion are not functions of human social institutions or behavior. And finally, once environmental scarcity becomes irreversible (as when a region's vital topsoil washes into the sea), then the scarcity is, by definition, an external influence on society. Even if enlightened reform of institutions and policies removes the original political and economic causes of the scarcity, it will be a continuing burden on society.

Scarcity-induced resource capture by Moors in Mauritania helped ignite violence over water and cropland in the Senegal River basin, producing tens of thousands of refugees. Expanding populations, land degradation, and drought spurred the rise of the Sendero Luminoso guerrillas in the southern highlands of Peru. In Haiti, forest and soil loss worsen a chronic economic crisis that generates strife and periodic waves of boat people. And land shortages in Bangladesh, exacerbated by fast population growth, have prompted millions of people to migrate to India – an influx that has, in turn, caused ethnic strife in the states of Assam and Tripura.

Close examination of such cases shows that severe environmental scarcity can reduce local food production, aggravate poverty of marginal groups, spur large migrations, enrich elites that speculate on resources, and undermine a state's moral authority and capacity to govern. These long-term, tectonic stresses can slowly tear apart a poor society's social fabric, causing chronic popular unrest and violence by boosting grievances and changing the balance of power among contending social groups and the state.

The violence that results is usually chronic and diffuse, and almost always sub-national not international. There is virtually no evidence that environmental scarcity causes major interstate war. Yet among international relations scholars, it has been conventional wisdom for some time that critical scarcities of natural resources can produce such war. During the 1970s, for example, Nazli Chourci and Robert North argued in their book *Nations in Conflict* that countries facing high resource demands and limited resource availability within their territories would seek the needed resources through trade or conquest beyond their boundaries. Although this "lateral pressure" theory helped explain some past wars, such as World War I, our more recent research highlights a number of the theory's errors. Most importantly, the theory makes no distinction between renewable and non-renewable resources.

There is no doubt that some major wars in this century have been motivated in part by one country's desire to seize another's non-renewable resources, such as fossil fuels or iron ore. For instance, prior to and during World War II, Japan sought to secure coal, oil, and minerals in China and Southeast Asia. But the story is different for renewables like cropland, forests, fish, and fresh water. It is hard to find clear examples from this century of major war motivated mainly by scarcities of renewables.

There are two possible explanations. First, modern states cannot easily convert cropland and forests seized from a neighbor into increased state power, whereas they can quickly use non-renewables like iron and oil to build and fuel the military machines of national aggression. Second, countries with economies highly dependent on renewables tend to be poor, and poor countries cannot easily buy large and sophisticated conventional armies to attack their neighbors. For both these reasons, the incentives and the means to launch resource wars are likely to be lower for renewables than for non-renewables.

The exception, some might argue, is water, in particular river water: adequate water supplies are needed for all aspects of national activity, including the production and use of military power, and rich countries are as dependent on water as poor countries (often, in fact, they are more dependent). Moreover, about 40 per cent of the world's population lives in the 214 river basins shared by more than one country. Thus at a meeting in Stockholm in August, 1995, Ismail Serageldin, the World Bank's Vice President for Environmentally Sustainable Development, declared that the "wars of the next century will be over water," not oil.

The World Bank is right to focus on the water crisis. Water scarcity and pollution are already hindering economic growth in many poor countries. With global water use doubling every 20 years, these scarcities – and the subnational social stresses they cause – are going to get much worse. But Mr. Serageldin is wrong to declare we are about to witness a surge of "water wars."

Wars over river water between upstream and downstream neighbors are likely only in a narrow set of circumstances: the downstream country must be highly dependent on the water for its national wellbeing; the upstream country must be able to restrict the river's flow; there must be a history of antagonism between the two countries; and, most importantly, the downstream country must be militarily much stronger than the upstream country. There are, in fact, very few river basins around the world where all these conditions hold. The most obvious example is the Nile: Egypt is wholly dependent on the river's water, has historically turbulent relations with its upstream neighbors Sudan and Ethiopia, and is vastly more powerful than either. And, sure enough, Egypt has several times threatened to go to war to guarantee an adequate supply of Nile waters.

But more common is the situation along the Ganges, where India has constructed a huge dam – the Farakka Barrage – with harsh consequences on downstream cropland, fisheries, and villages in Bangladesh. Bangladesh is so weak that the most it can do is plead with India to release more water. There is little chance of a water war here between upstream and downstream countries (although the barrage's effects have contributed to the migrations out of Bangladesh into India). The same holds true for other river basins where alarmists speak of impending wars, including the Mekong, Indus, Parana, and Euphrates.

The chronic, diffuse, subnational strife that environmental scarcity helps generate is exactly the kind of conflict that bedevils conventional military institutions. Around the world, we see conventional armies pinned down and often utterly impotent in the face of interethnic violence or attacks by ragtag bands of lightly armed guerrillas and insurgents. As yet, environmental scarcity is not a major factor behind most of these conflicts. But we can expect it to become a far more powerful influence in coming decades because of larger populations and higher resource consumption rates.

Globally, the human population is growing by 1.6 per cent a year; on average, real economic product per capita is also rising by 1.5 per cent a year. These increases combine to boost the earth's total economic product by about 3 per cent annually. With a doubling time of around 23 years, the current global product of $25 trillion should exceed $50 trillion in today's dollars by 2020.

A large component of this increase will be achieved through higher consumption of the planet's natural resources. Already, as the geographers R. Kates, B.L. Turner, and W.C. Clark write, "transformed, managed, and utilized ecosystems constitute about half of the ice-free earth; human-mobilized material and energy flows rival those of nature." Such changes are certain to grow, because of the rapidly increasing scale and intensity of our economic activity.

At the level of individual countries, these changes often produce a truly daunting combination of pressures. Some of the worst affected countries are "pivotal states" – to use the term recently coined in Foreign Affairs by historian Paul Kennedy. These include South Africa, Mexico, India, Pakistan, and China.

India deserves particularly close attention. Since independence, the country has often seemed on the brink of disintegration. But it has endured, despite enormous difficulties, and by many measures India has made real progress in bettering its citizens' lives. Yet, although recent economic liberalization has produced a surge of growth and a booming middle class (often estimated at 150 million strong), India's prospects are uncertain at best.

Population growth stubbornly remains around 2 per cent a year; the country's population of 955 million (of which about 700 million live in the countryside) grows by 17 million people annually, which means it doubles every 38 years and adds the equivalent of Indonesia to its population every 12. Demographers estimate that India's population will reach 1.4 billion by 2025. Yet, already, severe water scarcities and cropland fragmentation, erosion, and salinization are widespread. Fuelwood shortages, deforestation, and desertification also affect sweeping tracts of countryside.

Rural resource scarcities and population growth have combined with an inadequate supply of rural jobs and economic liberalization in cities to widen wealth differentials between countryside and urban areas. These differentials propel huge waves of rural–urban migration. The growth rates of many of India's cities are nearly twice that of the country's population, which means that cities like Delhi, Mumbai, and Bangalor double in size every 20 years. Their infrastructures are overtaxed: Delhi has among the worst urban air pollution in the world, power and water are regularly unavailable, garbage is left in the streets, and the sewage system can handle only a fraction of the city's waste-water.

India's rapidly growing population impedes further loosening of the state's grip on the economy: as the country's workforce expands by 6.5 million a year, and as resentment among the poor rises against those castes and classes that have benefited most from liberalization, left-wing politicians are able to exert strong pressure to maintain subsidies of fertilizers, irrigation, and inefficient industries and to keep statutory restrictions against corporate layoffs. Rapid population growth also leads to fierce competition for limited status and job opportunities in government and education. Attempts to hold a certain percentage of such positions for lower castes cause bitter inter-caste conflict. The right-wing Bharatiya Janata Party capitalizes on upper- and middle-caste resentment of encroachment on their privileges, mobilizing this resentment against minorities like Muslims.

These pressures are largely beyond the control of India's increasingly corrupt and debilitated political institutions. At the district and state levels, politicians routinely hire local gang leaders or thugs to act as political enforcers. At the national level, kickbacks and bribes have become common in an economic system still constrained by bureaucracy and quotas. The central government in Delhi and many state governments are widely seen as unable to manage India's rapidly changing needs, and as a result have lost much of their legitimacy. Furthermore, the 1996 national elections brought a dramatic decline in the Congress party, which has traditionally acted to aggregate the interests of multiple sectors of Indian society. The parties that gained at Congress's expense represent a profusion of narrow caste, class, religious, and regional interests.

The fast expansion of urban areas in poor countries like India may have the dual effect of increasing both the grievances and opportunities of groups challenging the state: people concentrated in slums can communicate more easily than those in scattered rural villages, which might reinforce incipient economic frustrations and, by reducing problems of co-ordination, also increase their power in relation to police and other authorities. There is, however, surprisingly little historical correlation between rapid urbanization and civil strife; and the exploding cities of the developing world have been remarkably quiescent in recent decades.

India shows that the record may be changing: the widespread urban violence in early 1993 was concentrated in the poorest slums. Moreover, although Western commentators usually described the rioting as strictly communal between Hindus and Muslims, in actual fact Hindus directed many of their attacks against recent Hindu migrants from rural areas. B.K. Chandrashekar, a sociology professor at the

Indian Institute of Management, says that "the communal violence was quite clearly a class phenomenon. Indian cities became the main battlegrounds because of massive migrations of the rural poor in the past decades."

Indian social institutions and democracy are now under extraordinary strain. The strain arises from a rapid yet incomplete economic transition, from widening gaps between the wealthy and the poor, from chronically weak political institutions, and – not least – from continued high levels of population growth and resource depletion. Should India suffer major internal violence as a result – or, in the worst case, should it fragment into contending regions – the economic, migration, and security consequences for the rest of the world would be staggering indeed.

Some people reading the preceding account of India will say "nonsense!" As long as market reforms and adequate economic growth continue, India should be able to solve its problems of poverty, population growth, and environmental stress.

The most rigorous representatives of this optimistic position are neo-classical economists. They generally claim that few if any societies face strict limits to population or consumption. Properly functioning economic institutions, especially markets, can provide incentives to encourage conservation, resource substitution, the development of new sources of scarce resources, and technological innovation. Increased global trade allows resource-rich areas to specialize in production of goods (like grain) that are derived from renewables. These optimists are commonly opposed by neo-Malthusians – often biologists and ecologists – who claim that finite natural resources place strict limits on the growth of human population and consumption both globally and regionally; if these limits are exceeded, poverty and social breakdown result.

The debate between these two camps is now thoroughly sterile. Each grasps a portion of the truth, but neither tells the whole story. Neo-classical economists are right to stress the extraordinary ability of human beings to surmount scarcity and improve their lot. The dominant trend over the past two centuries, they point out, has not been rising resource scarcity but increasing aggregate wealth. In other words, most important resources have become less scarce, at least in economic terms.

The optimists provide a key insight that we should focus on the supply of human ingenuity in response to increasing resource scarcity rather than on strict resource limits. Many societies adapt well to scarcity, without undue hardship to their populations; in fact they often end up better off than they were before. These societies supply enough ingenuity in the form of new technologies and new and reformed social institutions – like efficient markets, clear property rights, and rural development banks – to alleviate the effects of scarcity.

The critical question then is, what determines a society's ability to supply this ingenuity? The answer is complex: different countries, depending on their social, economic, political, and cultural characteristics, will respond to scarcity in different ways and, as a result, they will supply varying amounts and kinds of ingenuity.

Optimists often make the mistake of assuming that an adequate supply of the right kinds of ingenuity is always assured. However, in the next decades population growth, rising average resource consumption, and persistent inequalities in resource access guarantee that scarcities of renewables will affect many regions in the developing world with a severity, speed, and scale unprecedented in history. Resource substitution and conservation tasks will be more urgent, complex, and unpredictable, driving up the need for many kinds of ingenuity. In other words, these societies will have to be smarter – socially and technically – in order to maintain or increase their wellbeing in the face of rising scarcities.

Simultaneously, though, the supply of ingenuity will be constrained by a number of factors, including the brain drain out of many poor societies, their limited access to capital, and their chronically incompetent bureaucracies, corrupt judicial systems, and weak states. Moreover, markets in developing countries often do not work well: property rights are unclear; prices for water, forests, and other common resources do not adjust accurately to reflect rising scarcity; and thus incentives for entrepreneurs to respond to scarcity are inadequate.

Most importantly, however, the supply of ingenuity can be restricted by stresses generated by the very resource crises the ingenuity is needed to solve. In Haiti, for example, severe resource shortages – especially of forests and soil – have inflamed struggles among social groups, struggles that, in turn, obstruct technical and institutional reform. Scarcities exacerbate poverty in Haitian rural communities and produce significant profit opportunities for powerful elites. Both these changes deepen divisions and distrust between rich and poor and impede beneficial change. Thus, for example, the Haitian army has blocked reforestation projects by destroying tree seedlings, because the army and the notorious Tonton Macoutes fear such projects will bring disgruntled rural people together and threaten their highly profitable control of forest resource extraction.

Similar processes are at work in many places. In Bihar, India, which has some of the highest population growth rates and rural densities in the country, land scarcity has deepened divisions between land-holding and peasant castes, promoting intransigence on both sides that has brought land reform to a halt. In South Africa, scarcity-driven migrations into urban areas, and the resulting conflicts over urban environmental resources (such as land and water), encourage communities to segment along lines of ethnicity or residential status. This segmentation shreds networks of trust and debilitates local institutions. Powerful warlords, linked to Inkatha or the African National Congress, have taken advantage of these dislocations to manipulate group divisions within communities, often producing horrific violence and further institutional breakdown.

Societies like these may face a widening "ingenuity gap" as their requirement for ingenuity to deal with scarcity rises while their supply of ingenuity stagnates or drops. A persistent and serious ingenuity gap boosts dissatisfaction and undermines regime legitimacy and coercive power, increasing the likelihood of widespread and chronic civil violence. Violence further erodes the society's capacity to supply ingenuity, especially by causing human and financial capital to flee. Countries with a critical ingenuity gap therefore risk entering a downward and self-reinforcing spiral of crisis and decay.

A focus on ingenuity supply helps us rethink the neo-Malthusian concept of strict physical limits to growth. The limits a society faces are a product of both its physical context and the ingenuity it can bring to bear on that context. If a hypothetical society were able to supply infinite amounts of ingenuity, then that society's maximum sustainable population size and rate of resource consumption would be determined by biological and physical laws, such as the second law of thermodynamics. Since infinite ingenuity is never available, the resource limits societies face in the real world are more restrictive than this theoretical maximum. And since the supply of ingenuity depends on many social and economic factors and can therefore vary widely, we cannot determine a society's limits solely by examining its physical context, as neo-Malthusians do. Rather than speaking of limits, it is better to say that some societies are locked into a "race" between a rising requirement for ingenuity and their capacity to supply it.

In coming decades, some societies will win this race and some will lose. We can expect an increasing bifurcation of world into those societies that can adjust to population growth and scarcity – thus avoiding turmoil – and those that cannot. If several pivotal states fall on the wrong side of this divide, humanity's overall prospects will change dramatically for the worse.

26 VACLAV SMIL

"Some Contrarian Notes on Environmental Threats to National Security"

from *Canadian Foreign Policy* (1994)

As a natural scientist with a long-standing interest in interdisciplinary research I have welcomed the recent discovery of global environmental change by political scientists. Indeed, I have marveled at the speed with which the concerns about potential violent conflicts engendered by severe environmental degradation rose to become a leading contender to replace the threat of nuclear war as the ultimate global nightmare. Or, to look at it from a different angle, I have admired the adroitness with which many practitioners of the discipline, which has lost its main feeding (and funding) ground with the dissolution of the Soviet empire, repositioned themselves to forecast a new Apocalypse – and hence to attract reoriented granting largesse driven by a new global angst.

And I must confess that I have participated in this shift both directly – by trying, upon invitation, to find explicit links between environmental degradation and conflict and by speaking at fear-tinged meetings in far-flung places – and indirectly, by gathering and evaluating plenty of worrisome evidence on environmental decline in my writings.[1] Consequently, I approach the critique of links between population growth, environmental degradation and conflict as something of an insider. I believe that these concerns have been long overdue, and that this new field of inquiry needs a great deal of interdisciplinary research. But there are at least three major reasons why I cannot embrace this fashionable preoccupation with the zeal of a novice convert.

The first difficulty lies in the unmistakably catastrophic tilt of this new concern. Rather than being an impartial search for understanding, it appears to be – too often for my comfort – a quest for illustrating and affirming preconceived ideas about the dim future of civilization. This chant merely adds to a venerable chorus of environmental catastrophists who have been invoking images of inexorable famines, epidemics, economic collapse and social disintegration since the late 1960s.[2] When seen within this well-established perspective, the only major distinguishing tilt in the recent work of political scientists is their insistence on capping these declines with violent endings. Implausibilities of this linkage in many real world situations have already been analyzed in some detail.[3]

What places this new approach squarely within the old catastrophist paradigm is its insistence that current trends lead almost always to tomorrow's scary scenarios: too many people degrading the planet's environment and creating all sorts of scarcities can bring only decline, instability – and violence. These are very ahistorical and unbalanced views. Looking just five generations back at Europe and North America around 1900 it would have been inevitable to conclude that the continuation of trends then prevailing since the beginning of the nineteenth century will result in massive starvation (recycling of organic wastes was insufficient to provide nutrients for higher yields), farmland shortages (all accessible land was converted to cropping), virtual complete deforestation (rates of American tree cutting were, adjusted for population, faster in the nineteenth-century US than in twentieth-century Brazil), and unbearable air pollution from rising coal combustion.

Looking just at the last decade we can see how totally unforeseen socio-economic changes (privitization of Chinese agriculture and industries, the collapse of European communism) have helped to lower environmental impacts by drastically cutting energy intensities of reforming economies and by beginning to introduce more sensible resource pricing. In all of these cases larger populations, degradative trends and growing scarcities acted as useful stimuli for better solutions. Systematic appraisals of existing inefficiencies and malpractices show how huge are the reservoirs of waste and mismanagement which we can tap in our effort to manage even the most worrisome trends.[4] Rapid population growth, increasing scarcity of some environmental goods, and undermining of environmental services are more than just ingredients of catastrophic sermons: they have essential factors in finding adaptive solutions.

The second set of pitfalls in the current environment-security writings is in rushing to judgment on the basis of often exceedingly wobbly information. Prophets of new security fears have not studied the intricacies of environmental change long enough to appreciate many inherent weaknesses in our understanding of the biospheric realities. For example, I have recently participated in an international meeting on environment and security where the perils of desertification received much attention – and where none of the assembled political scientists seemed aware of the fact that desertification is such an ill-defined concept that impeccably documented papers and books have been written showing that the phenomenon is largely a creation of United Nations (UN) bureaucrats in search of a self-justifying mission.[5]

No less importantly, the new catastrophists searching for supporting numbers do not look beyond the readily available environmental data sets in standard compendia to discover the enormous weaknesses and dubious nature of many listed variables.[6] To mention just one signal example: standard statistics of the world's arable land are profoundly wrong. Recent cadastral measurements in the Nepali hills show four times as much farmland as is listed in government inventories, and the Chinese are now officially admitting that their arable land total is close to 130 million hectares, rather than the official figure of 95 million.[7] Clearly, disparities of such magnitude will have a profound effect on assessment of pending environmental scarcities.

Finally there is the overreaching – and yet at the same time constraining – embrace of new approaches to security which leads to inevitable attribution problems. The tempting line of underlying reasoning – new security concerns must include matters of environmental change because some of the degradative processes or scarcities are, or soon will be, causes of conflicts ranging from diplomatic disputes to mass violence – is hitched to symptoms, not causes. Unraveling and understanding the causes is much more challenging – and correspondingly more profitable.

Environmental change subsumes natural events (ranging from glaciation cycles unfolding over 103 years to fluctuation in solar activity discernible in decades) and human interferences. Very frequently we cannot be sure about their relative contributions, and hence about effective responses. Where we can make an unequivocal attribution to human action we find that such changes and scarcities are the consequences of economic strategies and political emphases springing from complex mixtures of material and ideological aspirations – not preordained responses to larger populations and increasing environmental scarcities.

Hence it would be much more appropriate – both in order to illuminate the roots of undesirable environmental changes and to offer effective solutions – if the new security concerns would embrace the matters of agricultural subsidies, budgetary policies, commodity pricing, consumer preferences, individual and corporate taxation, savings incentives, technical innovation and trade barriers. I suspect that such a redefinition would be too far reaching even for the most avid reformers of security studies.

Students of conflict should be encouraged to include environmental change in their long term perspectives. At the same time, they should eschew headline-generating catastrophism, and they should not perceive larger populations and resource scarcities as unmitigated agents of cataclysmic decline and impending conflict. And they should not overstate the link between environmental change and social conflict by misinterpreting the former on the basis of inadequate understanding and questionable data while exaggerating the latter by suggesting all too readily the possibility, even inevitability, of violent outcomes.

NOTES

1 V. Smil *Potential Environmental Conflicts Involving Countries of the North Pacific, North Pacific Cooperative Security Dialogue* (Toronto: York University, 1992); V. Smil "Environmental Change as a Source of Conflict and Economic Losses in China," Occasional paper series of the project on Environmental Change and Acute Conflict Vol. 2. (Cambridge, MA: American Academy of Arts and Sciences, 1992) pp. 5–39; V. Smil *China's Environment: An Inquiry into the Limits of National Development* (Armonk, NY: M.E. Sharpe, 1993); and V. Smil *Global Ecology, Environmental Change and Social Flexibility* (London: Routledge, 1993).

2 Reading the earliest wave of these prophesies, especially the writings of Paul Ehrlich, is perhaps most illuminating; all of us should have been dead by now! See Paul Ehrlich *The Population Bomb* (New York, NY: Ballantine, 1968) and Paul Ehrlich "Eco-Casatrophe!" *Ramparts* 8, 3 (1969) pp. 24–28.

3 Daniel Deudney, "The case against linking environmental degradation and national security" *Millennium* Vol. 19 (1990) pp. 461–476.

4 For numerous examples see V. Smil "How many people can the Earth feed?" *Population and Development Review* Vol. 20 pp. 255–292.

5 C.J. Tucker *et al.* "Expansion and contraction of the Sahara Desert from 1980 to 1990" *Science* No. 253 (1990) pp. 299–301; U. Hellden "Desertification: time for an assessment" *Ambio* Vol. 20 (1991) pp. 372–383; D.S.G. Thomas and N.J. Middleton *Desertification: Exploding the Myth* (Chichester: John Wiley & Sons, 1994).

6 This is especially true about most national totals, averages and rates listed in FAO's *Production Yearbook* and about many environmental statistics in *World Resources* compiled every two years by the World Resources Institute. G.J. Gill *O.K., The Data's Lousy, But its All We've Got (Being a Critique of Conventional Methods)* (London: HED, 1993).

7 F.W. Crook "Underreporting of China's cultivated land area: implications for world agricultural trade" *China Agriculture and Trade Report* RS-93, pp. 33–39.

27 GARETH PORTER

"Environmental Security as a National Security Issue"

from *Current History* (1995)

The distinguishing characteristic of post-cold world politics is the absence of what international security analyst Lawrence Freedman calls the "strategic imperative" – the motivation among the major states to compete for military power. As military threats have subsided or disappeared, other threats, especially environmental ones, have emerged with greater clarity. It has thus become possible to argue persuasively that environmental threats are an essential component of national or international security.

This idea, often expressed by the term "environmental security", has been adopted by the Clinton administration as part of United States national security doctrine. But it remains controversial, both conceptually and politically. And a strong isolationist trend brought into Washington by the new Republican controlled Congress threatens to reverse the progress already made in redefining United States national security.

NATIONAL SECURITY AND ENVIRONMENTAL SECURITY

The term "national security" has never had a precise definition, even during the Cold War. In the post-Cold War world divergent concepts of security have been advanced by theorists and statesmen, each of which can be categorized on the basis of three major dimensions:

- whether it assumes that security is based primarily on conflict or cooperation
- the unit of analysis (individual, national, or global)
- the threats with which it is concerned

The traditional concept of national security that evolved during the Cold War viewed security as a function of the successful pursuit of interstate power competition. It took the sovereign state as the exclusive unit of analysis, and was concerned only with military threats or those related to an "enemy." National security was also used to convey the idea that a particular set of problems was most important to the state, and required the mobilization of a high level of material and human resources.

Environmental security represents a significant departure from this approach to national security. It addresses two distinct issues: the environmental factors behind potentially violent conflicts, and the impact of global environmental degradation on the well-being of societies and economies. The idea that environmental degradation is a security issue when it is a cause of violent conflict appears to be consistent with the traditional definition of national security. However, proponents of environmental security emphasize that environmental degradation is the result of impersonal social and economic forces, and requires cooperative solutions. This focus on threats that do not involve an enemy state or political entity disturbs many theorists and practitioners of national security, for whom the only issues that should be viewed as "security" issues are those that revolve around conflict itself.

More broadly, environmental security is concerned with any threat to the well-being of societies and their populations from an external force that can be influenced by public policies. Proponents of environmental security argue that increasing stresses on the earth's life-support systems and renewable natural resources have profound implications for human health and welfare that are at least as serious as traditional military threats.

Whether environmental security is compatible or in conflict with an exclusive focus on the security of the nation-state is a question on which proponents have expressed different views. Some consider environmental threats within a framework of national security, although they also stress the inadequacy of traditional nation-state responses to global environmental challenges based on concerns with national sovereignty. Others argue that environmental security is inherently global rather than national in character, since environmental threats affect all humanity and require coordinated action on a global scale.[1]

Environmental security deals with threats that are not only the unintended consequences of social and economic activities, but that also develop very slowly compared with military threats. Thus the time horizons it requires for policy planning are extremely broad. While some programs aimed at reducing population growth rates can achieve significant results in a decade or two, it takes far longer for declining birth rates to affect natural resource management. A typical program to reverse the environmental degradation of an entire ecosystem and to rehabilitate that ecosystem can take as long as 50 years to produce the desired results. Policies to restore the ozone layer will take up to 10 years to take effect, and those to produce climate change could take even longer. These time horizons represent a major obstacle to integrating environmental security into policy making processes, since political systems are not organized to look that far ahead.

THE GLOBAL DIMENSIONS OF ENVIRONMENTAL THREATS

The case for environmental security rests primarily on evidence that there has been serious degradation of natural resources (freshwater, soils, forests, fishery resources, and biological diversity) and vital life support systems (the ozone layer, climate system oceans, and atmosphere) as a result of the recent acceleration of global economic activities. These global physical chances could have far-reaching effects in the long run.

The thinning of the stratospheric ozone layer because of the accumulation of certain man-made chemicals could have a severe impact on human health and nutrition. It is estimated that if the 1987 Montreal Protocol phasing out ozone-depleting chemicals had not been signed and strengthened by amendments, chlorine levels by the year 2020 would be six times higher than the level at which significant ozone depletion starts. A 10 per cent ozone loss over North America is expected by the year 2000, and ozone levels 20 to 35 per cent below normal have already been recorded over Siberia and Europe. Although research is still inadequate, there is some evidence that increased exposure to UV-B radiation as a result of the destruction of the ozone layer could damage crops and phytoplankton (the basis of the marine food chain) and reduce human immunity to infectious disease.

Climate warming, from increased concentrations of carbon dioxide and other gases that trap heat in the atmosphere, could alter the fundamental physical conditions of life on the planet. According to the Intergovernmental Panel on Climate Chance (IPCC), an international scientific body, a doubling of atmospheric concentrations of these gases (compared with those of the previous century) could increase average global temperatures by 1.5 to 4.5 degrees centigrade, or 2 to 9 degrees Fahrenheit. The high end of that estimate would be roughly the same as the total temperature rise since the peak of the last Ice Age. Such increases could raise sea levels by about one to one and a half feet by 2050, flooding coastal lowland plains and wetlands worldwide and increasing storm tides and the intrusion of saltwater into estuaries and groundwater. Among the other physical changes that could be triggered even by a modest warming of temperatures are increased frequency and severity of hurricanes, droughts, and flooding. And increased weather extremes that accompany climate warming may already be contributing to an increase in and geographical redistribution of vector-borne diseases.

Biological diversity is being lost at a rate estimated at 2 per cent to 10 per cent of all species per decade. This rate of loss is unparalleled since the last mass extinction of species 65 million years ago. Biological diversity is one of humankind's chief resources for coping with diseases and other unexpected natural chances: its loss would dramatically reduce the chances of discovering natural substances that might hold the cure for existing and future diseases. And genetic uniformity of the world's foodcrop varieties poses the risk that diseases or pests that develop resistance to pesticides could destroy a large proportion of the crops on which most of the world's population depends. The genes of relatives of those varieties that grow in the wild, which will be needed to respond to such threats to food security, are now threatened by deforestation and conversion of land to agriculture.

The health of the world economy itself depends on avoiding the depletion of renewable natural resources. The degradation of cultivated land threatens to reduce agricultural productivity in large areas of the developing world. It has been estimated that 11 per cent of the earth's total vegetated surface has already suffered moderate to extreme soil degradation because of deforestation, overgrazing, or unsound agricultural practices.

Developing countries have already suffered significant reductions in productivity because of soil loss, deforestation, and other forms of environmental degradation: Indonesia's loss has been estimated at 4 per cent of GDP and Nigeria's at nearly 18 per cent of GDP. If rates of economic loss from environmental degradation continue to rise in key developing countries in future decades, the health of the entire world economy will be affected.

Each of these environmental threats to global being is subject to significant empirical and scientific uncertainty: neither the actual increased exposure to UV-B from the thinning of the ozone layer nor degree of harm it will do to plants, animals, or humans is calculable; neither the eventual increase in global average temperatures from a given level of greenhouse gas emissions nor the consequences for weather patterns, disease, crops, or sea level rise can be known. The actual rate of species loss is still unknown and the impact of the loss of a given proportion of species cannot be easily gauged. Finally, there is no reliable global data on the actual rate of land degradation, nor can the impact of land degradation on future food production be predicted with any confidence.

The uncertainties associated with these environmental threats are comparable, however, to those associated with most military threats that national security establishments prepare for. Military planning is based on "worst-case" contingencies that are considered relatively unlikely to occur, yet military preparations for such contingencies are justified as a necessary insurance policy, or "hedge" against uncertainty. But in the United States, for example, the potential harm that global environmental degradation poses to the health and livelihoods of Americans is arguably worse than those posed by most military security threats for which the country is prepared.

ENVIRONMENTAL FLASHPOINTS

The relationship between scarce natural resources and international conflict is not a new issue. But unlike traditional national security thinking about such conflicts, which focus primarily on nonrenewable resources like minerals and petroleum, the environmental security approach addresses renewable resources – those that need not be depleted if managed sustainably.

Conflicts involving renewable natural resources are of two kinds: those in which resource depletion is the direct objective of the conflict, and those in which it is an indirect cause of the conflict. Freshwater resources and fish stocks are the clearest examples of renewable resources that have been the direct objective of potentially violent international conflicts.

Conflict over the shared waters of international rivers has long been of interest to national security planners. The United States intelligence community estimated in the mid-1980s that there are 10 places in the world – half in the Middle East – where war could break out because of dwindling freshwater supplies. Especially dangerous are the Jordan River, which is shared by Jordan, Israel, and Lebanon; the Nile, shared by Egypt, Ethiopia, and Sudan; and the Euphrates, shared by Iraq, Turkey, and Syria.

International conflicts over fishing grounds have been frequent in recent decades. Thirty such conflicts were reported last year alone, including several in which force was used. Without any international agreement on managing fish stocks that straddle the exclusive economic zones of states or that migrate between EEZs, or between coastal zones and the high seas, even normal fluctuations in stocks increase interstate competition over fishery resources. But with more than half the world's major maritime fisheries already in serious decline from overfishing and the rest exploited up to or beyond their natural limits, the potential for political and even military confrontation is growing. Coastal states, such as Canada, Chile, and Russia, whose fish catch in their own EEZs is reduced by the operations of distant fishing fleets in the adjoining high seas, have threatened to use force to stop ships that they find overfishing, even outside their EEZs.

Shared freshwater resources and maritime fisheries are good examples of issues that involve more than traditional competition for control over natural resources. Equitable sharing, of the Jordan River, for example, will not be enough to prevent Jordan and Israel from running short of water: it has been projected that, by the year 2000, Israel's demand for water will exceed available supply by one-third, while Jordan's demand will exceed its supply by one-fifth. Sustainable water-use plans for both states must be formulated as part of water sharing agreements, including provisions for greater efficiency in water use by eliminating water subsidies, choosing less water-intensive crops, reducing water losses in irrigation, and minimizing water pollution.

The primary reason for the decline in maritime fisheries is too many fishing boats with too much modern fishing technology, such as bigger nets, electronic fish detection equipment, and mechanized hauling gear. To protect the world's fish stocks from further depletion, the international community will have to establish strict limits on entry into the fishing industry; establish binding standards on capitalization of fishing fleets, excessive fleet size, and inappropriate fishing gear; and set a numerical limit on the total catch and the percentage of the total catch per entrant. Without such a tough, enforceable international treaty, traditional power tactics in pursuit of control over fisheries resources will do nothing to protect a state's interest in continued access to the resource.

The environmental security approach thus offers a clear alternative to traditional security thinking about international conflicts over renewable natural resources. It suggests that the key problem is to conserve the resource in order to maintain adequate supplies well into the future, rather than trying to control more of a resource that is being depleted. In the case of shared rivers, conservation efforts will involve two or three states; with maritime fisheries, it will require global agreement.

A distinctly different issue is the indirect effect of environmental degradation on violent domestic conflicts. This has been brought into focus by civil wars, the collapse of state structures, and major humanitarian crises in Africa. Eight African countries are already experiencing significant humanitarian crises (defined as putting at least 1 million people at risk) related to domestic strife, or are at risk of experiencing them. The annual costs to the United States of foreign disaster and humanitarian crises increased from less than $25 million in the latter half of the 1980s to nearly $1.8 billion in 1994 because of the growing frequency and intensity of such crises and the need to use military forces to prevent or reduce human suffering.

Both Somalia and Rwanda, according to some analysts, illustrate the role that environmental deterioration has played in civil violence in Africa. In Somalia the direct cause of the violence was a power struggle among clan leaders who were heavily armed with Western weapons. But it is argued that the conflict was also spurred by economic change that had depleted renewable resources. External assistance from the United States and the World Bank helped drive the process by supporting the production of bananas, sugar, and livestock for export, which depleted the soil in the river valleys and led to overgrazing and desertification in already arid lands. Unsustainable development, according to this argument, fueled conflict between herders and farmers over access to water and grazing land, which played into clan rivalries.[2]

While the direct cause of the genocidal violence in Rwanda in 1994 was a desperate regime exploiting ethnic fears in order to cling to power, the crisis also had a significant environmental dimension. One of the highest population growth rates in the world – 3.7 per cent annually by the 1970s – and relatively severe soil degradation contributed to reduced agriculture production and food availability, especially in areas with steep slopes or acidic soils. Agricultural decline was a key element in political protests by both Hutu and Tutsi farmers against the Hutu regime of President Juvenal Habyarimana in the early 1990s. The regime's response was to adopt a deliberate strategy of ethnic hatred against all Tutsi in order to rally Hutu behind the government.[3]

Some analysts contend that the problem of states dissolving in violence and chaos because of a combination of socioeconomic inequality and environmental degradation is not confined to Africa. Thomas Homer-Dixon, the coordinator of a research project on environment and violent conflict, has concluded, on the basis of a number of case studies that include China, the Philippines, and Peru, that conflicts fueled in part by the degradation of renewable resources (cropland water, forests, and fish), population growth, and unequal resource distribution are likely to become more frequent in future decades as more of these resources are depleted. He has suggested that a growing number of societies experiencing such conflicts will either fragment or become more authoritarian.[4]

THE UNITED STATES VIEW OF ENVIRONMENTAL SECURITY

In the wake of the Cold War's end, the United States has moved officially to redefine national security to encompass environmental threats. The Bush administration was the first to acknowledge environmental security as part of overall United States security. A 1991 presidential document summarizing United States national security policy defined United States national security objectives to include "assuring the sustainability and environmental security of the planet. . . ."

The Clinton administration has integrated environmental security even further into its national security policy. Official interest in the issue of "failed states" was spurred by a February 1994 article in *The Atlantic* by journalist Robert Kaplan, which popularized the idea that "chaos" will emerge as the main threat to global security in future decades. Weaving together personal reportage on West Africa and other developing regions with academic analyses, Kaplan declared that population growth and resource depletion would prompt mass migrations and incite group conflicts in Egypt and on the Indian subcontinent. The Kaplan article was read and discussed among Clinton administration officials, including Vice President Al Gore and President Bill Clinton himself.

In remarks to a forum on global issues last May, Clinton referred to civil wars in Africa and elsewhere that were "caused not only by historic conflicts but also by ... deterioration of not only the economy, but the environment in which those people live." And at a conference on global population in June, Clinton referred to Kaplan's article in describing a stark vision of a future world of overpopulated countries, depleted resources, and extreme divisions of wealth and poverty. Clinton called for a strategy of "sustainable development" as a "comprehensive approach to the world's future." Without uttering the phrase "national security" he appeared to invoke its essence, referring to the need to be "disciplined" and to "order our priorities" in addressing the interrelated global problems of population, health, environment, and equitable economic growth.

The Clinton administration explicitly adopted the concept of environmental security in its 1994 national security document, *A National Security Strategy of Engagement and Enlargement*, which asserts that increasing competition for dwindling renewable resources "is already a very real risk to regional stability around the world." The document also notes that "environmental degradation will ultimately block economic growth." It calls for partnerships between governments and nongovernmental organizations as well as between nations and between regions, and for a "strategically focused, long term policy for emerging environmental risks."

SOURCES OF OPPOSITION

The concept of environmental security has been opposed by some academics, national security specialists, and conservative Congressional leaders. An early criticism was that it muddies the concept of security, co-mingling threats that are related to conflict with those that are not. Similarly, it has been argued that including all forces that threaten well-being within the definition of national security would drain the term of its meaning.[5] Such arguments imply that the traditional definition of national security was intellectually coherent or useful. But proponents of environmental security would argue that the traditional definition of national security distorted perceptions of global realities as well as policy priorities.

Early critics of environmental security also argued that its adoption could result in the militarization of environmental issues, making the agenda vulnerable to manipulation by traditional national security constituencies, especially the military. Because it invokes conflictual images, some have argued, the term suggests that environmental threats are caused by enemies, thus raising the specter of an aggressive and even militaristic approach to environmental problems.

This assumes that environmental security lacks an internal logic that challenges the premises of traditional national security thinking. As suggested earlier, the concept of environmental security directs attention to policy responses that are cooperative, not conflictual, even when the focus is on environmental problems that are the subject of international conflicts. The Clinton administration's acceptance of the environmental security approach clearly has not led to the militarization of environmental policy issues.

Another criticism, raised by some officials and academics, is that the environmental security argument is mainly a means of leveraging changes in budgetary allocations. But one of the functions of the traditional concept of national security was to ensure that sufficient resources were committed to military programs as a matter of highest national priority. Indeed, the enormous disparity between the resources budgeted for military security ($250 billion) and those budgeted for global environment and other problems related to environmental security (less than $5 billion) in fiscal year 1994 makes it clear that there is nothing like a reasonable balance among components of security in the allocation of budgetary resources. So a concept that justifies reallocating some of these resources is quite legitimate.

A final objection directed at environmental security is that environmental degradation and population pressures are not the primary causes of such conflicts. Since environmental and natural resource degradation is always imbedded in larger socioeconomic and political causes of conflict, proponents of environmental security cannot prove that such issues are crucial to the resulting violence. But they can make a persuasive case that relatively modest investments in resource conservation and family planning are justified by the much higher costs of responding to the collapse of states and the resulting human suffering.

The election of a Republican Congress in 1994 brought this argument to the fore. Congressional leaders are now arguing that the preventive measures proposed by proponents of environmental security would not have made any difference in cases such as Somalia and Rwanda. Unfortunately, it is impossible to ascertain what would have happened in Somalia or Rwanda had adequate assistance for sustainable development been provided early enough – that is, 20 to 30 years before the violence. In fact, very little assistance was actually provided to either country to conserve resources or reduce population pressures.

The debate over whether development assistance could help prevent violent conflict could become irrelevant, however, because Republican leaders also assert that neither conflict in the developing world nor global environmental threats should be viewed as significant United States concerns. They argue that the countries likely to suffer violent conflict have no strategic importance to the United States and should not be recipients of United States foreign assistance (except for emergency relief). They also discount the idea of global environmental deterioration, arguing in some cases that it is a fraud promoted by environmentalists with an axe to grind. If the Republican Congress drastically reduces aid for sustainable development and withholds funding for efforts to reduce global environmental threats, it will reverse, in effect, the Clinton administration's embrace of environmental security.

NOTES

1. Jessica Tuchman Matthews, "Redefining Security," *Foreign Affairs*, Spring 1989; and Norman Myers, *Ultimate Security: The Environmental Basis of Political Stability* (New York: Norton, 1993).
2. "Africa's Greenwars: The Ecological Roots of the Starvation in Somalia and the Horn of Africa," Friends of the Earth, February 1993.
3. See Jennifer Olson, "Factors Behind the Recent Tragedy in Rwanda," Department of Geography and the Center for Advanced Study of International Development, Michigan State University, November 1994.
4. See Thomas F. Homer-Dixon, "Environmental Scarcities and Violent Conflict: Evidence from Cases," *International Security*, Summer 1994.
5. See Daniel Deudney, "The Case Against Linking Environmental Degradation and National Security," *Millennium*, Winter 1990; and Matthias Finger, "The Military, the Nation State and the Environment," *The Ecologist*, vol. 21 no. 5 (1991).

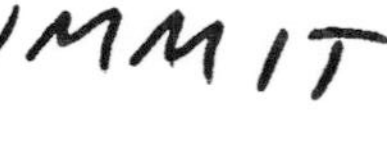

Cartoon 12 Earth summit (by Moir)
The consequences of "Northern" affluence reach worldwide as the spreading flows of pollution in this cartoon suggest. Clearly "the problem" may be in the "North" rather than the "South."
Source: Moir, Cartoon & Writers' Syndicate

28 MATTHIAS FINGER

"The Military, the Nation State and the Environment"

from *The Ecologist* (1991)

Throughout history, the military has viewed the environment as a tool to be used to deny resources to the enemy and as a potent weapon. In recent years, "environmental warfare" (defined by the leading authority on warfare and the environment, Arthur Westing, as "the manipulation of the environment for hostile military purposes") was carried out extensively by the US forces in Vietnam.[1] Herbicides were widely sprayed to destroy forest cover and enemy food crops, and apparently unsuccessful attempts were made to disturb regional weather patterns through cloud seeding.[2]

With the development of military technology and the spread of industrial artifacts such as chemical and nuclear plants, oil wells and large dams, the future potential for environmental warfare is vast. Westing speculates that asteroids could be diverted to strike enemy territory; the electrical properties of the ionosphere could be altered so as to disrupt enemy communications; the ozone layer above enemy territory could be destroyed; and wind, cloud and rainfall patterns could be altered.[3] Rivers could be diverted to deny the enemy access to essential water supplies, and both oceans and rivers could be poisoned with chemicals or nuclear materials. The acoustic or electromagnetic properties of the oceans could be altered and seismic sea waves could be used to destroy coastal and other near-shore facilities. The experience of the Gulf War shows some of the other possible mechanisms for environmental warfare; it also indicates that where the means for environmental warfare are available, they are likely to be used.

Even in peacetime, however, the impact of the military on the environment is considerable.[4] The direct consumption of oil by the US armed forces is about 3-4 per cent of the country's overall oil demand;[5] this percentage could easily triple if indirect consumption of oil is considered, for example in weapons' manufacturing. Michael Renner of the Worldwatch Institute estimates that the military sector's share of oil and energy use worldwide is also about 3–4 per cent and double this if indirect use is included. In some sectors the proportion of oil and energy use by the military is much greater, for example, it consumes about 25 per cent of all jet fuel worldwide.[6]

Non-fuel minerals are also heavily consumed by the military industrial complex. Renner gives an estimate for the use of steel and iron – "the backbones of any military machine" – of about 9 per cent of worldwide consumption.[7] The percentage of military use of other, more strategic minerals is between 5 and 15 per cent, but can rise to up to 40 per cent in the case of certain minerals used in high technology weapons. Renner concludes that, the "worldwide use of aluminium, copper, nickel and platinum for military purposes surpasses the entire Third World's demand for these minerals."[8]

Land and airspace constitute another form of resources in the service of the military. Renner believes that about 0.5–1 per cent of the planet's land mass is used for military bases alone, a percentage which would be increased if the territory occupied by the arms industry is included. This is comparable to the land area of Turkey or Indonesia.[9] This proportion increases still further if indirect land use for manoeuvres and flight exercises is taken into account. Twenty per cent of Canada and 25 per cent of West Germany are covered by such military exercises. These activities usually affect remote or uninhabited areas which are often explicitly set aside as natural wildlife reserves. During periods of war or crisis, whole countries potentially become arms training grounds. On the world's oceans, only coastal territorial limits are respected by naval ships and submarines.

GLOBAL MILITARY POLLUTION

In the opinion of Arthur Westing, "because about six per cent of the combined gross national products of the world's is devoted to military expenditures . . . roughly six per cent of the world's environmental pollution could be attributed to the military sector of the Global economy."[10] However, this is only part of the picture of military pollution; it neglects both the fact that military operations have considerably lower pollution standards than civilian activities, and that pollution from the military is of a qualitatively different nature than that from other sources.

It has been estimated that the operations of the armed forces may account for at least 6–10 per cent of global air pollution and that military-related activities may be responsible for 10–30 per cent of all global environmental degradation.[11] Renner states that the "total military-related carbon [dioxide] release in the US could be as high as 10 per cent."[12] Furthermore, the armed forces of the world are the largest producers of hazardous chemical and nuclear wastes.

Within the United States – where the best data are available – "the military is quite likely the largest generator of hazardous waste. . . . In recent years the Pentagon generated . . . more toxics than the top five US chemical companies combined."[13]

> Everything generates waste. The ships, planes, tanks, rocket launchers, barracks, maintenance yards and storage areas generate solid and liquid hazardous waste and, sometimes, radioactive waste. . . . In addition to the standard array of toxins, there are toxins that are unique to the military, such as propellant packs, explosives shells, explosives, obsolete chemical weapons, infectious waste from biological warfare experiments, and radioactive waste.[14]

Most military bases worldwide are probably heavily contaminated. The US Department of Defense has found almost 15,000 contaminated sites in about 1,600 military bases within the United States alone.[15] It is likely that the pollution problems are even worse on the 375 US bases abroad.

In the US, "99 per cent by volume of all high level radioactive waste and 75 per cent of low level radioactive waste . . . has come from nuclear reactors operated for military purposes, including ship and submarine pollution."[16] The US General Accounting Office admits that information about low level nuclear waste at its military bases is simply unavailable.[17] With regard to both nuclear and chemical waste, "the most severely poisoned areas could prove impossible to 'clean up' or otherwise rehabilitate."[18] Military nuclear pollution, of course, stems not only from the waste generated by nuclear reactors, but also from the mining and processing of nuclear materials.

Weapons tests and accidents have been the most significant military source of global radioactive pollution. From 1945 to 1989, more than 1,800 nuclear bombs were exploded in over 35 sites. Roughly one-quarter of the tests were conducted in the atmosphere.[19] About one-third of the US underground tests may have leaked radiation; the proportion may be higher for French and Soviet tests. In addition, more than 230 nuclear weapons accidents in the USA, the USSR and the UK took place between 1950 and 1988.[20]

THE SPECIAL NATURE OF MILITARY POLLUTION

Perhaps the most important factor differentiating the military from any other polluter is its special relationship with the nation state. The military has historically played a key role in the development of the nation state by securing access to natural resources for national industrial development. According to Westing "the rise of the State might not have occurred without a combination of natural resources limitations and the acceptance of war as an appropriate means for achieving societal aims."[21]

This relationship allows the military to shroud its polluting activities in secrecy and largely avoid environmental regulation and monitoring by national environmental agencies. Secrecy covers all military and military-related operations. The difficulty in finding relevant data for this article is an example of the privilege of secrecy granted to the military by the nation state.

The United States is one of the rare countries where environmental legislation does apply to military facilities and operations on its territory. In the name of national security, however, US military activities and facilities overseas are exempt from any environmental regulation. Even within the US,

> the military establishment has either ignored or obtained exemptions from laws such as the Resource Conservation and Recovery Act and the Clean Water Act that set environmental and public health and safety standards for private industries, individuals and municipalities in the United States.[22]

Even when environmental laws do apply to the military they often cannot be enforced. As Renner notes:

> The Justice Department has prevented the Environmental Protection Agency from suing other federal agencies, from imposing cleanup orders on them without their consent, or from fining them. And it has gone to court several times to preclude state agencies from fining federal installations. In consequence, EPA has had to settle for negotiating "voluntary compliance agreements" of doubtful value with the military.[23]

Even privately operated defence contractors can receive environmental immunity by obtaining a "national security exemption."[24]

In times of crisis or war the few environmental regulations that have gradually come to be applied in limited areas are rapidly waived. During the build-up to the Gulf War, the White House exempted the Pentagon from the legal requirement to carry out environmental assessments of its projects, thus allowing the military to test new weapons and carry out new activities at its bases without the elaborate public review normally required.[25] Obviously, such considerations apply to all nation states; the US has simply been chosen because of the availability of information.

REDEFINING SECURITY NOT SOCIETY

With the end of the Cold War, one might have expected the power of the military to decrease. This has not been the case. On the contrary, both national governments and the military have seized on public concerns over environmental degradation to give the military a new *raison d'être*, primarily by defining environmental degradation as a threat to national security.

Literature about "environmental" or "ecological" security has proliferated since 1987 when the UN General Assembly first introduced the concept.[26] But the phrase can be and has been interpreted in two very different ways. For those in the peace and development movements, the reference point is the individual. Environmental degradation – like unemployment, poverty, racism, authoritarian power structures and the military – are all, it is said, threats to the "security" of individuals.

For others, however, security is defined solely with regard to the nation state.[27] In the past, the argument goes, states defined their security in military terms. Now, however, states must recognize that they are all dependent upon the biosphere: the term 'national security' must be enlarged to include 'environmental security." According to Renner:

> National security is a rather meaningless concept if it does not encompass the preservation of liveable conditions on the Earth. Indeed, environmental degradation may imperil a nation's most fundamental aspects of security by eroding the natural support systems on which all human activity depends.[28]

Historically, this approach is inspired by the

threat of nuclear war. As Joe Clark, Canadian Secretary of State for External Affairs, stated before the 44th Session of the UN General Assembly:

> The environment is emerging as the most important international challenge of the remainder of this century and the next. In a very few years the environment will be seen as a threat to human existence in the same way as nuclear war has been regarded in the past. It is now a challenge to *national survival* (emphasis added).[29]

What has to be managed, according to this approach, is not so much environmental change and degradation, but rather the risks they pose to the nation state system. Threats to environmental security are thus only addressed when they threaten the core of national security. Concrete examples of "international environmental risk management" include the Partial Test Ban Treaty, the Intermediate Nuclear Forces Treaty, the Montreal Protocol on Substances that Deplete the Ozone Layer, and the Basel Convention on the Control of Transboundary Movements of Hazardous Waste and their Disposal.

The implicit model behind this approach is military: states collaborate in reaction to threats and combat them in a military-like manner. The Club of Rome, for example, has proposed the creation of a "UN Ecological Security Council." It is conceivable that such security councils would use military force or other coercive means to force recalcitrant states or other bodies to comply with international risk management agreements.

The basic weakness of this model is that it only becomes applicable once the environmental problem in question has become sufficiently urgent to pose a security threat to more than one state. In addition, it assumes that the common security threat can be isolated in time and space, and that identifiable causes for it can be found. As a result, it tends to deal with symptoms, rather than with fundamental causes. The military, the state, science and technology are not considered as part of the problem: on the contrary, they are considered to be effective tools with which to fight the common environmental security threat.

TWO DIFFERENT TYPES OF SECURITY

The term environmental security seeks to overcome the distinction between the interests of the individual and the interests of the nation state. Security, for the individual, is a matter of perception; it is subjective but nonetheless absolute at a certain moment of a person's life and in a certain socio-cultural context. The individual can feel more or less secure, and this feeling of security can depend upon family relationships, economic factors and social, cultural and environmental conditions. The extent to which the nation state is responsible for providing the individual with security varies from one country to another. For instance, in socialist countries "job security" is, or at least until recently was, considered a part of the nation state's responsibility towards the individual.

The security of individuals – even when provided by states – is epistemologically different from the security of nation states. States derive their security from their perceived relationship with other states; their security is relative not absolute. It is therefore perfectly conceivable to have an absolute increase in the threats to nation states as a whole (for example from global environmental degradation), but if these threats are equally distributed, and do not affect the equilibrium of the nation state system, this absolute increase in threats will not translate into a decrease in national security.

THE IDEOLOGICAL FUNCTION OF "ENVIRONMENTAL SECURITY"

Facing a common enemy - earlier this was another nation state, but now, we are told, the common enemy is the degradation of the environment - individuals and states supposedly have common security interests. In Renner's view, "military, economic and ecological developments increasingly seem to dictate a global community of interests."[30] And according to Buzan, "the concept of security binds together individuals, states and the international system so closely that it demands to be treated in a holistic perspective."[31]

This identification of the security interest of

the individual with those of the state is intellectually flawed when applied to global environmental change and degradation. There are at least three reasons for this:

- It assumes that individuals and states are affected by environmental degradation in the same way, which is demonstrably not the case. It therefore suggests that states and their citizens have the same interests in addressing global environmental change and therefore can and must collaborate in order to do so.
- It is based on the "American model of democracy," where the interest of the majority of the citizens is believed to be reflected by state policy. But other political systems do not conceive democracy as the articulation of individual interests. How do the national security interests of a military dictatorship, for example, reflect the interests of the country's citizens?
- It implicitly assumes that a worldwide coalition of individuals against environmental degradation would be identical with a worldwide coalition of nation states pursuing the same purpose. Again, because of the different ways environmental degradation affects individual security and national security, coalitions of individuals and coalitions of states – at least in environmental matters – are two different things.

The idea of "environmental security" – the context within which global environmental degradation is currently approached on an international level – blurs this difference of interests, deliberately ignores the different epistemological nature of individual and state security interests, and treats environmental degradation in a conceptual framework of *military* defence *against environmental* threats.

GLOBAL ENVIRONMENTAL RESOURCES MANAGEMENT

Framing environmental politics within the general context of global environmental security implies that environmental politics must be global in nature, which automatically leads to nation states being cast as the major actors in any solution to the global environmental crisis. Just as self-serving to the interests of the state is the new accent on global environmental resource management. For humanity to have a decent future, the argument goes, development must be sustained and for this to be possible the management of resources must become more efficient and be moved to a global level.

This rather idealistic approach is taken by the World Commission on Environment and Development.[32] The WCED conceives of global environmental resources management as a collective endeavour within an organizational structure that is probably best qualified as a "superstate," that is as a nation state on a global level. This global state is modelled after the American model of democracy, where every individual and every collective actor is supposed to have the right and the possibility to lobby for his, her or its interests. It is assumed that national governments, NGOs, corporations, scientists and individuals all have a common interest in managing the worldwide pool of resources; and that all their interests can be satisfied by "sustainable development."

There are several problems with this approach, the most important being that the military is largely ignored and the nation state is reduced to a simple actor comparable to individuals, multinationals and NGOs. Indeed it is likely that global environmental resources management as conceived by WCED would only "work" if (1) resources are available to be exploited; (2) nations states are not restricted in their national development; and (3) the militaries of the world are not threatened either as polluters or as consumers of strategic resources. The more important the military-industrial complex is within a country, the more likely it is that that nation state will act as a protector of its military rather than as a protector of the biosphere. For instance, the United States, with its huge military-industrial complex, has either vetoed or substantially watered down every major international environmental agreement. As Sand reports,

> starting with the 1983 Cartagena Convention for the Protection and Development of the Marine Environment of the Wider Caribbean Region, the US State Department introduced a new variety of dispute-settlement clauses in all UNEP conventions that reserve each party's right to block third

> party adjudication while leaving open an option to waive the veto right upon signing the treaty.[33]

The main justification cited for this is "national security."

Unless the military and its special relationship to the nation state are explicitly addressed, global environmental resources management can only be successful as long as some sort of global economic growth – and therefore profits for all parties – results from it. But, economic growth in its present form is clearly unsustainable.

ENVIRONMENTAL DEGRADATION AND MILITARIZATION

As the responses to "the global environmental crisis" described above cannot address its root causes, global environmental degradation will progress, and environmental threats to human society will continue to grow. At certain times, nation states are likely to collaborate in order to manage (in a military fashion and with the help of the military, such as at Chernobyl) specific threats as far as they can be addressed with conventional problem-solving approaches using science, technology and "rational management."

According to Thomas Homer-Dixon, international conflicts will become more likely as increased environmental stress and scarcity make states more unstable.[34] Other researchers make similar arguments. Janet Welsh-Brown of the World Resources Institute believes that the "accelerating deterioration of the resource base, combined with rapid growth of populations that depend more directly than most on natural systems, threatens the economic and political stability of countries vital to US interests."[35] Others believe that the creation of "winners and losers," as environmental change and degradation affect the economic and political stability of different nation states to differing degrees, will be likely to lead to conflicts.[36]

Daniel Deudney, on the other hand, does not believe that environmental change and degradation will cause international conflicts.[37] Although he agrees with Homer-Dixon that declining domestic living standards have the potential to lead to international conflict because they alter the relative power of states, Deudney argues that with modern (especially nuclear) weapons a country can be poor and still have a strong military capability. If it is true that the military of each nation state can be maintained at a relatively low cost, then it is highly likely that some sort of balance of power can be maintained despite continuous environmental degradation. In other words, the nation state system can function relatively smoothly without states necessarily perceiving a need to address environmental change and degradation unless forced to do so by their own citizens. Therefore the present high degree of global militarization can and will be maintained.

Continued militarization will prevent the global environmental crisis from being addressed other than by "international environmental risk management," where the military can appear environmentally "useful." Thus, environmental degradation, whether or not it leads to more conflict between states, will increase the relative importance of the military-industrial complex within each state, which in turn will perpetuate military pollution, which will raise global environmental security concerns and so further strengthen the military.

Global environmental change and degradation can only be effectively addressed if this vicious circle is broken. The military must be addressed as a cause and not a cure of global environmental problems. In the long run, the industrial-military complex must be dismantled. This is a *sine qua non* for effectively dealing with the entire global environmental crisis. If we delay taking action on this, worldwide militarization will progress, thus diminishing our future options for finding a way out of the crisis.

NOTES

1 Westing, A. "Environmental Warfare: An Overview", in A. Westing (ed.) *Environmental Warfare: A Technical, Legal and Policy Appraisal*, Taylor and Francis, London, 1984, p. 3.

2 Stanford Biology Study Group. "The Destruction of Indochina", *Instant Research on Peace and Violence* 2, 1972, pp. 2–8; Westing, A.

Warfare In a Fragile World. Military Impact on the Human Environment, Taylor and Francis, London, 1980, p. 360.

3 Westing, op. cit. 1, pp. 4–9.

4 Information on military resource consumption and pollution is difficult to obtain; most of what is available refers only to the US military. In gathering information on this topic I have benefited from the precious assistance given by Dan Cook Hoffman and Zoltan Bedy whom I would like to thank.

5 Renner, M. "Assessing the Military's War on the Environment", in Lester Brown *et al. State of the World 1991*, Norton, New York, 1991, pp. 137–138.

6 The military consumes 27 per cent of jet fuel in the USA, 34 per cent in the USSR and 50 per cent in Germany.

7 Renner, op. cit. 5, p. 140.

8 Ibid., loc. cit.

9 Westing, op. cit. 5, p. 143.

10 Westing, op. cit. 1, p. 349.

11 Quoted in Renner, op. cit. 5, p. 143.

12 Ibid., loc. cit.

13 Renner, op. cit. 5, p. 143.

14 Bloom, S. "Military Toxic Waste: The Pentagon's Pandora's Box", *The Mobilizer/Weapons Facilities Network Bulletin* Fall 1989, p. 11.

15 Renner, op. cit. 5, p. 143.

16 *The Defense Monitor* 18, 6, 1989, p. 3.

17 GAO/RCED-90-96, p. 16.

18 Renner, op. cit. 5, p. 151.

19 Renner, op. cit. 5, p. 149.

20 Gregory, S. and Edwards, A. "The Hidden Costs of Deterrence: Nuclear Weapons Accidents 1950–1988", *Bulletin of Peace Proposals* 20, 1, 1989.

21 Westing, A. "Constraints on Military Disruption of the Biosphere: An Overview", in A. Westing (ed.) *Cultural Norms. War and the Environment*, Oxford University Press, Oxford, 1988, p. 4.

22 *The Defense Monitor* 8, 6, 1989, p. 1.

23 Renner, op. cit. 5, p. 151.

24 *The Defense Monitor* 8, 6, 1989, p. 7.

25 Schneider, K. *New York Times*, 31 January 1991.

26 Kakonen, J. "The Concept of Security: From Limited to Comprehensive", paper presented at the 25th Annual International Peace Research Conference, Groningen, 3–7 July 1990, p. 15.

27 Renner, M. "Forging Environmental Alliances", *WorldWatch*, November/December 1989, pp. 8–15; Renner, M. "National Security: The Economic and Environmental Dimensions", *Worldwatch Paper* 89, Worldwatch Institute, Washington, DC, 1989; Renner, M. "From Military Security to Environmental Security: The Challenge of Transboundary Perils", *PAWSS Perspectives* 1, 1, April 1990, pp. 9–12; Myers, N. "Environment and Security", *Foreign Policy* 74, 1989, pp. 23–41.

28 Renner, *Worldwatch Paper* 89, ibid., p 6.

29 Quoted in Hampson, F.O. "Peace, Security and New Forms of International Govemance", in C. Mungall and D. McLaren (eds) *Planet Under Stress*, Oxford University Press, Toronto, 1990.

30 Renner, op. cit. 23, p. 62.

31 Buzan, B. *People, Stares and Fear: The National Security Problem in International Relations*, North Carolina Press, Durham, 1983, p. 245.

32 World Commission on Environment and Development, *Our Common Future*, Oxford University Press, 1987.

33 Sand, P. *Lessons Learned in Global Environmental Governance*, World Resources Institute, Washington, 1990, p. 21.

34 Homer-Dixon, T. "Environmental Change and Violent Conflict", *Emerging Issues Occasional Paper Series*, American Academy of Arts and Sciences, Cambridge, MA, 1990.

35 Brown, J. (ed.) *In the US Interest: Resources, Growth, and Security in the Developing World*, Westview Press, Boulder, CO, 1990.

36 Glaniz, M. "On Assessing Winners and Losers in the Context of Global Warming", Report of the Workshop, Environmental and Societal Impacts Group, National Center for Atmospheric Research, Boulder, CO, 1990.

37 Deudney, D. "The Case Against Linking Environmental Degradation and National Security", *Millennium* 19, 3, Winter 1990, pp. 461–476.

Cartoon 13 "Yo! Amigo!!" (by Scott Willis)
"Northern" environmentalists have often argued that rainforests in the "South" are important to prevent climate changes caused by the enhanced Greenhouse effect. But the question that this cartoon poses so powerfully is about who is responsible for most of the greenhouse gases and who should take action to deal with the problem.
Source: Scott Willis

29 VANDANA SHIVA

"The Greening of Global Reach"

from *Global Ecology: A New Arena of Political Conflict* (1993)

The green movement grew out of local awareness and local efforts to resist environmental damage. The crisis of deforestation in the Himalayas was a concern first voiced by the local peasant women of Garhwa. The crisis of toxic hazards was first recognized by the affected residents of Love Canal.

The pattern that emerged over the 1970s and 1980s was the recognition that major environmental threats were posed by globally powerful institutions, such as multinational corporations, and multilateral development banks such as the World Bank, whose operations reach every city, village field and forest worldwide.

In recent years, two decades of the green movement are being erased. The local has disappeared from environmental concern. Suddenly, it seems, only "global" environmental problems exist, and it is taken for granted that their solution can only be "global".

In this chapter I shall look more closely at what the concept of the "global" conceals and projects, how it builds power relations around environmental issues, and how it transforms the environmental crisis from being a reason for change into a reason for strengthening the status quo.

THE "GLOBAL" AS A GLOBALIZED LOCAL

Unlike what the term suggests, the global as it emerged in the discussions and debates around the UN Conference on Environment and Development (UNCED) – eventually held in June 1992 – was not about universal humanism or about planetary consciousness. The life of all people, including the poor of the Third World, or the life of the planet, is not at the centre of concern in international negotiations on global environmental issues.

The "global" in the dominant discourse is the political space in which a particular dominant local seeks global control, and frees itself of local, national and international restraints. The global does not represent the universal human interest, it represents a particular local and parochial interest which has been globalized through the scope of its reach. The seven most powerful countries, the G-7, dictate global affairs, but the interests that guide them remain narrow, local and parochial. The World Bank is not really a Bank that serves the interests of all the world's communities. It is a Bank where decisions are based on the voting power weighted by the economic and political power of donors, and in this decision-making it is the communities who pay the real price and the real donors (such as the tribals of Narmada Valley whose lives are being destroyed by a Bank financed mega-dam) but have no say. The "global" of today reflects a modern version of the global reach of a handful of British merchant adventurers who, as the East India Company, later the British Empire, raided and looted large areas of the world. Over the past 500 years of colonialism, whenever this global reach has been threatened by resistance, the language of opposition has been co-opted, redefined and used to legitimize future control.

The independence movement against colonialism had revealed the poverty and deprivation caused by the economic drain from the colonies to the centres of economic power. The post-war world order which saw the emergence of independent political states in the South, also saw the emergence of the Bretton Woods institutions such as the World Bank and IMF which took over the language of underdevelopment and poverty, removed these independent political states' history, and made them the reason for a new bondage based on development financing and debt burdens.

The environment movement revealed the environmental and social costs generated by maldevelopment, conceived of and financed by such institutions as the World Bank. Now, however, the language of the environment is itself being taken over and made the reason for strengthening such "global" institutions and increasing their global reach.

In addition to the legitimacy derived from co-opting the language of dissent is the legitimization that derives from a false notion that the globalized "local" is some form of hierarchy that reflects geographical and democratic spread, and to which lower order hierarchies should somehow be subservient. Operationalizing undemocratic development projects was based on a similar false notion of "national interest", and every local interest felt morally compelled to make sacrifices for what seemed the larger interest. It was this moral compulsion that led each community to make way for the construction of mega-dams in post-independence India. Only during the 1980s, when the different "local" interests met nationwide, did they realize that what was projected as the "national interest" was, in fact, the electoral interests of a handful of politicians financed by a handful of contractors, such as J.P. and Associates who benefit from the construction of all dams, such as Tehri and the Narmada Valley projects. Against the narrow and selfish interest that had been elevated to the status of "national" interest, the collective effort of communities engaged in resistance against large dams began to emerge as the real though subjugated national interest.

In a similar way the World Bank's Tropical Forest Action Plan (TFAP) was projected as responding to a global concern about the destruction of tropical forests. When rainforest movements formed a worldwide coalition under the World Rainforest Movement, however, it became clear that TFAP reflected the narrow commercial interests of the World Bank and multinational forestry interests such as Shell, Jaako Poyry and others, and that the global community best equipped to save tropical forests were forest dwellers themselves and farming communities dependent on forests.

GLOBAL ENVIRONMENT OR GREEN IMPERIALISM?

Instead of extending environmental concern and action, the recent emergence of a focus on "global" environmental problems has in fact narrowed the agenda.

The multiple environmental concerns that emerged from the grassroots, including the forest, and the water crises, toxic and nuclear hazards and so on have been marginalized. Thus the Global Environmental Facility (GEF) set up at the World Bank addresses only four environmental issues: (1) a reduction in greenhouse gas emissions; (2) protection of biodiversity; (3) a reduction in pollution of international waters; and (4) a reduction in ozone layer depletion.

The exclusion of other concerns from the global agenda is spurious, since, for example, the nuclear and chemical industries operate globally, and the problems they generate in every local situation are related to their global reach.

"Global environmental problems" have been so constructed as to conceal the fact that globalization of the local is responsible for destroying the environment which supports the subjugated local peoples. The construction becomes a political tool not only to free the dominant destructive forces operating worldwide from all responsibility for all the destruction on to the communities that have no global reach.

Consider the case of ozone depletion. CFCs, which are a primary cause of ozone depletion, are manufactured by a handful of transnationals, such as Dupont, with specific locally identifiable manufacturing plants. The rational mechanism to control CFC production and use was to control these plants. That such substances as CFC are produced by particular companies in particular plants is totally ignored when ozone depletion becomes transformed into a "global" environmental problem. The producers of CFCs are apparently blameless and the blame laid instead on the potential use of refrigerators and air-conditioners by millions of people in India and China. Through a shift from present to future, the North gains a new political space in which to control the South. "Global" concerns thus create the moral base for green imperialism.

It also creates the economic base, since through conventions and protocols, the problem is reduced to technology and aid transfer. Dupont then becomes essential to the problem it has created, because it has patented CFC substitutes, for which a market must be found. The financial resources that go into the Montreal Protocol Fund for transfer of technology are in effect subsidies for Dupont and others, not for the Third World.

The erosion of biodiversity is another area in which control has been shifted from the South to the North through its identification as a global problem. Biodiversity erosion has occurred because of habitat destruction in diversity rich areas, by dams, mines and highways financed by the World Bank for the benefit of transnational corporations (TNCs), and by replacing diversity-based agriculture and forest systems with monocultures of "green revolution" wheat and rice and eucalyptus plantations, which were also supported and planned by the World Bank, in order to create markets for seed and chemical industries.

The most important step in biodiversity conservation is to control the World Bank's planned destruction of biodiversity. Instead, by treating biodiversity as a global resource, the World Bank emerges as its protector through the GEF (Global Environmental Facility) and the North demands free access to the South's biodiversity through the proposed Biodiversity Convention. But biodiversity is a resource over which local communities and nations have sovereign rights. Globalization becomes a political means to erode these sovereign rights, and a means to shift control over and access to biological resources from the gene-rich South to the gene-poor North. The "global environment" thus emerges as the principal weapon to facilitate the North's worldwide access to natural resources and raw materials on the one hand, and on the other, to enforce a worldwide sharing of the environmental costs it has generated, while retaining a monopoly on benefits reaped from the destruction it has wreaked on biological resources. The North's slogan at UNCED and the other global negotiation fora seems to be: "What's yours is mine. What's mine is mine".

The notion of "global" facilitates this skewed view of a common future. The construction of the global environment narrows the South's options, while increasing the North's. Through its global reach, the North exists in the South, but the South exists only within itself, since it has no global reach. Thus the South can *only* exist locally, while only the North exists globally.

Solutions to the global environmental problems can come only from the global, that is the North. Since the North has abundant industrial technology and capital, if it has to provide a solution to environmental problems, they must be reduced to a currency that the North dominates. The problem of ecology is transformed into a problem of technology transfer and finance. What is absent from the analysis is that the assumption that the South needs technology and finances from the North is a major cause of the environmental crisis, and a major reason for the drain of resources from South to North. While the governments of the South demand "new and additional sources of finance" for the protection of the environment, they ignore the reverse transfer of $50 billion per year of capital from the poor South to the affluent North. The old order does not change through the environmental discussions, rather it becomes more deeply entrenched.

THE PROBLEM OF FALSE CAUSALITY

With the masking-out of the role of the globalized local in local environmental destruction worldwide, the multiple facets of destruction are treated as local causes of problems with global impact. Among the many simultaneously occurring impacts of maldevelopment and colonialism are: the rise of poverty; the increase of environmental degradation; the growth of population; polarization; and conflict between men and women, and between ethnic communities.

Extraction of surplus and the exploitation and destruction of resources have left people without livelihoods. Lacking access to resources for survival, the poor have been forced to generate economic security by having large families. The collapse of social cohesion and economic stability has provided the ground for ethnic conflict.

Instead of identifying the cause of these multifaceted problems as global domination of certain narrow interests of the North, however, these problems are selectively transformed from consequence to cause. Poverty and population are identified as *causes* of environmental degradation. Diversity is seen as a defect and identified as a *cause* of ethnic conflict.

False causality is applied to explain false connections. Thus some UNCED documents went to the extent of pointing to population growth as a *cause* of the explosive growth of toxic chemicals. A problem caused by an irresponsible chemical industry is converted into a problem caused by fertility rates in the poor countries of the South. The 1991 cyclone in Bangladesh was similarly linked causally to the number of babies in Bangladesh.

THE "GLOBAL" IS NOT PLANETARY

The visual image of planet Earth used in the discourse on global ecology disguises the fact that at the ethical level the global as construct does not symbolize planetary consciousness. The global reach by narrow and selfish interests is not based on planetary or Gaian ethics. In fact, it abstracts the planet and peoples from the conscious mind, and puts global institutions in their place. The planet's security is invoked by the most rapacious and greedy institutions to destroy and kill the cultures which employ a planetary consciousness to guide their concrete daily actions. The ordinary Indian woman who worships the tulsi plant worships the cosmic as symbolized in the plant. The peasants who treat seeds as sacred, see in them the connection to the universe. Reflexive categories harmonize balance from planets to plants to people. In most sustainable traditional cultures, the great and the small have been linked so that limits, restraints, responsibilities are always transparent and cannot be externalized. The great exists in the small and hence every act has not only global but cosmic implications. To tread gently on the earth becomes the natural way to be. Demands in a planetary consciousness are made on the self, not on others.

The moral framework of the global reach, however, is quite the opposite. There are no reflexive relationships. The G-7 can demand a forest convention that imposes international obligations on the Third World to plant trees. But the Third World cannot demand that the industrialized countries reduce the use of fossil fuels and energy. The "global" has been so structured, that the North (as the globalized local) has all rights and no responsibility, and the South has no rights, but all responsibility. "Global ecology" at this level becomes a moralization of immorality. It is devoid of any ethics for planetary living; and based on concepts not of universal brotherhood but of universal bullying.

DEMOCRATIZING "GLOBAL" INSTITUTIONS

The creation of new mechanisms for responding to the global ecological crisis was one of UNCED's agendas. Problematizing the "global" through collective articulation of all local concerns and interests, in all their diversity, is the creative intervention in the global/local conflicts as they are emerging.

To democratize the "global" is the next step. What at present exists as the global is not the democratic distillation of all local and national concerns worldwide, but the imposition of a narrow group of interests from a handful of nations on a world scale. But if genuine democracy is to exist at local and national levels it is essential for international interests to become democratized.

The roots of the ecological crisis at the

institutional level lie in the alienation of the rights of local communities to actively participate in environmental decisions. The reversal of ecological decline involves strengthening local rights. Every local community equipped with rights and obligations, constitutes a new *global* order for environmental care.

The current trend in global discussions and negotiations, however, is to move rights further upwards towards more distant, non-local centralization in such agencies as the World Bank.

Multilateralism in a democratic set-up must mean a lateral expansion of decision-making based on the protection of local community rights where they exist, and the institutionalization of rights where they have been eroded. Two central planks of local environmental rights include: (1) the right to information; and (2) right to prior consent; that is, any activity with a potential impact on the local environment should be subject to consent by local people.

Basing an environmental order on globally institutionalized local rights also avoids the impracticable issue of representation and the terrible bungling resulting from international NGOs "selecting" national NGOs to "select" local NGOs to represent "people" at global negotiations.

The "global" must accede to the local, since, the local exists with nature, while the "global" exists only in offices of World Bank/IMF and headquarters of multinational corporations. The local is everywhere. The real ecological space of global ecology is to be found in the integration of all locals. The "global" in global reach is a political not an ecological space.

Institutionally, we should not be concerned about how to enable the last tribal to be present at World Bank decisions in Washington. What we need to ensure is that no World Bank decision affecting the tribals' resources is taken without their prior informed consent.

Whether the local as global and the global as local will exist in a way different from the imperialistic order of the last 500 years depends on this process of democratization. The imperialistic category of global is disempowering at the local level. Its coercive power comes from abolishing limits for the forces of domination and destruction and imposing restrictions on the forces of conservation.

The ecological category of global is an empowering one at the local level because it charges every act, every entity, with the largeness of the cosmic and the planetary and adds meaning to it. It is also empowering because precisely by embodying the planetary in the local, it creates conditions for local autonomy and local control.

An Earth democracy cannot be realized as long as global domination is in the hands of undemocratic structures. Neither can it be realized on an anthropocentric basis – the rights of non-human nature cannot be ignored. And it cannot be realized if the need to ensure the survival of the planet is made the reason for denying the right to survival of those who today are poor and marginalized because they have borne the accumulated burden of centuries of subjugation.

Cartoon 14 "OK. You hand us . . ." (by Skauge)
As Visvanathan's reading makes especially clear, technical expertise of "environmental science" which turns the world into resources often ignores the experience of poor people in "Southern" states.
Source: Skauge, Cartoon & Writers' Syndicate

30

SHIV VISVANATHAN

"Mrs Brundtland's Disenchanted Cosmos"

from *Alternatives* (1991)

I

The whole world loves a storyteller. There is magic to the act of storytelling, a warmth of waiting as someone begins, "Once there was a ..." The storyteller unravels the struggle of good and evil, gods and demons, humans and animals. The beauty of a good story is that it can be told again and again. But in the modern world, storytelling is dying out. The battle between good and evil is no longer embodied in myths, fables, anecdotes, or parables. The struggles of humankind are now sought to be captured in the grids of social science, and the classic narrative of social science is the bureaucratic report.

A report is the end of storytelling. It is not told; it is authored. It is generally chaired by a person heading a committee. A report is too impersonal to have the warmth of a story, and yet the story of the world, its fate, is caught in reports – the Brandt Report, the Brundtland Report, the Report of the Club of Rome. They are all stories of the world, but they do not belong to the world of storytelling. Dry as dust, they reduce even the hell of Dante and its horrific circles to sanitized departments, each headed by a bureaucrat. Yet these narratives – unreadable – and opaque as they are – must be taken seriously. They represent new charters of conquest. They "speak" (if a report can speak at all) with forked tongues. Reports as narratives mimic modern violence. They are bloodless and antiseptic, but with one stroke of a file, a world can die; with one erasure, a man can cease to be a citizen.

These reports capture the new styles of control and surveillance. The usual methods of modern control – the factory, the school, the prison – were all modes of vigilance. They are best embodied in the Benthamite ideas of the Panopticon. The English philosopher, Bentham, elaborated a system of vigilance where the poor, vagrant, alcoholic, and orphan were made to work in an inspection house, where every phase of their work could be examined by a central eye. It was an all encompassing plan for surveillance, which served as a general model of control. Prisons, factories, even schools are panopticons.

Today, the Benthamite model of vigilance, embodied in concrete structures, is no longer enough. One does not need specific groups panopticonized. The goals are bigger as the whole world needs to be panopticonized. The new modes of surveillance are more subtle. There is little blatant aggression, no group of colonials sitting around a table and carving colonies like a steak. Open aggression is too uneconomical.

Why kill, when you can co-opt? Why destroy, when you can absorb the world through Keynesian strategies? Why police blatantly, when the expert advertisement can make the victim compliant?

The new epidemic of reports uses the style of concern to control: it restates certain problems to erase people's memory of them. But the entire act is performed within the antiseptic confines of the club. One can comfortably invite a few Third Worlders, even to write the foreword to the report. They pose no dangers. In fact, they outdo the West in their need to retain membership. But note, the entire act of violence is sanitized. There is no Cortez or Shakha here. It is killing through concepts, through coding, by creating grammars that decide which sentence can be spoken and which cannot. It is from such a perspective that the Brundtland Report – well intentioned as it is – must be seen not as a statement of intention, but in terms of the logic of the world it seeks to create and impose.

II

Every contemporary report needs a key word or a slogan to keep it alive long after the report itself gathers dust. The key words of the Brundtland Report are sustainable development. There is no great contradiction in terms. Sustainability and development belong to different, almost incommensurable worlds. We were told in catechism class that even God cannot square a circle. Sustainable development is another example of a similar exercise.

Sustainability is about care and concern; it speaks the ethics of self-restraint. It exudes the warmth of locality, of Earth as home. Development is a genocidal act of control. It represents a contract between two major agents, between the modern nation-state and modern Western science. The first is deemed to be the privileged form of politics, the second claims to be the universal form of knowledge. One cannot conceive of a nation-state without a science policy program. Development is a compact between the nation-state and modern Western science to reduce all forms of difference – all ethnic forms, all ethnic knowledges – to create a flatland called modernity. Within such a Hobbesian world, dams displace people, forest bills turn ecocidal, and nuclear energy becomes a reason for the state. If differences exist between modern and peasant/tribal, such differences are reduced through a time series. The tribal peasant or folk are labeled premodern and therefore must be driven into modernity. Every act of protest is heresy. What legitimates this violence is the doctrine of progress, which imposes a linearity to this world and justifies any violence done by modernizing elites on allegedly backward sectors. Here, traditions are neither privileged ways of looking or being, but only an obsolescent world to be developed or museumized. Real autonomy is granted to no world view other than development. All history, all biography, all memory is aligned to facilitate this long march to modernity.

Consider just two cases. In 1974, the government of Paraguay was charged with the genocide of the Ache Indians. The charges included enslavement, torture, and deliberate withholding of food and medicines. The response of the Paraguayan minister to the UN was to deny intent. He admitted that there were victims and victimizers, but no intent. In a similar way, Brazil was accused of genocide against the Indians of the Amazon. Brazil, too, accepted that the Indians had been eliminated and their land forfeited. But it argued that it was done for purely economic reasons. In both cases, genocide was seen as an accidental by-product of development. Today, a hydroelectric dam or a hamburger can create more refugees than war. It is a system of nonresponsibility for all cultures that do not fit into the grids of progress.

III

But the insurrection of local groups across the world is challenging this regime of development. A thousand Copernican revolutions are threatening this Ptolemaic mandate. What does the mandarin world of development do? It convenes a secretariat of international civil servants to invent a few epicycles. To the general notions of GNP, GDP, its money indicators, and market models, it adds a few concepts like pollution, diversity, recycling, and sustainability. It is like adding a few more blind alleys to the labyrinth of development. However, the way this new scholasticism creates this world is important. Its bloodless methods need to be understood.

The styles of conquest of this new scholasticism embody a combination of four tactics. We can call them the clock, the map, the dictionary, and the file. Through each of these, history and geography are rewritten to suit the needs of power.

If the first waves of modernity sought to caricature the past, the second wave seeks to control the future. Note the use of the singular. It is not a promise of multiple futures. It is the future. We thought the future was a place of dreams, a dream of possibility. Freedom was essentially the freedom to dream differently, and have different languages for interpreting our dreams. Today, a group of experts tells us what to dream. The future is suddenly no longer fiction or fantasy. It is being colonized by an oracle of international civil servants who have mapped it with cybernetics and systems theory. The future has become a territory of surveillance; a group of grammarians has moved in before the poet has uttered a word. They have already decided that the future is a different country, where all of us must behave alike. The future is not a carnival time, where dreams spoof the pomposities of the present.

In fact, bureaucrats don't dream. They only extrapolate. They are so conventional that they merely replicate their life-world in the future. The future becomes organized into a large secretariat, with many departments. So there are resources for the future, energy for the future, populations for the future, cities for the future. Before man arrives, the "file-ariat" of the UN already wants to be in place.

Remember that these new colonials lack the fabulous mythology of their crass predecessors. There is no myth of the Orient, no Burtons or Lawrence ready for the new Arabia. This new class is too puritan, too antiseptic for the old and fabulous excesses of the Orient. Heidegger once said "abstraction is conceptual rape," and our new international file-ariat rape through abstraction. These new utopias are merely divided according to the levels of systems theory. There are only systems and subsystems, all disciplined into a hierarchy. They don't talk, they feed back. It is a cybernetic world where the future is already in equilibrium.

Before we forget, the bureaucrats remind us that there is a common future. It is a notion of commonness that emphasizes unity, order, and uniformity. There is no place of plurality, or difference, or multiplicity. For us, globalization is the dull ideology of bureaucratic uniformity. It is a dismal civics. To conquer, the bureaucrats need a language, a "new" vernacular. Who needs dialogue or translation or even the laughter of misunderstanding, the celebration of ambiguity when a cold formal language can be constructed. The key words of this language are still the old clichés of efficiency, ecology, diversity, and underdevelopment.

IV

The entire discourse of bureaucrats is still written in the language of a monetized economy. Consider the word "ecology." In the bureaucrat's world, it becomes bastardized because money and market dissolve real ecologies. Ecology – true ecology – should be an attempt to liberate the imagination of democracy from the constraints that "big science," the nation-state, and development have imposed. It seeks a notion of a good life, an idea of restraint and self-limits that cannot be reduced to economic audits and goes beyond a world that values obsolescence as godhead. In Brundtland, ecology is merely a search for managerial efficiency.

As a theory of both culture and nature, such an "ecology" is a misnomer. First, it never acknowledges nature for itself. Nature is never seen as a dwelling. It is only a resource or a toolshed. It is not an *oikos*, a word that denotes prudence or care, the world of the housewife or tribe. Its jargon stems from the accountant's office, where a being *is* only if it is monetized. The world of nature is reduced to the world of commodities.

Here the role of nature's economy is never highlighted. Its task in maintaining the hydrological cycle, preserving soils, or sustaining genetic diversity is never celebrated. Nature to the bureaucrats is an outdated craftsman, to be retired with the advent of the scientist; nature has to be preserved until scientific techniques improve. It is a world where the moral economy of the forest is irrelevant. What it needs is not the forest but the forest reduced to a park, reserve, or plantation, where selected aspects of nature are maintained until genetic engineering can take over. As the Brundtland Report remarks: "It would be a grim irony indeed if just as the new genetic engineering techniques begin to let us peer into life's diversity and use genes more effectively to better the human condition, we looked and found this treasure sadly depleted." Here the word "treasure" smacks of the language of piracy. We feel the need to go beyond words like park, reserve, and gene bank to empathize with nature. The language of nature in these reports is crudely Cartesian. Nature is still outside, objectified, analyzed. One would suggest that this perspective had much to learn from the world of farmers and tribes where the ego and nature are constructed differently. Here, to paraphrase Levinas, "I care for the other, therefore I am." It is not so much a world of otherness but of togetherness.

The language of parks and reserves only disguises the language of mining, which is so central to the Western, technocratic discourse on nature. At least the old colonials were more blatant. For them nature was "raw material" or "primary product."

It is also interesting to note the puritanical strain of these documents. One feels that nature, the tropics and the Third World are often synonymous and all suffer from excess. All need to be disciplined, and nature as a celebration is now corseted in a "park" or "reserve." Notice the sheer restraint connoted by these terms. It is as if raw energy has to be sublimated or dammed for industrial use. "Excess" is transformed into "reserve" as the disciplinary grid of science begins to operate.

Even in cultural terms, the word resource is myopic. As a concept, it is so obsessed with capital and labor that many sources of work and well-being are ignored. And if recognized, they are immediately disembodied from locality and context. As the acerbic wisdom of Wolfgang Sachs stated it,

> Numerous things which had so far been taken for granted as part of ordinary life acquire new and dramatic significance. They change into valuable resources. Cowdung for example, kindled by the Senegalese peasants to heat water for a cooking pot, suddenly becomes an energy resource, Kenyan women cultivating village fields are discovered to be a human resource for boosting food production.

Sachs utters a wise warning, that the language of resources is the language of economic inputs.

When nature and human beings become natural resources and human resources, the process of abstraction and exploitation has already begun. A coconut is no longer celebrated as a coconut, a forest is no longer a forest when it is treated as a resource. Rather, the magic cover that existed for them, the totemic field that protected them, is stripped off and, as resources, they begin the long journey into the world economy.

V

The world of Brundtland is still the mechanical world of Jevons and Walras. Its systems vocabulary does not eliminate the still mechanical mind-set of its experts. The fundamental mind-set of the machine is clock time. Clock time is empty, clock time is reversible. Yet the clock is the icon of modern urgency. The *Bulletin of the Atomic Scientists* has a clock on its cover. The clock at twelve denotes apocalypse. Yet only mechanical time allows for reversibility – nature does not. By speaking the language of growth, the Brundtland Report gets caught in contradictions. It believes that time can be reversed and speeded up. Growthmanship involves both. It believes that you can virtually consume as before. It believes that the pace of technological innovation can still be sustained. Its ideal citizen is still the consumer with the big mouth and a faith in science. In this Cartesian–consumerist world, there is no real place for *satyagrah* or any form of enlightened self-restraint.

Finally, the language and philosophy of the Brundtland Report is still the language of what David Ehrenfeld calls universalist–humanism. Such a perspective is anthropocentric. Maybe if parts of the report were written from the perspective of a microbe, a tree, or a spider, the report might have created more empathy. It is also ethnocentric. The industrial consumerman is still the privileged citizen. Such a humanism, Ehrenfeld showed, still believes that all problems are solvable and that technology and management can solve all problems. In this, the expert and his technology are always part of the solution and not part of the problem. Apropos of this, Rene Dubos remarks,

> Developing counter technologies to correct new kinds of damage constantly being created by technological innovations is a policy of despair. If we follow this course, we shall increasingly behave like hunted creatures, fleeing from one protective device to another, each more costly, each more complex, and more undependable than before, we shall be chiefly sheltering ourselves from environmental dangers while sacrificing the values that make life worth living.

Deep down the Brundtland Report still believes that the expert and the World Bank can save the world. All it needs is the application of better technology and management. What it fails to understand is that a club of experts, whether Brandt, Brundtland, or Rome, is an inadequate basis for society. What one needs is not a common future but the future as a commons.

A commons is the plurality of life worlds to which all citizens have access. It is not merely the availability of nature as being but of alternative imaginations, skills that survival in the future might require. But it is on such a world that Brundtland misses out.

Unfortunately, the Western technocentric man has officially only two notions of Earth. The first is the "life boat world" of Garrett Hardin. In his "Life Boat Ethics," Hardin pictures the world as an ocean where poor swim or drift, while rich sit snugly in a life boat. The rules of the technocracy claim that the rich should not help the poor, as the latter lack the discipline to survive. In this scenario, the future is a place where all those who have nothing in common with you are eliminated.

The second view is the model of spaceship Earth. To call Earth a spaceship is to reduce Earth to a complex but constraining machine. In fact, it reminds one of Buckminster Fuller's statement that the space capsule is the truly perfect environment perfected for man. Such a view reduces Earth to a space module, which is "the most deadly, defunctionalized environment that the mind of man has yet conceived, compared to which the most backward, Neolithic village was a paradise of creativity and autonomy." We need myths of Earth that go beyond metaphor of systems and machine.

The Earth is a celebration that cannot be reduced to the puritanism of machine philosophy. The expert world of Brundtland is too repressed, too provincial to dance the dance of Earth. Instead of classifying species and economizing them, it is time to talk to the Earth like St. Francis, inviting brother spider and sister sparrow. But it is just such a view that Brundtland misses. Brundtland's is a world without animals and therefore contents itself with stuffed toys. It is so autistic that it can neither sing nor dance nor tell stories, yet without it you cannot talk of the Earth. Brundtland seeks a co-optation of the very groups that are creating a new dance of politics, where democracy is not merely order and discipline, where Earth is a magic cosmos, where life is still a mystery to be celebrated. Nowhere in Brundtland, even in the chapter on species diversity, is there an understanding of this.

We spoke of the insurrection of local groups across the world – groups against dams, groups against war, groups wanting to farm without herbicides, groups for the rights of man. The experts of the global state would love to co-opt them turning them into a secondary, second rate bunch of consultants, a lower order of nurses and paramedics still assisting the surgeon and physician. It is this that we seek to resist not by mimicking their experts but by creating an explosion of imaginations that this club of experts seeks to destroy with its cries of lack and excess. The world of official science and the nation-state is not only destroying soils and silting up lakes, it is freezing the imagination. The wisdom of voluntarism realizes that the lakes and the imagination have to be desilted simultaneously. For this the large dams of expertise reified as the Narmada or Aswan Dam or as the World Bank and IMF have to be destroyed together. We have to see the Brundtland Report as a form of published illiteracy and say a prayer for the energy depleted and the forests lost in publishing the report. And finally, a little prayer, an apology to the tree that supplied the paper for this document. Thank you, tree.

PART 5

Anti-geopolitics

PART 5

INTRODUCTION

Paul Routledge

As the previous sections attest, geopolitical knowledge tends to be constructed from positions and locations of political, economic, and cultural power and privilege. Hence the histories of geopolitics have tended to focus upon the actions of states and their elites, understating rebellion and overemphasizing statesmanship. However, the geopolitical policies enacted by states, and the discourses articulated by their policy makers, have rarely gone without some form of contestation by those who have faced various forms of domination, exploitation and/or subjection which result from such practices. As Foucault has noted, "there are no relations of power without resistances ... like power, resistance is multiple and can be integrated in global strategies" (1980: 142).

Indeed, myriad alternative stories can be recounted that frame history from the perspective of those who have engaged in resistance to the state and the practices of geopolitics. These histories keep alive the memory of people's resistances, and in doing so suggest new definitions of power that are not predicated upon military strength, wealth, command of official ideology and cultural control (Zinn, 1980). These histories of resistance can be characterized as a 'geopolitics from below' emanating from subaltern (i.e. dominated) positions within society that challenge the military, political, economic and cultural hegemony of the state and its elites. These challenges are counter-hegemonic struggles in that they articulate resistance to the coercive force of the state – in both domestic and foreign policy – as well as withdrawing popular consent to be ruled "from above." They are expressions of what we would term "anti-geopolitics."

Anti-geopolitics can be conceived as an ethical, political and cultural force within civil society – i.e. those institutions and organizations that are neither part of the processes of material production in the economy, nor part of state-funded or state-controlled organizations (e.g. religious institutions, the media, voluntary organizations, educational institutions and trades unions) – that challenges the notion that the interests of the state's political class are identical to the community's interests. Anti-geopolitics represents an assertion of permanent independence from the state *whoever is in power*, and articulates two interrelated forms of counter-hegemonic struggle. First, it challenges the *material* (economic and military) geopolitical power of states and global institutions; and second, it challenges the *representations* imposed by political elites upon the world and its different peoples that are deployed to serve their geopolitical interests.

Anti-geopolitics can take myriad forms, from the oppositional discourses of dissident intellectuals to the strategies and tactics of social movements (although the former may frequently be speaking on behalf of the latter). While anti-geopolitical practices are usually located within the political boundaries of a state, with the state frequently being the principal opponent, this is not to suggest that anti-geopolitics is necessarily localized. For example, with the intensity of the processes of globalization, social movements are increasingly operating across regional, national and international scales, as they challenge elite international institutions and global structures of domination.

COLONIAL ANTI-GEOPOLITICS

At the end of World War II, the world became reconfigured into a bi-polar political order characterized by the geopolitical, military and ideological competition between the US and the USSR. The "Third World" soon became a battleground upon which this competition was played out as the superpowers waged an ideological struggle for the hearts and minds of non-Western peoples who were liberating themselves from colonialism. During the Cold War, economic aid and development served as a means to encourage capitalist market economies, thereby providing conditions under which Western-style "democracy" could flourish. Economic development programs were constructed as a strategy to bring Third World states into the geopolitical orbit of the US and its allies. It was a process by which the "colonial world" was reconfigured into the "developing world" (Peet and Watts, 1993; Sachs, 1992). Concurrently, the Soviet Union also provided aid to those states that had nascent revolutionary or communist governments, such as Cuba and Vietnam.

Independence from colonialism was aided by the emergence of the US and the USSR as two self-proclaimed anti-imperialist superpowers; the weakening of the economic and military strength of the imperial powers (such as Britain and France) due to the debilitating effects of World War II; and the development of powerful nationalist political movements within the colonized countries whose aim was to secure the independence for their countries and their effective sovereignty in world affairs. Although the character of these nationalist movements varied widely according to local contexts, armed struggle and guerrilla warfare were frequently employed against the colonizing forces in order to defeat and remove them as a force of occupation (e.g. the Mau Mau rebellion in Kenya against the British, and the Algerian resistance against the French).

However, what frequently defined and decided the successful outcome of independence struggles was the social struggle of the population at large in combination with the armed struggle. For example, armed struggle by a guerrilla army was frequently used in concert with non-violent sanctions such as strikes and civil disobedience conducted by the population at large (see Sharp, 1973). Together, these sanctions effectively withdrew popular consent from the colonizing ruling power. The power of the guerrilla army, then, lay both in its military capabilities and in the fact that it represented the embodiment of the collective will of a colonized people to resist. The use of non-violent sanctions enabled widespread popular participation in the struggle against colonialism which, in turn, frequently enabled the development of a national consciousness to develop amongst the colonized. The strategy of non-violent resistance was probably most effective in the Indian independence movement led by Mohandas K. Gandhi and the Indian National Congress against the British.

One of the principal legitimizations of colonial exploitation had been that empire building was, in part, an unselfish, even noble act. Colonialism was further legitimized through the (mis)representation of other cultures and places as primitive, savage and uneducated, in need of Western civilization and enlightenment. Edward Said – a Professor of Comparative Literature at Columbia University and former member of the Palestinian National Council – argues that such representations were "imaginative geographies," or fictional realities, that shaped the West's perception and experience of other places and cultures. Such representations designated geographical space into familiar and unfamiliar spaces, dramatizing the distance and difference between the West and its others, separating the occident from the orient, the colonizers from the colonized, and the developed from the underdeveloped (Reading 31). These representations were constructed around essentialist conceptions of (non-Western) *others* that equated difference with inferiority, and served to inform and legitimate geopolitical strategies of control and colonization by the Western countries, as they subjected other territories to military conquest and commercial exploitation. Said exposes and articulates the ideological and political purposes of imaginative geographies for the purposes of imperialism and notes that such distortions are not confined to the colonial era, but are continually deployed to this day because they serve geopolitical ends. Hence in the Cold War, American policy makers such as Kennan could refer to the inherent deception of the Russian mind when explaining Soviet foreign policy, while more recently the demonization of Islam in general, and the Palestinian cause in particular, have served to legitimate the Israeli occupation of the West Bank. Said's work poses a challenge to the representation of others by Western "experts" articulating an intellectual project against such deformed and self-interested representations of the world. For Said the role of the dissident intellectual is to articulate an oppositional consciousness to dominant (Western or elitist) representations of others.

Colonialism was invariably a violent process, constructed upon and maintained by a profound alienation of colonized peoples, and premised upon a geographical, political, and cultural division of the colonized world. This is described by Frantz Fanon – a medical doctor and psychiatrist who worked for the *Front de Liberation Nationale* (FLN): the principal nationalist organization involved in the anti-colonial war against French occupation of Algeria (Reading 32). Fanon articulates how the indigenous inhabitants of colonial societies are "othered" by the colonizing culture, which constructs a Manichean world of colonizer and colonized. Such a division occurs on both physical and representational levels as the colonized are spatially separated from the colonizers, and their culture depicted in negative terms relative to the colonizer. As Fanon explains, the colonized are dehumanized by the colonizers in order to legitimize their control and exploitation. Writing from within the turmoil of decolonization, Fanon speaks with the voice of the heterogeneous peoples who comprised the colonized: those who were silenced and (mis)represented by the West. The constant state of emergency that exists within the colonized world, Fanon explains, also becomes a state of emergence for the colonized, to throw off both the colonizer's material occupation and their appropriation of the colonized's right to speak for themselves and to represent themselves.

Moreover, as Fanon notes, the decolonization process was a global phenomenon, influenced both by other anti-colonial struggles (e.g. the French army's defeat at Dien Bien Phu in Vietnam) and by the geopolitics of the Cold War. Both the US and the USSR attempted to support and control independence movements as part of broader geopolitical strategies against one another. For Fanon, decolonization entails both the physical removal of the occupier from one's territory and what Ngugi (1986) has termed a decolonization of the mind. This involves opposition to Western ways of representing and organizing the world and the peoples in it – a struggle over who decides and controls how different cultures are interpreted and represented.

Decolonization had mixed results. For example, independence left many states with borders that had been established arbitarily by colonial rulers. As a result, many newly independent states were forced to manage a poisonous legacy of colonialism, as certain ethnic groups were divided by states while other ethnic groups were empowered by state geography to divide and rule competing ethnic groups. The disjuncture between state geography and ethnic geography has been a major source of tension and conflict in post-colonial Africa. Moreover, independence often resulted in the replacement of white colonial rule by indigenous elites dedicated to perpetuating ethnic divisions as a means of staying in power and continuing the exploitation of the country's peoples and natural resources. This process often took place with the active support of Western states and institutions, who through so-called "development programmes" helped establish what Kwame Nkrumah of Ghana, a leading advocate of pan-Africanism, called *neocolonialism*. An example of this process was the regime of General Joseph Mobutu established in Zaire. Covertly supported by the Western powers, Mobutu (who was on the CIA's payroll) came to power in 1965 by means of a military *coup d'état*. With the goodwill and support of states like France and the United States, who viewed him as a stabilizing force and anti-communist ally, Mobutu established a corrupt and repressive regime that plundered the country's resources and impoverished the country's people. As the living standards of average Zaireans fell, Mubuto enriched himself and his entourage, purchasing luxury homes in the south of France and in Switzerland.

COLD WAR ANTI-GEOPOLITICS

Decolonization took place within a global geopolitical map of satellites and spheres of influence, dominated by the two superpowers. In addition to Western Europe, where it had deployed troops and military bases during the course of World War II, the US wielded considerable influence in Central and Latin America. Over the past century and a half, in order to protect its geopolitical and geo-economic interests in its "backyard" – maintaining its southern neighbours as a rich source of resources and profits – the US has engaged in direct military intervention (e.g. Dominican Republic 1965), the threat of force, the use of surrogate troops (e.g. the *contras* in Nicaragua, 1981–1989), clandestine "destabalizing" operations against radical regimes (e.g. Chile 1973) and economic blockades and sanctions (e.g. Cuba 1962–present).

Although American foreign policy was constantly cloaked in the rhetoric of anticommunism, it was, nevertheless, intimately connected with its economic interests in the region. As Jenny Pearce (1981: 2) argues:

> The U.S. dominates the economies of the region, shaping them to its needs through investment and trading policies in a way that has left a lasting legacy of dependency and underdevelopment. At the same time the U.S. has ensured that only a small minority of the population of these countries can benefit from its involvement. The history of United States foreign policy in the region is also the history of its support for local elites favourable to its interests. The close alliance between those who control political and economic power within the region and the military and economic might of the U.S. has resulted in some of the most extreme forms of exploitation and repression anywhere in the world. Such attempts by one country to dominate others are usually called "imperialism".

There were both intellectual and material challenges to this state of affairs. Prominent among the intellectual challenges were a group of scholars known as "dependency theorists." They sought to analyze the extent to which the political economy of developing countries was influenced by a global economy dominated by the advanced capitalist countries. The analysis of dependency included a variety of related theories including those formulated by the Economic Commission for Latin America (see Love, 1980; Frank, 1967, 1978; Dos Santos, 1970; Cardoso, 1972; Emmanuel, 1972; Amin, 1976; and Cardoso and Faletto, 1979).

Dependency theory focused on the unequal economic and political exchange that took place between the advanced capitalist countries (the "core") and the developing countries (the "periphery"). The economies of the periphery were seen as conditioned by, and dependent upon, the development and expansion of economies in the core. The process of development was seen as selective, reinforcing the accumulation of wealth in the core at the expense of the periphery. Dependency was seen as both the relationship between states – an industrialized core and an impoverished periphery – and also the unequal relationship between groups and classes within states.

The most important material challenge was the Cuban Revolution of 1959, led by Fidel Castro and Che Guevara, which overthrew the US-supported Batista regime. Supported by Soviet aid, Cuba withstood repeated attempts by the US at destabilization, and instituted land reforms, literacy, housing, and public health improvements. Inspired by the success of Cuba, numerous peasant guerrilla movements emerged throughout Central and Latin America in attempts to challenge authoritarian regimes and alleviate poverty (e.g. the Farabundo Marti National Liberation Front in El Salvador, and the Sandanista National Liberation Front in Nicaragua, which overthrew the Somoza dictatorship in 1979). The response of the US to these resistances was to raise the bogeyman of communism and intervene indirectly through the provision of military training, financing and hardware to its puppet regimes in order to establish brutal counter-insurgency programs, euphemistically termed "low intensity conflict" (Galeano, 1973, 1995).

Outside of the Americas, the US adopted similar methods when faced with newly independent revolutionary regimes, following the defeat of the colonial powers. Such events were interpreted within the logics of Cold War geopolitics, demanding US intervention to counter potential Soviet or Chinese influence. This is dramatically illustrated by the involvement of the US in Vietnam. Following the establishment of the Democratic Republic of Vietnam by Ho Chi Minh in 1945, the country was plunged into an eight year war with France who opposed the existence of a communist-led government in its former colony. Following the defeat of the French at Dien Bien Phu, the Geneva Accords of 1954 saw the partition of Vietnam: the north remaining under the control of the Vietnam Communist Party, while the south was controlled by the anti-communist government of Ngo Dinh Diem.

The US immediately sought to buttress Diem's government by forming the Southeast Asia Treaty Organization (SEATO) – a military treaty specifically aimed at containing communist expansion in the region. Furthermore, the US sent military advisors and military and economic aid to South Vietnam and supported Diem's decision to withhold the scheduled 1956 elections. By 1960 a communist-led insurgency against Diem's regime, supported by the North, threatened to topple the government. As a result, the US gradually began to escalate its involvement in the country, sending military advisors, and subsequently troops, to South Vietnam which totalled half a million during the height of the war (1966–1967).

Framed within the discourse of communist containment, US policy actively sought to prevent a reunification of Vietnam under the control of Ho Chi Minh. Such a geopolitical strategy determined that the US militarily support an unpopular, authoritarian regime in the South irrespective of the cost in human lives. However, the US was unable to defeat the National Liberation Front, who waged an effective guerrilla war, mobilizing the entire population of the North against the joint US and South Vietnamese forces. Moreover, as evidence of US atrocities in the war became known (e.g. the My Lai massacre) and as the number of US casualties mounted, so public pressure began to mount within the US to bring an end to the war.

As the anti-war movement within the US gathered momentum, it became linked to the civil rights movement under the leadership of Martin Luther King, which had been agitating for an end to racial segregation and discrimination. In 1967, King publicly voiced his opposition to the war in Vietnam (Reading 33). As a counter-hegemonic discourse, King's address articulates a moral challenge to the state's right to make war against others for its own geopolitical interests. He provides a powerful denunciation of the violence inherent in his government's foreign policy – a violence that included mass bombings, the use of napalm against civilian populations, the poisoning of food and water resources and over a million Vietnamese casualties. In so doing, King attempts to give voice to the Vietnamese, demystifying their negative representation by his government.

King also makes links between the issues of imperialism abroad with racism and poverty at home, noting the large percentage of black conscripts to the US army in Vietnam, and the deleterious impact that foreign wars had upon the domestic cultural and economic life of the poor. In contrast to the armed conflict occurring in Vietnam, King calls for domestic non-violent resistance to the war in the form of conscientious objection – a withdrawal of popular consent to be conscripted. His critique of the ongoing interventionist character of US foreign policy – which necessitated a domestic resistance that continued generation after generation – proved prophetic given subsequent domestic opposition to US foreign policy in Nicaragua and El Salvador during the 1980s.

While the Soviet Union sought to support certain revolutionary movements in the Third World for its own geopolitical purposes *contra* the US, within Eastern Europe, Soviet hegemony enacted what Václav Havel (Reading 34) termed a "post-totalitarian" politics. This characterized a political system that, while it exhibited many of the features of classic dictatorships, differed in terms of the nature of power. First, the Soviet system was not localized but extended over an entire bloc – despite some local variations – and had developed over a period of 70 years (40 years in Eastern Europe). Each country within the Soviet bloc was penetrated in varying ways by Soviet mechanisms of control, and subordinated to the interests of Soviet communism. The rule of the Communist Party in the different states of Eastern Europe and the Soviet Union was facilitated economically by state ownership and central direction of the means of production, and politically by an elaborate network of spies and secret police. Second, cultural conformity and ideological consent were promoted and fostered by equating communist dogma with state truth, and state truth with political truth. All opposition to communism was, as a consequence, opposition to the state and to its political truths (and vice versa). Opposition to any was opposition to all and was dubbed variously as "anti-patriotic," "pro-imperialist" and "counter-revolutionary."

Despite the Soviet's dominance within Eastern Europe, popular uprisings against Soviet occupation and control periodically surfaced within its "satellites" – in the German Democratic Republic in 1953, in Hungary in 1956, in Czechoslovakia in 1968 and in Poland in 1981. Although these expressions of opposition proved unsuccessful, they were indicative of broader counter-hegemonic currents within the Soviet bloc. As Havel – a writer who was imprisoned for nine years by the Czech regime – argues, the power of the communist regime lay as much in its ability to produce consent from its citizens as from its ability to coerce them through force. That people either accepted the ideology of the system, or at least behaved as if they did, Havel terms "living within a lie." To engage in effective resistance to Soviet hegemony required challenges to the systems ideological manipulation, or "living within the truth." This involved freedom of political and artistic expression in all its myriad forms. What first came to the notice of the West as "dissent" – articulated by dissidents such as Andrei Sakharov (in the USSR), and Havel himself – was symptomatic of the development within the Soviet Union and the Warsaw Pact countries of various independent initiatives that emerged "from below." These sought to extend the space available within society for autonomous action out of the control and discipline of state political culture, articulating a "second culture." Moreover, such dissent set up parallel – and frequently underground – organizational forms that challenged the state's claims to truth and sought to strengthen the development of an independent civil society.

One of the many organizations that emerged throughout the Soviet bloc was the Czechoslovakia-based Charter 77 which was established by dissident writers and other professionals including Havel himself. Originally a human rights group attempting to ensure that Czechoslovakian politicians observe international law, the organization also issued position papers proposing alternative perspectives on economic, political, environmental, and social problems. The actions of such groups laid the groundwork for increased popular participation in challenging the Soviet bloc regimes in open protests as Gorbachev's policy of *glasnost* began to be implemented. For example, Charter 77 activists helped to organize the Civic Forum in 1989 – the umbrella organization that coordinated the Velvet revolution that ousted the communist regime in Czechoslovakia, and saw Havel himself become President of the Czech and Slovak Federative Republic.

The dissident movements in Eastern Europe also forged links with what proved to be the largest popular resistance against the Cold War itself – the peace movement – which opposed the deployment of Cruise and Pershing missiles in Europe by NATO and SS20s by the Soviet Union. The movement comprised a variety of anti-nuclear and anti-militarist groups, including the Nuclear Freeze in the US, the Campaign for Nuclear Disarmament (CND) in Britain, and the European Nuclear Disarmament (END) movement (see Reading 13). This movement emerged at a time of heightened tension between the superpowers, and ascendant US rearmament under the Reagan presidency. Although clearly differentiated on a state by state basis, END was non-statal, and sought to evolve mechanisms for transnational solidarity and identity, in an attempt to revitalize democracy within Europe. Intellectuals within the peace movement – such as historian E.P.Thompson (Reading 35) – articulated a theoretical critique of the Cold War, voicing opposition to the superpower arms race and the division of Europe into ideological and militarized blocs.

Thompson argues that the expansionist ideologies of the US and the USSR were the driving force of the Cold War, each legitimated through the threat of a demonized *other* (communism and capitalism respectively) that served to define an approved national self-image against that of one's ideological opponent. This ideology permeated both the state and civil society in many areas (e.g. the media), conflating the interests of the state with the public interest, thus compromising the integrity and autonomy of civil society. Thompson argues that the Cold War should be seen as a means by which the dominant states within each bloc controlled and disciplined their own citizens, populations, and clients, and by which those who stood to benefit from increased arms production and political anxiety (e.g. financial, commercial, military and political interests) promoted the rivalry. Hence the Cold War served as legitimation for both US and USSR intervention in other states, their appropriation of vast resources for military purposes, and for keeping powerful elites in both blocs in power. Within this geopolitical regime, nuclear weapons served to suppress the political process, substituting the threat of annihilation for the negotiated resolution of differences.

However, Thompson also notes that in reality the principle threat of the Cold War was not the demonized other but rather was within each of the superpower blocs - i.e. the peace movements of the Western bloc and the dissident movements of the Eastern bloc. These movements articulated both material challenges to superpower militarism – through direct action, underground organizations, etc. – and also an intellectual challenge to the geopolitical othering that the Cold War was predicated upon. Their calls for international solidarity, rather than antagonism, were seen as a threat to the power of political elites within each bloc to determine geopolitical spheres of influence. Moreover, by attempting to revitalize spaces of public autonomy, these movements challenged each superpower's ability to control public opinion.

George Konrad, a Hungarian writing during the Cold War, terms such resistances antipolitics (Reading 36). Antipolitics, as conceived by Konrad, is a moral force within civil society that articulates a distrust and public rejection of the power monopoly of the political class within the state – a power that is wielded against domestic populations through repressive legislation (e.g. censorship) and against others through the threat or prosecution of war. It is a practice that does not seek to overthrow the state, but opposes the political power that is exerted over people. The Cold War articulated a particularly dangerous manifestation of this power, since politicians within the NATO and Warsaw Pact blocs had the power to unleash weapons of mass destruction. Konrad critiques Cold War geopolitics as a form of terrorism, since it enables the powerful to keep the masses dominated by the threat of nuclear annihilation. The threat of war, Konrad argues, is synonymous with the absence of democracy. He critiques the complicity of intellectuals with the state, whose role is to manufacture ideological justifications for the prosecution of geopolitical power. In response, the project of antipolitics seeks to expose the propaganda that equates the preparation, threat or waging of war with patriotism. Moreover, it attempts to develop an internationalist solidarity, premised upon the notions of mutual co-existence, that transgresses state borders in an attempt to undermine the politics of othering upon which Cold War discourse is constructed.

NEW WORLD ORDER ANTI-GEOPOLITICS

With the revolutions of 1989, the demise of the Soviet Union and the Gulf War of 1990–1991, the geopolitical discourse of the US in particular, and the West in general, has shifted from that of the Cold War to that of the new world order. As discussed in Part 3, the Gulf War provided the rationale for this new discourse, which has geopolitical and geo-economic dimensions. The geopolitical dimension involves the maintenance of the US national security state and the legitimation of (continued) US military and economic intervention around the world in order to ensure "freedom" and "democracy."

However, as Abouali Farmanfarmaian – an Iranian writer – notes, this legitimation is constructed upon racialized and sexualized representations of (non-American) others (Reading 37). Farmanfarmaian argues that such representations are part of a constructed international division of attributes, which arose out of the colonial experience, that posit the world beyond the boundaries of the US and the West as uniformly characterized by chaos, irrationality and violence, including sexual violence. Such representations of others serve to establish and reinforce the construction of American self-identity, particularly at a time when the US economy and values are being internally and globally questioned.

Farmanfarmaian argues that the discourse surrounding the Gulf War drew upon these constructs in order to legitimate the US military response to Iraq. Framed within a sexualized discourse that sought to display American "virility," Farmanfarmaian argues that the war was prosecuted in order to reassert the country's military prowess following its perceived impotence in Vietnam and during the Iran hostage crisis. Moreover, he notes that the boundaries that define US identity – and inform its foreign policy – continue to depend on the representations of others as inferior. Such representations serve to legitimize geopolitical practices that are characterized by excessive military violence, rather than negotiated settlement, when dealing with non-Western states such as Iraq.

The geo-economic dimension of the new world order involves the doctrine of transnational liberalism or neoliberalism. The fundamental principal of this doctrine is economic liberty for the powerful, that is that an economy must be free from the social and political "impediments," "fetters" and "restrictions" placed upon it by states trying to regulate in the name of the public interest. These "impediments" – which include national economic regulations, social programs and class compromises (i.e. national bargaining agreements between employers and trade unions, assuming these are allowed) – are considered barriers to the free flow of trade and capital and the freedom of transnational corporations to exploit labor and the environment in their best interests. Hence, the doctrine argues that national economies should be deregulated (e.g. through the privatization of state enterprises) in order to promote the allocation of resources by "the market" which, in practice, means by the most powerful. As a result of the power of international organizations like the International Monetary Fund (IMF) and the World Bank to enforce the doctrine of neoliberalism upon developing states desperately in need of the liquidity controlled by these organizations, there has been a drastic reduction in government spending on health, education, welfare and environmental protection across the world. This has occurred as states strive to reduce inflation and satisify demands to open their markets to transnational corporations and capital inflows from abroad. Transnational liberalism celebrates capital mobility and "fast capitalism," the decentralization of production away from developed states and the centralization of control of the world economy in the hands of transnational corporations and their allies in key government agencies (particularly those of the seven most powerful countries, the G-7), large international banks, and institutions like the World Bank, the IMF and the World Trade Organization. As transnational corporations have striven to become "leaner and meaner" in this highly competitive global environment, they have engaged in massive cost-cutting and "downsizing," reducing the costs of wages, health care provisions and environmental protections in order to make production more competitive.

Transnational liberalism has been institutionalized through various international free trade agreements, such as the North American Free Trade Agreement (NAFTA) between the US, Canada and Mexico. These agreements are based upon the doctrine that each country and region should produce goods and services in which they have a competitive advantage, and that barriers to trade between countries (such as tariffs) should be reduced. However, such agreements are more concerned with removing the barriers to the movement of capital, to enable transnational corporations to operate without government interference or regulation, and to exploit the competitive advantage in cheap labor, lax environmental regulations and natural resources.

The resulting global competition for jobs and investment has resulted in the pauperization and marginalization of indigenous peoples, women, peasant farmers, and industrial workers, and a reduction in labor, social and environmental conditions - what Brecher and Costello (Reading 39) term "the race to the bottom" or "downward levelling." However, such processes have not occurred without challenges by their victims. One of the most prominent recent examples has been that of the *Ejercito Zapatista Liberacion National* – the EZLN or the Zapatistas – in Chiapas, Mexico, which has articulated resistance to the NAFTA and the Mexican state.

Coinciding the emergence of their rebellion with the coming into effect of the NAFTA, the Zapatistas, a predominantly indigenous (Mayan) guerrilla movement, demanded the democratization of Mexican civil society and an end to NAFTA which they argued was a "death certificate for the ethnic peoples of Mexico." Although they initially engaged in a guerrilla insurgency by occupying the capital of Chiapas and several other prominent towns in the state, the Zapatistas were more concerned to globalize their resistance. The appearance of an armed insurgency, at a moment when the Mexican economy was entering into a free trade agreement, enabled the Zapatistas to attract national and international media attention. Through their spokesperson, Subcommandante Marcos, the Zapatistas engaged in a war of words, fought primarily with communiqués rather than bullets, giving voice to the victims of neoliberalism (Reading 38). The particular importance of the Zapatista struggle has lain in its ability, with limited resources and personnel, to disrupt international financial markets, and their investments within Mexico, while exposing the inequities on which development and transnational liberalism are predicated.

Marcos articulates an alternative geography of the Mexican state of Chiapas, highlighting the exploitation of the economy, culture, and environment of the indigenous Mayan peoples and peasants to enrich national and international markets. Marcos considers Chiapas to be an internal colony within the modernizing, industrializing Mexican state. The "wind from above" – that of neoliberalism, culminating in the NAFTA accords – includes the amendment to Article 27 of the Mexican constitution which enabled all *ejidos* (individual and communally held peasant lands) to be sold to powerful corporate interests, and has been accompanied by the impoverishment of Chiapas' indigenous people and the militarization of the state. The response to this has been what Marcos terms the "wind from below" – an increasing number of peasant rebellions throughout Chiapas, and subsequently Mexico, demanding democratic and economic rights.

The Zapatistas are but one example of resistances to transnational liberalism that have proliferated across the world during the past fifteen years. These have involved leftist guerrillas, social movements, non-government organizations, human rights groups, environmental organizations, and indigenous peoples movements. Frequently coalitions have formed across national borders and across different political ideologies in order to oppose transnational institutions and agreements. Such resistances are frequently responses to local conditions that are in part the product of global forces, and resistance to these conditions has taken place at both the local and the global level. In contrast to official political discourse about the global economy, these challenges articulate a "globalization from below" that comprises a "geopolitics from below" – an evolving international network of groups, organizations and social movements.

Jeremy Brecher, a historian, and Tim Costello, a labor activist (Reading 39), articulate a normative agenda for such resistance, to counter and transform what they term the corporate agenda of transnational liberalism. This includes revitalizing democratic practices and public institutions, promoting economic and environmental sustainability, encouraging grassroots economic development, and holding transnational corporations accountable to enforceable codes of conduct. Such an agenda has to be enacted across local, national, and global scales if it is to be aimed at empowering popular collective action. However, while the discursive logics of the Cold War effectively prevented such an agenda in the past, the contemporary material and ideological hegemony of transnational liberalism is already marginalizing any alternatives to itself. Just as it was crucial to imagine a world beyond the geopolitics of the Cold War in order to challenge that hegemony, so Brecher and Costello argue that we must begin to imagine and create alternatives to the geo-economics of the new world order to establish an economically just and environmentally sustainable future.

REFERENCES AND FURTHER READING

On colonial anti-geopolitics

Fanon, F. (1965) *A Dying Colonialism*, New York: Grove Press.

Guha, R. and Spivak, G.C. (1988) *Selected Subaltern Studies*, New York: Oxford University Press.

Memmi, A. (1965) *The Colonizer and the Colonized*, Boston: Beacon Press.

Ngugi, wa Thiong'o (1986) *Decolonising the Mind*, London: Heinemann.

Said, E. (1978) *Orientalism*, New York: Vintage Books.

——(1993) *Culture and Imperialism*, London: Vintage Books.

Scott, J.C. (1985) *Weapons of the Weak*, New Haven: Yale University Press.

——(1990) *Domination and the Arts of Resistance*, New Haven: Yale University Press.

Wolf, E.R. (1969) *Peasant Wars of the Twentieth Century*, New York: Harper and Row.

On Cold War anti-geopolitics

Amin, S. (1976) *Unequal Development: An Essay on the Social Formations of Peripheral Capitalism*, New York: Monthly Review Press.

Cardoso, F.H. (1972) "Dependency and Development in Latin America," *New Left Review*, 74 (July–August): 83–95.

Cardoso, F.H. and Faletto, E. (1979) *Dependency and Development*, Berkeley: University of California Press.

Chailand, G. (1982) *Guerrilla Strategies*, London: Penguin.

——(1989) *Revolution in the Third World*, London: Penguin.

Dos Santos, T. (1970) "The Structure of Dependence," *American Economic Review*, 60 (May): 231–236.

Emmanuel, A. (1972) *Unequal Exchange: A Study of the Imperialism of Trade*, New York: Monthly Review Press.

Frank, A.G. (1967) *Capitalism and Underdevelopment in Latin America*, New York: Monthly Review Press.

——(1978) *Dependent Accumulation and Underdevelopment*, London: Macmillan.

Galeano, E. (1973) *The Open Veins of Latin America*, New York: Monthly Review Press.

——(1995) *Memory of Fire*, London: Quartet Books.

Katziaficas, G. (1987) *The Imagination of the New Left: A Global Analysis of 1968*, Boston: South End Press.

Love, J.L. (1980) "Raul Prebisch and the Origins of the Doctrine of Unequal Exchange," *Latin American Research Review*, 15(3): 45–72.

Pearce, J. (1981) *Under the Eagle*, London: Latin American Bureau.

Sharp, G. (1973) *The Politics of Nonviolent Action* (3 vols), Boston: Porter Sargent.

Smith, D. and Thompson, E.P. (eds) (1987) *Prospectus for a Habitable Planet*, London: Penguin.

Zinn, H. (1980) *A People's History of the US*, New York: Harper and Row.

On new world order anti-geopolitics

Churchill, W. (1993) *Struggle for the Land*, Monroe, ME: Common Courage Press.

Ekins, P. (1992) *A New World Order: Grassroots Movements for Global Change*, New York: Routledge.

Escobar, A. and Alvarez, S.E. (eds) (1992) *The Making of Social Movements in Latin America*, Boulder: Westview Press.

Esteva, G. (1987) "Regenerating People's Space," *Alternatives*, 12: 125–152.

Foucault, M. (1980) *Power/Knowledge*, New York: Pantheon Books.

Gedicks, A. (1993) *The New Resource Wars*, Boston: South End Press.

Ghai, D. and Vivian, J. (1992) *Grassroots Environmental Action*, London: Routledge.

Gramsci, A. (1971) *Prison Notebooks*, New York: International Publishers.

Haraway, D. (1991) "Situated Knowledges: The

Science Question in Feminism and the Privilege of Partial Perspective," in Haraway, D. *Simians, Cyborgs, and Women*, London: Free Association Books.

Hecht, S. and Cockburn, A. (1990) *The Fate of the Forest*, New York: Harper Collins.

Mendlovitz, S.H. and Walker, R.B.J. (eds) (1987) *Towards a Just World Peace: Perspectives from Social Movements*, London: Butterworths.

Peet, R. and Watts, M. (1993) "Development Theory and Environment in an Age of Market Triumphalism," *Economic Geography*, 69(3): 227–253.

Ross, J. (1995) *Rebellion from the Roots*, ME: Common Courage Press.

Sachs, W. (ed.) (1992) *The Development Dictionary*, London: Zed Books.

31 EDWARD SAID

"Orientalism Reconsidered"

from *Europe And Its Others, Volume 1* (1984)

There are two sets of problems that I'd like to take up, each of them deriving from the general issues addressed in *Orientalism*.

As a department of thought and expertise Orientalism of course refers to several overlapping domains: firstly, the changing historical and cultural relationship between Europe and Asia, a relationship with a 4000 year old history; secondly, the scientific discipline in the West according to which, beginning in the early nineteenth century, one specialised in the study of various Oriental cultures and traditions; and, thirdly, the ideological suppositions, images and fantasies about a currently important and politically urgent region of the world called the Orient. The relatively common denominator between these three aspects of Orientalism is the line separating Occident from Orient and this, I have argued, is less a fact of nature than it is a fact of human production, which I have called imaginative geography.

This is, however, neither to say that the division between Orient and Occident is unchanging nor is it to say that it is simply fictional. It is to say – emphatically – that . . . the Orient and the Occident are facts produced by human beings, and as such must be studied as integral components of the social, and not the divine or natural, world. And because the social world includes the person or subject doing the studying as well as the object or realm being studied, it is imperative to include them both in any consideration of Orientalism for, obviously enough, there could be no Orientalism without, on the one hand, the Orientalists, and on the other, the Orientals.

Yet, and this is the first set of problems I want to consider, there is still a remarkable unwillingness to discuss the problems of Orientalism in the political or ethical or even epistemological contexts proper to it. This is as true of professional literary critics who have written about my book, as it is of course of the Orientalists themselves. Since it seems to me patently impossible to dismiss the truth of Orientalism's political origin and its continuing political actuality, we are obliged on intellectual as well as political grounds to investigate the resistance to the politics of Orientalism, a resistance that is richly symptomatic of precisely what is denied.

If the first set of problems is concerned with the problems of Orientalism reconsidered from the standpoint of local issues, like who writes or studies the Orient, in what institutional or discursive setting, for what audience, and with what ends in mind, the second set of problems takes us to a wider circle of issues. These are the issues raised initially by methodology and then considerably sharpened by questions as to how the production of knowledge best serves communal, as opposed to factional, ends, how knowledge that is non-dominative and non-coercive can be produced in a setting that is deeply inscribed with the politics, the considerations, the positions and the strategies of power. In these methodological and moral re-considerations of Orientalism I shall quite consciously be alluding to similar issues raised by the experiences of feminism or women's studies, black or ethnic studies, socialist and anti-imperialist studies, all of which take for their point of departure the right of formerly un- or mis-represented human groups to speak for and represent themselves in domains defined, politically and intellectually as normally excluding them, usurping their signifying and representing functions, overriding their historical reality. In short, Orientalism reconsidered in this wider and libertarian optic entails nothing less than the creation of new objects for a new kind of knowledge.

Certainly there can be no doubt that – in my own rather limited case – the consciousness of being an Oriental goes back to my youth in colonial Palestine and Egypt, although the impulse to resist its accompanying impingements was nurtured in the heady atmosphere of the post-World War II period of independence when Arab nationalism, Nasserism, the 1967 War, the rise of the Palestine national movement, the 1973 War, the Lebanese Civil War, the Iranian Revolution and its horrific aftermath, produced that extraordinary series of highs and lows which has neither ended nor allowed us a full understanding of its remarkable revolutionary impact.

The interesting point here is how difficult it is to try to understand a region of the world whose principal features seem to be, first, that it is in perpetual flux, and second, that no one trying to grasp it can by an act of pure will or of sovereign understanding stand at some Archimedean point outside the flux. That is, the very reason for understanding the Orient generally, and the Arab world in particular, was first that it prevailed upon one, beseeched one's attention urgently, whether for economic, political, cultural or religious reasons, and second, that it defied neutral, disinterested, or stable definition.

... [E]ven so relatively inert an object as a literary text is commonly supposed to gain some of its identity from its historical moment interacting with the attentions, judgements, scholarship and performances of its readers. But, I discovered, this privilege was rarely allowed the Orient, the Arabs, or Islam, which separately or together were supposed by mainstream academic thought to be confined to the fixed status of an object frozen once and for all in time by the gaze of Western percipients.

Far from being a defence either of the Arabs or Islam – as my book was taken by many to be – my argument was that neither existed except as "communities of interpretation" which gave them existence, and that, like the Orient itself, each designation represented interests, claims, projects, ambitions and rhetorics that were not only in violent disagreement, but were in a situation of open warfare. So saturated with meanings, so overdetermined by history, religion and politics are labels like "Arab" or "Muslim" as subdivisions of "The Orient" that no one today can use them without some attention to the formidable polemical mediations that screen the objects, if they exist at all, that the labels designate.

I do not think it is too much to say that the more these observations have been made by one party, the more routinely they are denied by the other; this is true whether it is Arabs or Muslims discussing the meaning of Arabism or Islam, or whether an Arab or Muslim disputes these designations with a Western scholar. Anyone who tries to suggest that nothing, not even a simple descriptive label, is beyond or outside the realm of interpretation, is almost certain to find an opponent saying that science and learning are designed to transcend the vagaries of interpretation, and that objective truth is in fact attainable. This claim was more than a little political when used against Orientals who disputed the authority and objectivity of an Orientalism intimately allied with the great mass of European settlements in the Orient.

The challenge to Orientalism and the colonial era of which it is so organically a part, was a challenge to the muteness imposed upon the Orient as object. Insofar as it was a science of incorporation and inclusion by virtue of which the Orient was constituted and then introduced into Europe, Orientalism was a scientific movement whose analogue in the world of empirical politics was the Orient's colonial accumulation and acquisition by Europe. The Orient was therefore not Europe's interlocutor, but its silent Other. From roughly the end of the eighteenth century, when in its age, distance and richness the Orient was re-discovered by Europe, its history had been a paradigm of antiquity and originality, functions that drew Europe's interests in acts of recognition or acknowledgement but *from* which Europe moved as its own industrial, economic and cultural development seemed to leave the Orient far behind. Oriental history – for Hegel, for

Marx, later for Burkhardt, Nietzsche, Spengler, and other major philosophers of history – was useful in protraying a region of great age, and what had to be left behind.

Here, of course, is perhaps the most familiar of Orientalism's themes – they cannot represent themselves, they must therefore be represented by others who know more about Islam than Islam knows about itself. Now it is often the case that you can be known by others in different ways than you know yourself, and that valuable insights might be generated accordingly. But that is quite a different thing than pronouncing it as immutable law that outsiders *ipso facto* have a better sense of you as an insider than you do of yourself. Note that there is no question of an *exchange* between Islam's views and an outsider's: no dialogue, no discussion, no mutual recognition. There is a flat assertion of quality, which the Western policy-maker, or his faithful servant, possesses by virtue of his being Western, white, non-Muslim.

Now this, I submit, is neither science, nor knowledge, nor understanding: it is a statement of power and a claim for relatively absolute authority. It is constituted out of racism, and it is made comparatively acceptable to an audience prepared in advance to listen to its muscular truths . . . for whom Islam is not a culture, but a nuisance . . . associate[d] with the other nuisances of the 1960s and the 1970s – blacks, women, post-colonial Third World nations that have tipped the balance against the US in such places as UNESCO and the UN.

. . . Orientalism's large political setting, which is routinely denied and suppressed . . . comprises two other elements, about which I'd like to speak very briefly, namely the recent (but at present uncertain) prominence of the Palestinian movement, and secondly, the demonstrated resistance of Arabs in the United States and elsewhere against their portrayal in the public realm.

As for the Palestinian issue . . . the Israeli occupation of the West Bank and Gaza, the destruction of Palestinian society, and the sustained Zionist assault upon Palestinian nationalism have quite literally been led and staffed by Orientalists. Whereas in the past it was European Christian Orientalists who supplied European culture with arguments for colonising and suppressing Islam, as well as for despising Jews, it is now the Jewish national movement that produces a cadre of colonial officials whose ideological theses about the Islamic or Arab mind are implemented in the administration of the Palestinian Arabs, an oppressed minority within the white-European-democracy that is Israel. . . . Hebrew University's Islamic studies department has produced every one of the colonial officials and Arab experts who run the Occupied Territories.

Underlying much of the discussion of Orientalism is a disquieting realisation that the relationship between cultures is both uneven and irremediably secular. This brings us to the point I alluded to a moment ago, about recent Arab and Islamic efforts, well-intentioned for the most part, but sometimes motivated by unpopular regimes, who in attracting attention to the shoddiness of the Western media in representing the Arabs or Islam divert scrutiny from the abuses of their rule and therefore make efforts to improve the so-called image of Islam and the Arabs. Parallel developments have been occurring . . . in UNESCO where the controversy surrounding the world information order – and proposals for its reform by various Third World and Socialist governments – has taken on the dimensions of a major international issue. Most of these disputes testify, first of all, to the fact that the production of knowledge, or information, of media images is unevenly distributed: its locus, and the centers of its greatest force are located in what, on both sides of the divide, has been polemically called the metropolitan West. Secondly, this unhappy realisation on the part of weaker parties and cultures has reinforced their grasp of the fact that although there are many divisions within it, there is only one secular and historical world, and that neither nativism, nor divine intervention, nor regionalism, nor ideological smokescreens can hide societies, cultures and peoples from each other, especially not from those with the force and will to penetrate others for political as well as economic ends. But, thirdly, many of these disadvantaged post-colonial states and their loyalist intellectuals have, in my opinion, drawn the wrong set of conclusions, which in practice is that one must either attempt to impose control upon the production of knowledge at the source, or, in the worldwide media economy, to attempt to improve, enhance, ameliorate the images currently in circulation without doing anything to change the political situation from which they emanate and on which to a certain extent they are based.

The failings of these approaches strike me as obvious, and here I don't want to go into such matters as the squandering of immense amounts of petro-dollars for various short-lived public relations scams, or the increasing repression, human-rights abuses, outright gangsterism that has taken place in many formerly colonial countries, all of them occurring in the name of national security and fighting neo-imperialism. What I do want to talk about is the much larger question of what, in the context recently provided by such relatively small efforts as the critique of Orientalism, is to be done, and on the level of politics and criticism how we can speak of intellectual work that isn't merely reactive or negative.

I come finally now to the second and, in my opinion, the more challenging and interesting set of problems that derive from the reconsideration of Orientalism. One of the legacies of Orientalism, and indeed one of its epistemological foundations, is historicism, that is, the view propounded by Vico, Hegel, Marx, Ranke, Dilthey and others, that if humankind has a history it is produced by men and women, and can be understood historically as, at each given period, epoch or moment, possessing a complex, but coherent unity. So far as Orientalism in particular and the European knowledge of other societies in general have been concerned, historicism meant that the one human history uniting humanity either culminated in or was observed from the vantage point of Europe, or the West. What was neither observed by Europe nor documented by it was therefore "lost" until, at some later date, it too could be incorporated by the new sciences of anthropology, political economics and linguistics. It is out of this later recuperation of what Eric Wolf has called people without history, that a still later disciplinary step was taken, the founding of the science of world history, whose major practitioners include Braudel, Wallerstein, Perry Anderson and Wolf himself.

But along with the greater capacity for dealing with – in Ernst Bloch's phrase – the non-synchronous experiences of Europe's Other, has gone a fairly uniform avoidance of the relationship between European imperialism and these variously constituted, variously formed and articulated knowledges. What, in other words, has never taken place is an epistemological critique at the most fundamental level of the connection between the development of a historicism which has expanded and developed enough to include antithetical attitudes such as ideologies of Western imperialism and critiques of imperialism on the one hand, and on the other, the actual practise of imperialism by which the accumulation of territories and population, the control of economies, and the incorporation and homogenisation of histories are maintained. If we keep this in mind we will remark, for example, that in the methodological assumptions and practice of world history – which is ideologically anti-imperialist – little or no attention is given to those cultural practices like Orientalism or ethnography affiliated with imperialism, which in genealogical fact fathered world history itself; hence the emphasis in world history as a discipline has been on economic and political practices, defined by the processes of world historical writing, as in a sense separate and different from, as well as unaffected by, the knowledge of them which world history produces. The curious result is that the theories of accumulation on a world scale, or the capitalist world state, or lineages of absolutism (a) depend on the same displaced percipient and historicist observer who had been an Orientalist or colonial traveller three generations ago; (b) depend also on a homogenising and incorporating world historical scheme that assimilated non-synchronous developments, histories, cultures and peoples to it; and (c) block and keep down latent epistemological critiques of the institutional, cultural and disciplinary instruments linking the incorporative practice of world history with partial knowledges like Orientalism on the one hand, and on the other, with continued "Western" hegemony of the non-European, peripheral world.

In fine, the problem is once again historicism

and the universalising and self-validating that has been endemic to it ... in a whole series of studies produced in a number of both interrelated and frequently unrelated fields, there has been a general advance in the process of, as it were, breaking up, dissolving and methodologically as well as critically re-conceiving the unitary field ruled hitherto by Orientalism, historicism, and what could be called essentialist universalism.

I shall be giving examples of this dissolving and decentering process in a moment. What needs to be said about it immediately is that it is neither purely methodological nor purely reactive in intent. You do not respond, for example, to the tyrannical conjuncture of colonial power with scholarly Orientalism simply by proposing an alliance between nativist sentiment buttressed by some variety of native ideology to combat them. This, it seems to me, has been the trap into which many Third World and anti-imperialist activists fell in supporting the Iranian and Palestinian struggles, and who found themselves either with nothing to say about the abominations of Khomeini's regime or resorting, in the Palestine case, to the time-worn clichés of revolutionism and, if I might coin a deliberately barbaric phrase, rejectionary armed-strugglism after the Lebanese debacle. Nor can it be a matter simply of re-cycling the old Marxist or world historical rhetoric which only accomplishes the dubiously valuable task of re-establishing intellectual and theoretical ascendancy of the old, by now impertinent and genealogically flawed, conceptual models. No: we must, I believe, think both in political and above all theoretical terms, locating the main problems in what Frankfurt theory identified as domination and division of labor, and along with those, the problem of the absence of a theoretical and utopian as well as libertarian dimension in analysis. We cannot proceed unless therefore we dissipate and re-dispose the material of historicism into radically different objects and pursuits of knowledge, and we cannot do that until we are aware clearly that no new projects of knowledge can be constituted unless they fight to remain free of the dominance and professionalised particularism that comes with historicist systems and reductive, pragmatic or functionalist theories.

These goals are less grand and difficult than my description sounds. For the re-consideration of Orientalism has been intimately connected with many other activities of the sort I referred to earlier, and which it now becomes imperative to articulate in more detail. Thus, for example, we can now see that Orientalism is a praxis of the same sort, albeit in different territories, as male gender dominance, or patriarchy, in metropolitan societies: the Orient was routinely described as feminine, its riches as fertile, its main symbols the sensual woman, the harem and the despotic – but curiously attractive – ruler. Moreover, Orientals like Victorian housewives were confined to silence and to unlimited enriching production. Now much of this material is manifestly connected to the configurations of sexual, racial and political asymmetry underlying mainstream modern Western culture, as adumbrated and illuminated respectively by feminists, by black studies critics and by anti-imperialist activists.

What I want to do in conclusion is to try to draw ... the larger enterprise of which the critique of Orientalism is a part. Firstly, we note a plurality of audiences and constituencies; none of the works and workers I have cited claims to be working on behalf of One audience which is the only one that counts, or for one supervening, over-coming Truth, a truth allied to Western (or for that matter Eastern) reason, objectivity, science. On the contrary, we note here a plurality of terrains, multiple experiences and different constituencies, each with its admitted (as opposed to denied) interest, political desiderata, disciplinary goals. All these efforts work out of what might be called a decentered consciousness, not less reflective and critical for being decentered, for the most part non- and in some cases anti-totalizing and anti-systematic. The result is that instead of seeking common unity by appeals to a center of sovereign authority, methodological consistency, canonicity and science, they offer the possibility of common grounds of assembly between them. They are therefore planes of activity and praxis, rather then one topography commanded by a geographical and historical vision locatable in a known centre of metropolitan power. Secondly, these activities and praxes are consciously secular, marginal and oppositional with reference to the mainstream, generally authoritarian systems from which they emanate, and against which they now agitate. Thirdly, they are political and practical in as much as they intend – without necessarily succeeding in implementing the end of dominating, coercive systems of knowledge. I do not think it too much to say that the political meaning of analysis, as carried out in all these fields, is uniformly and programmatically libertarian by virtue of the fact that, unlike Orientalism, it is not based on the finality and closure of antiquarian or curatorial knowledge, but on investigative open models of analysis, even though it might seem that analyses of this sort – frequently difficult and abstruse – are in the final count paradoxically quietistic. I think we must remember the lesson provided by Adorno's negative dialectics, and regard analysis as in the fullest sense being *against* the grain, deconstructive, utopian.

But there remains the one problem haunting all intense, self-convicted and local intellectual work, the problem of the division of labor, which is a necessary consequence of that reification and commodification first and most powerfully analysed in this century by George Lukacs. This is the problem sensitively and intelligently put by Myra Jehlen for women's studies, whether in identifying and working through anti-dominant critiques, subaltern groups – women, blacks, and so on – can resolve the dilemma of autonomous fields of experience and knowledge that are created as a consequence. A double kind of possessive exclusivism could set in: the sense of being an excluding insider by virtue of experience (only women can write for and about women, and only literature that treats women or Orientals well is good literature), and secondly, being an excluding insider by virtue of method (only Marxists, anti-Orientalists, feminists can write about economics, Orientalism, women's literature).

This is where we are now, at the threshold of fragmentation and specialisation, which impose their own parochial dominations and fussy defensiveness, or on the verge of some grand synthesis which I for one believe could very easily wipe out both the gains and the oppositional consciousness provided by these counter-knowledges hitherto. Several possibilities propose themselves, and I shall conclude simply by listing them. A need for greater crossing of boundaries, for greater interventionism in cross-disciplinary activity, a concentrated awareness of the situation – political, methodological, social, historical – in which intellectual and cultural work is carried out. A clarified political and methodological commitment to the dismantling of systems of domination which since they are collectively maintained must, to adopt and transform some of Gramsci's phrases, be collectively fought, by mutual siege, war of manoeuvre and war of position. Lastly, a much sharpened sense of the intellectual's role both in the defining of a context and in changing it, for without that, I believe, the critique of Orientalism is simply an ephemeral pastime.

32

FRANTZ FANON

"Concerning Violence"

from *The Wretched of the Earth* (1963)

Decolonization is the meeting of two forces, opposed to each other by their very nature, which in fact owe their originality to that sort of substantification which results from and is nourished by the situation in the colonies. Their first encounter was marked by violence and their existence together – that is to say the exploitation of the native by the settler – was carried on by dint of a great array of bayonets and cannons.

In decolonization, there is therefore the need of a complete calling in question of the colonial situation. If we wish to describe it precisely, we might find it in the well known words: "The last shall be first and the first last." Decolonization is the putting into practice of this sentence. The naked truth of decolonization evokes for us the searing bullets and bloodstained knives which emanate from it. For if the last shall be first, this will only come to pass after a murderous and decisive struggle between the two protagonists.

The colonial world is a world cut in two. The dividing line, the frontiers, are shown by barracks and police stations. In the colonies it is the policeman and the soldier who are the official, instituted go-betweens, the spokesmen of the settler and his rule of oppression. In capitalist societies the educational system, whether lay or clerical, the structure of moral reflexes handed down from father to son, the exemplary honesty of workers who are given a medal after fifty years of good and loyal service, and the affection which springs from harmonious relations and good behavior – all these aesthetic expressions of respect for the established order serve to create around the exploited person an atmosphere of submission and of inhibition which lightens the task of policing considerably. In the capitalist countries a multitude of moral teachers, counselors and "bewilderers" separate the exploited from those in power. In the colonial countries, on the contrary, the policeman and the soldier, by their immediate presence and their frequent and direct action, maintain contact with the native and advise him by means of rifle butts and napalm not to budge. It is obvious here that the agents of government speak the language of pure force. The intermediary does not lighten the oppression, nor seek to hide the domination; he shows them up and puts them into practice with the clear conscience of an upholder of the peace; yet he is the bringer of violence into the home and into the mind of the native.

The zone where the natives live is not complementary to the zone inhabited by the settlers. The two zones are opposed, . . . they both follow the principle of reciprocal exclusivity.

The settlers' town is a strongly built town, all made of stone and steel. It is a brightly lit town; the streets are covered with asphalt, and the garbage cans swallow all the leavings, unseen, unknown and hardly thought about. The settler's feet are never visible, except perhaps in the sea; but there you're never close enough to see them. His feet are protected by strong shoes although the streets of his town are clean and even, with no holes or stones. The settler's town is a well-fed town, an easygoing town; its belly is always full of good things. The settlers' town is a town of white people, of foreigners.

The town belonging to the colonized people, or at least the native town, the Negro village, the *medina*, the reservation, is a place of ill fame, peopled by men of evil repute. They are born there, it matters little where or how; they die there, it matters not where, nor how. It is a world without spaciousness; men live there on top of each other, and their huts are built one on top of the other. The native town is a hungry town, starved of bread, of meat, of shoes, of coal, of light. The native town is a crouching village, a town on its knees, a town wallowing in the mire. It is a town of niggers and dirty Arabs.

In the colonies, the foreigner coming from another country imposed his rule by means of guns and machines. In defiance of his successful transplantation, in spite of his appropriation, the settler still remains a foreigner. It is neither the act of owning factories, nor estates, nor a bank balance which distinguishes the governing classes. The governing race is first and foremost those who come from elsewhere, those who are unlike the original inhabitants, "the others."

To break up the colonial world does not mean that after the frontiers have been abolished lines of communication will be set up between the two zones. The destruction of the colonial world is no more and no less than the abolition of one zone, its burial in the depths of the earth or its expulsion from the country.

The colonial world is a Manichean world. It is not enough for the settler to delimit physically, that is to say with the help of the army and the police force, the place of the native. As if to show the totalitarian character of colonial exploitation the settler paints the native as a sort of quintessence of evil. Native society is not simply described as a society lacking in values. It is not enough for the colonist to affirm that those values have disappeared from, or still better never existed in, the colonial world. The native is declared insensible to ethics; he represents not only the absence of values, but also the negation of values. He is, let us dare to admit, the enemy of values, and in this sense he is the absolute evil. He is the corrosive element, destroying all that comes near him; he is the deforming element, disfiguring all that has to do with beauty or morality; he is the depository of maleficent powers, the unconscious and irretrievable instrument of blind forces.

At times this Manicheism goes to its logical conclusion and dehumanizes the native, or to speak plainly, it turns him into an animal. In fact, the terms the settler uses when he mentions the native are zoological terms. He speaks of the yellow man's reptilian motions, of the stink of the native quarter, of breeding swarms, of foulness, of spawn, of gesticulations.

The native knows all this, and laughs to himself every time he spots an allusion to the animal world in the other's words. For he knows that he is not an animal; and it is precisely at the moment he realizes his humanity that he begins to sharpen the weapons with which he will secure its victory.

In the colonial context the settler only ends his work of breaking in the native when the latter admits loudly and intelligibly the supremacy of the white man's values. In the period of decolonization, the colonized masses mock at these very values, insult them and vomit them up.

The immobility to which the native is condemned can only be called in question if the native decides to put an end to the history of colonization – the history of pillage – and to bring into existence the history of the nation – the history of decolonization.

The uprising of the new nation and the breaking down of colonial structures are the result of one of two causes: either of a violent struggle of the people in their own right, or of action on the part of surrounding colonized peoples which acts as a brake on the colonial regime in question.

A colonized people is not alone. In spite of all that colonialism can do, its frontiers remain open to new ideas and echoes from the world outside. It discovers that violence is in the atmosphere, that it here and there bursts out, and here and there sweeps away the colonial regime – that same violence which fulfills for the native a role that is not simply informatory, but also operative. The great victory of the Vietnamese people at Dien Bien Phu is no longer, strictly speaking, a Vietnamese victory. Since July, 1954, the question which the colonized peoples have asked themselves has been, "What must be done to bring about another Dien Bien Phu? How can we manage it?" Not a single colonized individual could ever again doubt the possibility of a Dien Bien Phu; the only problem was how best to use the forces at their disposal, how to organize them, and when to bring them into action. This encompassing violence does not work upon the colonized people only; it modifies the attitude of the colonialists who become aware of manifold Dien Bien Phus. This is why a veritable panic takes hold of the colonialist governments in turn. Their purpose is to capture the vanguard, to turn the movement of liberation toward the fight, and to disarm the people: quick, quick, let's decolonize. Decolonize the Congo before it turns into another Algeria.... To the strategy of Dien Bien Phu, defined by the colonized peoples, the colonialist replies by the strategy of encirclement – based on the respect of the sovereignty of states.

... [T]he reconstruction of the nation continues within the framework of cutthroat competition between capitalism and socialism.

This competition gives an almost universal dimension to even the most localized demands. Every meeting held, every act of repression committed, reverberates in the international arena ... each act of sedition in the Third World makes up part of a picture framed by the Cold War. Two men are beaten up in Salisbury, and at once the whole of a bloc goes into action, talks about those two men, and uses the beating-up incident to bring up the particular problem of Rhodesia, linking it, moreover, with the whole African question and with the whole question of colonized people. The other bloc however is equally concerned in measuring by the magnitude of the campaign the local weaknesses of its system. Thus the colonized peoples realize that neither clan remains outside local incidents. They no longer limit themselves to regional horizons, for they have caught on to the fact that they live in an atmosphere of international stress.

When every three months or so we hear that the Sixth or Seventh Fleet is moving toward such-and-such a coast; when Khrushchev threatens to come to Castro's aid with rockets; when Kennedy decides upon some desperate solution for the Laos question, the colonized person or the newly independent native has the impression that whether he wills it or not he is being carried away in a kind of frantic cavalcade. In fact, he is marching in it already.

Strengthened by the unconditional support of the socialist countries, the colonized peoples fling themselves with whatever arms they have against the impregnable citadel of colonialism. If this citadel is invulnerable to knives and naked fists, it is no longer so when we decide to take into account the context of the Cold War. In this fresh juncture, the Americans take their role of patron of international capitalism very seriously. Early on, they advise the European countries to decolonize in a friendly fashion. Later on, they do not hesitate to proclaim first the respect for and then the support of the principle of "Africa for the Africans." The United States is not afraid today of stating officially that they are the defenders of the right of all peoples to self-determination.

... [W]e understand why the violence of the native is only hopeless if we compare it in the abstract to the military machine of the oppressor. On the other hand, if we situate that violence in the dynamics of the international situation, we see at once that it constitutes a terrible menace for the oppressor. Persistent *jacqueries* and Mau-Mau disturbance unbalance the colony's economic life but do not endanger the mother country. What is more important in the eyes of imperialism is the opportunity for socialist propaganda to infiltrate among the masses and to contaminate them. This is already a serious danger in the Cold War; but what would happen to that colony in case of real war, riddled as it is by murderous guerrillas? Thus capitalism realizes that its military strategy has everything to lose by the outbreak of nationalist wars.

Again, within the framework of peaceful coexistence, all colonies are destined to disappear, and in the long run neutralism is destined to be respected by capitalism. What must at all costs be avoided is strategic insecurity: the breakthrough of enemy doctrine into the masses and the deep-rooted hatred of millions of men. The colonized peoples are very well aware of these imperatives which rule international political life; for this reason even those who thunder denunciations of violence take their decisions and act in terms of this universal violence.

The mobilization of the masses, when it arises out of the war of liberation, introduces into each man's consciousness the ideas of a common cause, of a national destiny, and of a collective history. In the same way the second phase, that of the building-up of the nation, is helped on by the existence of this cement which has been mixed with blood and anger. Thus we come to a fuller appreciation of the originality of the words used in these underdeveloped countries. During the colonial period the people are called upon to fight against oppression; after national liberation, they are called upon to fight against poverty, illiteracy, and underdevelopment. The struggle, they say, goes on. The people realize that life is an unending contest.

33 MARTIN LUTHER KING

"A Time to Break Silence"

from *A Testament of Hope: The Essential Writings of Martin Luther King, Jr.* (1986)

THE IMPORTANCE OF VIETNAM

Since I am a preacher by trade, I suppose it is not surprising that I have ... major reasons for bringing Vietnam into the field of my moral vision. There is at the outset a very obvious and almost facile connection between the war in Vietnam and the struggle I, and others, have been waging in America. A few years ago there was a shining moment in that struggle. It seemed as if there was a real promise of hope for the poor – both black and white – through the poverty program. There were experiments, hopes, new beginnings. Then came the buildup in Vietnam and I watched the program broken and eviscerated as if it were some idle political plaything of a society gone mad on war, and I knew that America would never invest the necessary funds or energies in rehabilitation of its poor so long as adventures like Vietnam continued to draw men and skills and money like some demonic destructive suction tube. So I was increasingly compelled to see the war as an enemy of the poor and to attack it as such.

Perhaps the more tragic recognition of reality took place when it became clear to me that the war was doing far more than devastating the hopes of the poor at home. It was sending their sons and their brothers and their husbands to fight and to die in extraordinarily high proportions relative to the rest of the population. We were taking the black young men who had been crippled by our society and sending them eight thousand miles away to guarantee liberties in Southeast Asia which they had not found in southwest Georgia and East Harlem. So we have been repeatedly faced with the cruel irony of watching Negro and white boys on TV screens as they kill and die together for a nation that has been unable to seat them together in the same schools. So we watch them in brutal solidarity burning the huts of a poor village, but we realize that they would never live on the same block in Detroit. I could not be silent in the face of such cruel manipulation of the poor.

My third reason moves to an even deeper level of awareness, for it grows out of my experience in the ghettos of the North over the last three years – especially the last three summers. As I have walked among the desperate, rejected and angry young men I have told them that Molotov cocktails and rifles would not solve their problems. I have tried to offer them my deepest compassion while maintaining my conviction that social change comes most meaningfully through nonviolent action. But they asked – and rightly so – what about Vietnam? They asked if our own nation wasn't using massive doses of violence to solve its problems, to bring about the changes it wanted. Their questions hit home, and I knew that I could never again raise my voice against the violence of the oppressed in the ghettos without having first spoken clearly to the greatest purveyor of violence in the world today – my own government. For the sake of those boys, for the sake of this government, for the sake of the hundreds of thousands trembling under our violence, I cannot be silent.

For those who ask the question, "Aren't you a civil rights leader?" and thereby mean to exclude me from the movement for peace, I have this further answer. In 1957 when a group of us formed the Southern Christian Leadership Conference, we chose as our motto: "To save the soul of America." We were convinced that we could not limit our vision to certain rights for black people, but instead affirmed the conviction that America would never be free or saved from itself unless the descendants of its slaves were loosed completely from the shackles they still wear.

Now, it should be incandescently clear that no one who has any concern for the integrity and life of America today can ignore the present war. If America's soul becomes totally poisoned, part of the autopsy must read Vietnam. It can never be saved so long as it destroys the deepest hopes of men the world over. So it is that those of us who are yet determined that America *will* be are led down the path of protest and dissent, working for the health of our land.

... This I believe to be the privilege and the burden of all of us who deem ourselves bound by allegiances and loyalties which are broader and deeper than nationalism and which go beyond our nation's self-defined goals and positions. We are called to speak for the weak, for the voiceless, for victims of our nation and for those it calls enemy, for no document from human hands can make these humans any less our brothers.

STRANGE LIBERATORS

And as I ponder the madness of Vietnam and search within myself for ways to understand and respond to compassion my mind goes constantly to the people of that peninsula. I speak now not of the soldiers of each side, not of the junta in Saigon, but simply of the people who have been living under the curse of war for almost three continuous decades now. I think of them too because it is clear to me that there will be no meaningful solution there until some attempt is made to know them and hear their broken cries.

They must see Americans as strange liberators. The Vietnamese people proclaimed their own independence in 1945 after a combined French and Japanese occupation, and before the Communist revolution in China. They were led by Ho Chi Minh. Even though they quoted the American Declaration of Independence in their own document of freedom, we refused to recognize them. Instead, we decided to support France in its reconquest of her former colony.

Our government felt then that the Vietnamese people were not "ready" for independence, and we again fell victim to the deadly Western arrogance that has poisoned the international atmosphere for so long. With that tragic decision we rejected a revolutionary government seeking self-determination, and a government that had been established not by China (for whom the Vietnamese have no great love) but by clearly indigenous forces that included some Communists. For the peasants this new government meant real land reform, one of the most important needs in their lives.

For nine years following 1945 we denied the people of Vietnam the right of independence. For nine years we vigorously supported the French in their abortive effort to recolonize Vietnam.

Before the end of the war we were meeting eighty per cent of the French war costs. Even before the French were defeated at Dien Bien Phu, they began to despair of the reckless action, but we did not. We encouraged them with our huge financial and military supplies to continue the war even after they had lost the will. Soon we would be paying almost the full costs of this tragic attempt at recolonization.

After the French were defeated it looked as if independence and land reform would come again through the Geneva agreements. But instead there came the United States, determined that Ho should not unify the temporarily divided nation, and the peasants watched again as we supported one of the most vicious modern dictators – our chosen man, Premier Diem. The peasants watched and cringed as Diem ruthlessly routed out all opposition, supported their extortionist landlords and refused even to discuss reunification with the north. The peasants watched as all this was presided over by US influence and then by increasing numbers of US troops who came to help quell the insurgency that Diem's methods had aroused. When Diem was overthrown they may have been happy, but the long line of military dictatorships seemed to offer no real change – especially in terms of their need for land and peace.

The only change came from America as we

increased our troop commitments in support of governments which were singularly corrupt, inept and without popular support. All the while the people read our leaflets and received regular promises of peace and democracy – and land reform. Now they languish under our bombs and consider us – not their fellow Vietnamese – the real enemy. They move sadly and apathetically as we herd them off the land of their fathers into concentration camps where minimal social needs are rarely met. They know they must move or be destroyed by our bombs. So they go – primarily women and children and the aged. They watch as we poison their water, as we kill a million acres of their crops. They must weep as the bulldozers roar through their areas preparing to destroy the precious trees. They wander into the hospitals, with at least twenty casualties from American firepower for one "Vietcong"-inflicted injury. So far we may have killed a million of them – mostly children. They wander into the towns and see thousands of the children, homeless, without clothes, running in packs on the streets like animals. They see the children degraded by our soldiers as they beg for food. They see the children selling their sisters to our soldiers, soliciting for their mothers.

What do the peasants think as we ally ourselves with the landlords and as we refuse to put any action into our many words concerning land reform? What do they think as we test out our latest weapons on them, just as the Germans tested out new medicine and new tortures in the concentration camps of Europe? Where are the roots of the independent Vietnam we claim to be building? Is it among these voiceless ones? We have destroyed their two most cherished institutions: the family and the village. We have destroyed their land and their crops. We have cooperated in the crushing of the nation's only non-Communist revolutionary political force – the unified Buddhist church. We have supported the enemies of the peasants of Saigon. We have corrupted their women and children and killed their men. What liberators!

Now there is little left to build on – save bitterness. Soon the only solid physical foundations remaining will be found at our military bases and in the concrete of the concentration camps we call fortified hamlets. The peasants may well wonder if we plan to build our new Vietnam on such grounds as these. Could we blame them for such thoughts? We must speak for them and raise the questions they cannot raise. These too are our brothers. Perhaps the more difficult but no less necessary task is to speak for those who have been designated as our enemies. What of the National Liberation Front – that strangely anonymous group we call VC or Communists? What must they think of us in America when they realize that we permitted the repression and cruelty of Diem which helped to bring them into being as a resistance group in the south? What do they think of our condoning the violence which led to their own taking up of arms? How can they believe in our integrity when now we speak of "aggression from the north" as if there were nothing more essential to the war? How can they trust us when now we charge them with violence after the murderous reign of Diem and charge them with violence while we pour every new weapon of death into their land? Surely we must understand their feelings even if we do not condone their actions. Surely we must see that the men we supported pressed them to their violence. Surely we must see that our own computerized plans of destruction simply dwarf their greatest acts.

How do they judge us when our officials know that their membership is less than twenty-five per cent Communist and yet insist on giving them the blanket name? What must they be thinking when they know that we are aware of their control of major sections of Vietnam and yet we appear ready to allow national elections in which this highly organized political parallel government will have no part? They ask how we can speak of free elections when the Saigon press is censored and controlled by the military junta. And they are surely right to wonder what kind of new government we plan to help form without them – the only party in real touch with the peasants. They question our political goals and they deny the reality of a peace settlement from which they will be excluded. Their questions are frighteningly relevant. Is our nation planning to build on political myth again and then shore it up with the power of new violence?

Here is the true meaning and value of compassion and non-violence when it helps us to see the enemy's point of view, to hear his questions, to know his assessment of ourselves. For from his view we may indeed see the basic weaknesses of our own condition, and if we are mature, we may learn and grow and profit from the wisdom of the brothers who are called the opposition.

So, too, with Hanoi. In the north, where our bombs now pummel the land, and our mines endanger the waterways, we are met by a deep but understandable mistrust. To speak for them is to explain this lack of confidence in Western words, and especially their distrust of American intentions now. In Hanoi are the men who led the nation to independence against the Japanese and the French, the men who sought membership in the French commonwealth and were betrayed by the weakness of Paris and the willfulness of the colonial armies. It was they who led a second struggle against French domination at tremendous costs, and then were persuaded to give up the land they controlled between the thirteenth and seventeenth parallel as a temporary measure at Geneva. After 1954 they watched us conspire with Diem to prevent elections which would have surely brought Ho Chi Minh to power over a united Vietnam, and they realized they had been betrayed again.

When we ask why they do not leap to negotiate, these things must be remembered. Also it must be clear that the leaders of Hanoi considered the presence of American troops in support of the Diem regime to have been the initial military breach of the Geneva agreements concerning foreign troops, and they remind us that they did not begin to send in any large number of supplies or men until American forces had moved into the tens of thousands.

Hanoi remembers how our leaders refused to tell us the truth about the earlier North Vietnamese overtures for peace, how the president claimed that none existed when they had clearly been made. Ho Chi Minh has watched as America has spoken of peace and built up its forces, and now he has surely heard of the increasing international rumors of American plans for an invasion of the north. He knows the bombing and shelling and mining we are doing are part of traditional pre-invasion strategy. Perhaps only his sense of humor and of irony can save him when he hears the most powerful nation of the world speaking of aggression as it drops thousands of bombs on a poor weak nation more than eight thousand miles away from its shores.

At this point I should make it clear that while I have tried in these last few minutes to give a voice to the voiceless on Vietnam and to understand the arguments of those who are called enemy, I am as deeply concerned about our troops there as anything else. For it occurs to me that what we are submitting them to in Vietnam is not simply the brutalizing process that goes on in any war where armies face each other and seek to destroy. We are adding cynicism to the process of death, for they must know after a short period there that none of the things we claim to be fighting for are really involved. Before long they must know that their government has sent them into a struggle among Vietnamese, and the more sophisticated surely realize that we are on the side of the wealthy and the secure while we create a hell for the poor.

PROTESTING THE WAR

Meanwhile we in the churches and synagogues have a continuing task while we urge our government to disengage itself from a disgraceful commitment. We must continue to raise our voices if our nation persists in its perverse ways in Vietnam. We must be prepared to match actions with words by seeking out every creative means of protest possible.

As we counsel young men concerning military service we must clarify for them our nation's role in Vietnam and challenge them with the alternative of conscientious objection. I am pleased to say that this is the path now being chosen by more than seventy students at my own *alma mater*, Morehouse College, and I recommend it to all who find the American course in Vietnam a dishonorable and unjust one. Moreover I would encourage all ministers of draft age to give up their ministerial exemptions and seek status as conscientious objectors. These are the times for real choices and not false ones. We are at the moment when our lives must be placed on the line if our nation is to survive its own folly. Every man of humane convictions must decide on the protest that best suits his convictions, but we must all protest.

There is something seductively tempting about stopping there and sending us all off on what in some circles has become a popular crusade against the war in Vietnam. I say we must enter the struggle, but I wish to go on now to say something even more disturbing. The war in Vietnam is but a symptom of a far deeper malady within the American spirit, and if we ignore this sobering reality we will find ourselves organizing clergy and laymen-concerned committees for the next generation. They will be concerned about Guatemala and Peru. They will be concerned about Thailand and Cambodia. They will be concerned about Mozambique and South Africa. We will be marching for these and a dozen other names and attending rallies without end unless there is a significant and profound change in American life and policy.

In 1957 a sensitive American official overseas said that it seemed to him that our nation was on the wrong side of a world revolution. During the past ten years we have seen emerge a pattern of suppression which now has justified the presence of US military "advisors" in Venezuela. This need to maintain social stability for our investments accounts for the counter-revolutionary action of American forces in Guatemala. It tells why American helicopters are being used against guerrillas in Colombia and why American napalm and green beret forces have already been active against rebels in Peru. It is with such activity in mind that the words of the late John F. Kennedy come back to haunt us. Five years ago he said, "Those who make peaceful revolution impossible will make violent revolution inevitable."

Increasingly, by choice or by accident, this is the role our nation has taken – the role of those who make peaceful revolution impossible by refusing to give up the privileges and the pleasures that come from the immense profits of overseas investment.

I am convinced that if we are to get on the right side of the world revolution, we as a nation must undergo a radical revolution of values. We must rapidly begin the shift from a "thing-oriented" society to a "person-oriented" society. When machines and computers, profit motives and property rights are considered more important than people, the giant triplets of racism, materialism and militarism are incapable of being conquered.

A true revolution of values will soon cause us to question the fairness and justice of many of our past and present policies. On the one hand we are called to play the good Samaritan on life's roadside; but that will be only an initial act. One day we must come to see that the whole Jericho road must be transformed so that men and women will not be constantly beaten and robbed as they make their journey on life's highway. True compassion is more than flinging a coin to a beggar; it is not haphazard and superficial. It comes to see that an edifice which produces beggars needs restructuring. A true revolution of values will soon look uneasily on the glaring contrast of poverty and wealth. With righteous indignation, it will look across the seas and see individual capitalists of the West investing huge sums of money in Asia, Africa and South America, only to take the profits out with no concern for the social betterment of the countries, and say: "This is not just." It will look at our alliance with the landed gentry of Latin America and say: "This is not just." The Western arrogance of feeling that it has everything to teach others and nothing to learn from them is not just. A true revolution of values will lay hands on the world order and say of war: "This way of settling differences is not just." This business of burning human beings with napalm, of filling our nation's homes with orphans and widows, of injecting poisonous drugs of hate into veins of peoples normally humane, of sending men home from dark and bloody battlefields physically handicapped and psychologically deranged, cannot be reconciled with wisdom, justice and love. A nation that continues year after year to spend more money on military defense than on programs of social uplift is approaching spiritual death.

THE PEOPLE ARE IMPORTANT

These are revolutionary times. All over the globe men are revolting against old systems of exploitation and oppression and out of the wombs of a frail world new systems of justice and equality are being born. We in the West must support these revolutions. It is a sad fact that, because of comfort, complacency, a morbid fear of communism, and our proneness to adjust to injustice, the Western nations that initiated so much of the revolutionary spirit of the modern world have now become the arch anti-revolutionaries. This has driven many to feel that only Marxism has the revolutionary spirit. Therefore, Communism is a judgment against our failure to make democracy real and follow through on the revolutions that we initiated. Our only hope today lies in our ability to recapture the revolutionary spirit and go out into a sometimes hostile world declaring eternal hostility to poverty, racism and militarism.

34

VÁCLAV HAVEL

"The Power of the Powerless"

from *The Power of the Powerless* (1985)

A spectre is haunting eastern Europe: the spectre of what in the West is called 'dissent'. This spectre has not appeared out of thin air. It is a natural and inevitable consequence of the present historical phase of the system it is haunting. It was born at a time when this system, for a thousand reasons, can no longer base itself on the unadulterated, brutal, and arbitrary application of power, eliminating all expressions of nonconformity. What is more, the system has become so ossified politically that there is practically no way for such nonconformity to be implemented within its official structures.

Between the aims of the post-totalitarian system and the aims of life there is a yawning abyss: while life, in its essence, moves towards plurality, diversity, independent self-constitution and self-organization, in short, towards the fulfillment of its own freedom, the post-totalitarian system demands conformity, uniformity, and discipline.

Ideology, in creating a bridge of excuses between the system and the individual, spans the abyss between the aims of the system and the aims of life. It pretends that the requirements of the system derive from the requirements of life. It is a world of appearances trying to pass for reality. The post-totalitarian system touches people at every step, but it does so with its ideological gloves on. This is why life in the system is so thoroughly permeated with hypocrisy and lies . . .

[. . .]

Individuals need not believe all these mystifications, but they must behave as though they did, or they must at least tolerate them in silence, or get along well with those who work with them. For this reason, however, they must *live within a lie*. They need not accept the lie. It is enough for them to have accepted their life with it and in it. For by this very fact, individuals confirm the system, fulfill the system, make the system, are the system.

Revolt . . . steps out of living within the lie . . . [and] is an attempt to *live within the truth* . . .

When I speak of living within the truth, I naturally do not have in mind only products of conceptual thought, such as a protest or a letter written by a group of intellectuals. It can be any means by which a person or a group revolts against manipulation: anything from a letter by intellectuals to a workers' strike, from a rock concert to a student demonstration, from refusing to vote in the farcical elections, to making an open speech at some official congress, or even a hunger strike, for instance. If the suppression of the aims of life is a complex process, and if it is based on the multifaceted manipulation of all expressions of life then, by the same token, every free expression of life indirectly threatens the post-totalitarian system politically, including forms of expression to which, in other social systems, no one would attribute any potential political significance, not to mention explosive power.

Undeniably, the most important political event in Czechoslovakia after the advent of the Husak leadership in 1969 was the appearance of Charter 77. The spiritual and intellectual climate surrounding its appearance, however, was not the product of any immediate political event. That climate was created by the trial of some young musicians associated with a rock group called "The Plastic People of the Universe". Their trial was not a confrontation of two differing political forces or conceptions, but two differing conceptions of life. On the one hand, there was the sterile Puritanism of the post-totalitarian establishment and, on the other hand, unknown young people who wanted no more than to be able to live within the truth, to play the music they enjoyed, to sing songs that were relevant to their lives, and to live freely in dignity and partnership. These people had no past history of political activity. They were not highly motivated members of the opposition with political ambitions, nor were they former politicians expelled from the power structures. They had been given every opportunity to adapt to the status quo, to accept the principles of living within a lie and thus to enjoy life undisturbed by the authorities. Yet they decided on a different course. Despite this, or perhaps precisely because of it, their case had a very special impact on everyone who had not yet given up hope. Moreover, when the trial took place, a new mood had begun to surface after the years of waiting, of apathy and of skepticism towards various forms of resistance. Many groups of differing tendencies which until then had remained isolated from each other, reluctant to co-operate, or which were committed to forms of action that made co-operation difficult, were suddenly struck with the powerful realization that freedom is indivisible. Everyone understood that an attack on the Czech musical underground was an attack on a most elementary and important thing, something that in fact bound everyone together: it was an attack on the very notion of "living within the truth", on the real aims of life. The freedom to play rock music was understood as a human freedom and thus as essentially the same as the freedom to engage in philosophical and political reflection, the freedom to write, the freedom to express and defend the various social and political interests of society. People were inspired to feel a genuine sense of solidarity with the young musicians and they came to realize that not standing up for the freedom of others, regardless of how remote their means of creativity or their attitude to life, meant surrendering one's own freedom. This was the climate, then, in which Charter 77 was created. Who could have foreseen that the prosecution of one or two obscure rock groups would have such far-reaching consequences?

I think that the origins of Charter 77 illustrate very well what I have already suggested above: that in the post-totalitarian system, the real background to the movements that gradually assume political significance does not usually consist of overtly political events of confrontations between different forces or concepts that are openly political. These movements for the most part originate elsewhere, in the far broader area of the "pre-political", where "living within a lie" confronts "living within the truth", that is, where the demands of the post-totalitarian system conflict with the real aims of life. These real aims can naturally assume a great many forms. Sometimes they appear as the basic material or social interests of a group or an individual; at other times, they may appear as certain intellectual and spiritual interests; at still other times, they may be the most fundamental of existential demands, such as the simple longing of people to live their own lives in dignity. Such a conflict acquires a political character, then, not because of the elementary political nature of the aims demanding to be heard but simply because, given the complex system of manipulation on which the post-totalitarian system is founded and on which it is also dependent, every free human act or expression, every attempt to live within the truth, must necessarily appear as a threat to the system and, thus, as something which is political *par excellence*. Any eventual political articulation of the movements that grow out of this "pre-political" hinterland is secondary. It develops and matures as a result of a subsequent confrontation with the system, and not because it started off as a political program, project or impulse.

Charter 77 would have been unimaginable without that powerful sense of solidarity among widely differing groups, and without the sudden realization that it was impossible to go on waiting any longer, and that the truth had to be spoken loudly and collectively, regardless of the virtual certainty of sanctions and the uncertainty of any tangible results in the immediate future.

If some of the most important political impulses in Soviet bloc countries in recent years have come initially – that is, before being felt on the level of actual power – from mathematicians, philosophers, physicians, writers, historians, ordinary workers and so on, more frequently than from politicians, and if the driving force behind the various "dissident movements" comes from so many people in "non-political" professions, this is not because these people are more clever than those who see themselves primarily as politicians. It is because those who are not politicians are also not so bound by traditional political thinking and political habits and therefore, paradoxically, they are more aware of genuine political reality and more sensitive to what can and should be done under the circumstances.

If the basic job of the "dissident movements" is to serve truth, that is, to serve the real aims of life ... then another stage of this approach, perhaps the most mature stage so far, is what Vaclav Benda has called the development of parallel structures. When those who have decided to live within the truth have been denied any direct influence on the existing social structures, not to mention the opportunity to participate in them, and when these people begin to create what I have called the independent life of society, this independent life begins, of itself, to become structured in a certain way.

What are these structures? Ivan Jirous was the first in Czechoslovakia to formulate and apply in practice the concept of a "second culture". Although at first he was thinking chiefly of non-conformist rock music and only certain literary, artistic or performance events close to the sensibilities of those non-conformist musical groups, the term "second culture" very rapidly came to be used for the whole area of independent and repressed culture, that is, not only for art and its various currents but also for the humanities, the social sciences, and philosophical thought. This "second culture", quite naturally, has created elementary organizational forms: *samizdat* editions of books and magazines, private performances and concerts, seminars, exhibitions and so on. Culture, therefore, is a sphere in which the "parallel structures" can be observed in their most highly developed form. Benda, of course, gives thought to potential or embryonic forms of such structures in other spheres as well: from a parallel information network to parallel forms of education (private universities), parallel trade unions, parallel foreign contacts, to a kind of hypothesis on a parallel economy.

These parallel structures, it may be said, represent the most articulated expressions so far of "living within the truth". One of the most important tasks the "dissident movements" have set themselves is to support and develop them. After all, the parallel structures do not grow *a priori* out of a theoretical vision of systemic changes but from the aims of life and the authentic needs of real people. In fact, all eventual changes in the system, changes we may observe here in their rudimentary forms, have come about as it were *de facto*, from "below".

The primary purpose of the outward direction of these movements is always to have an impact on society, not to affect the power structure, at least not directly and immediately. Independent initiatives address the hidden sphere; they demonstrate that living within the truth is a human and social alternative and they struggle to expand the space available for that life; they help to raise the confidence of citizens; they shatter the world of "appearances" and unmask the real nature of power. They do not assume a messianic role; they are not a social "avant-garde" or "elite" that alone knows best, and whose task it is to "raise the consciousness" of the "unconscious" masses.... Nor do they want to lead anyone. They leave it up to each individual to decide what he or she will or will not take from their experience and work.

These movements, therefore, always affect the power structure as such indirectly, as a part of society as a whole, for they are primarily addressing the hidden spheres of society, since it is not a matter of confronting the regime on the level of actual power.

The relationship of the post-totalitarian system – as long as it remains what it is – and the independent life of society – as long as it remains the locus of a renewed responsibility for the whole and to the whole – will always be one of either latent or open conflict.

In this situation there are only two possibilities: either the post-totalitarian system will go on developing (that is, will be *able* to go on developing), thus inevitably coming closer to some dreadful Orwellian vision of a world of absolute manipulation, while all the more articulate expressions of living within the truth are definitively snuffed out; or the independent life of society (the parallel polis), including the "dissident movements", will slowly but surely become a social phenomenon of growing importance, taking a real part in the life of society with increasing clarity and influencing the general situation.

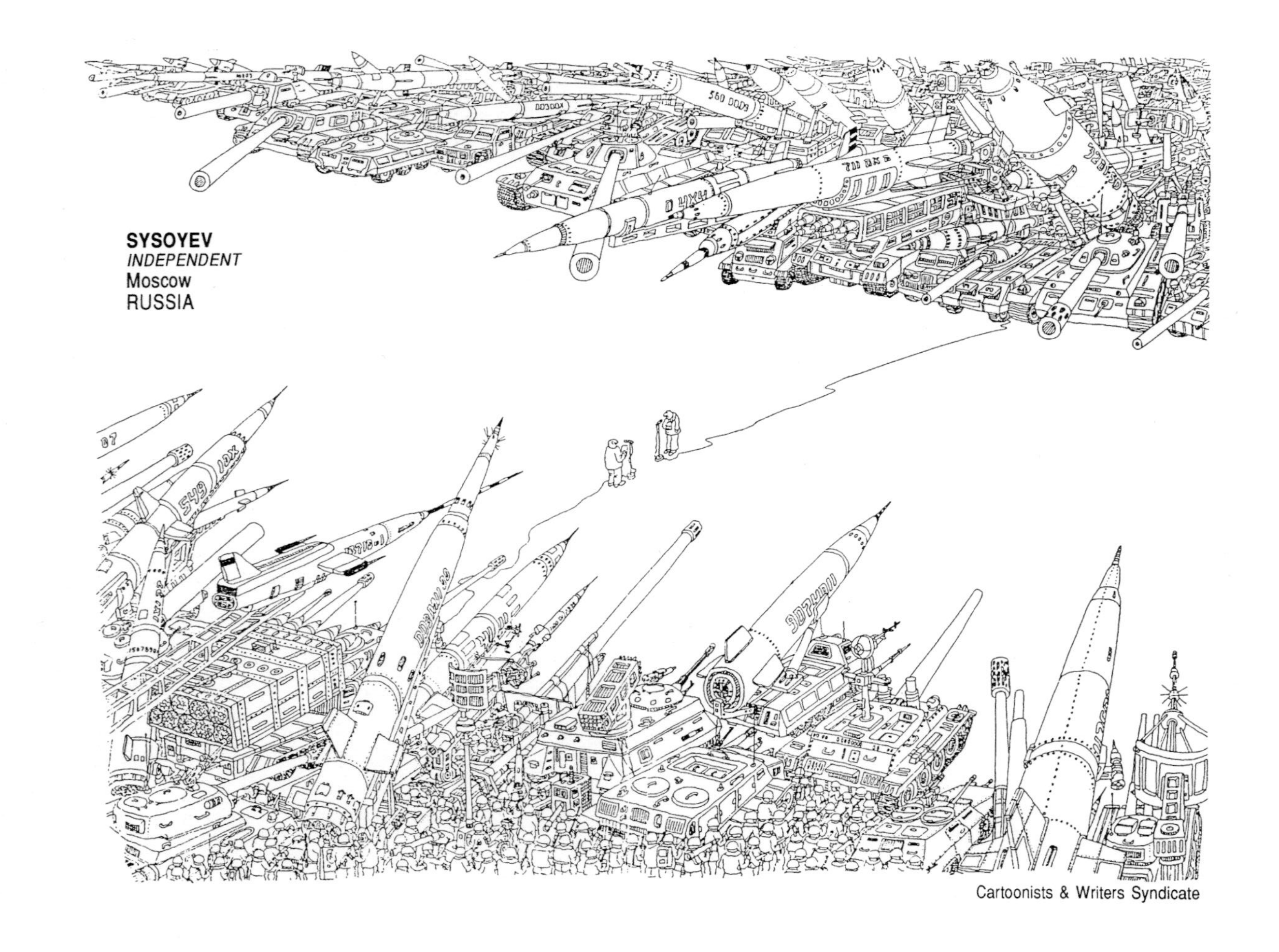

Cartoon 15 Cold War diplomacy (by Sysoyev)
During the Cold War, two military-industrial war machines faced each other. They dwarfed the human art of diplomacy and substituted the threat of annihilation for the negotiated resolution of differences.
Source: Sysoyev, Cartoon & Writers' Syndicate

35 E.P. THOMPSON

"America and the War Movement"

from *The Heavy Dancers* (1985)

There are two ways into the disarmament argument. One is from the globe itself: the threat to the species, the ecological imperative of survival. The other is from the injury done to people by the deformations, whether economic or cultural, of their own war-directed societies. And the problem is this: both arguments are being won, yet none of the structures of power has been shifted an inch by the argument, and not one missile has yet been stopped in its tracks.

There can no longer be any doubt that the procurers of nuclear weapons threaten the human species and most mammalian species besides. Yet the procurers continue with their business, and simply enlarge their public relations staffs to handle meddlesome scientists, bishops, doctors and peace movements.

Since the crisis is now global, then the shared perception and the strategies of resistance must increasingly become international. This is where the American peace movement (whose work is crucial to civilizations survival) faces quite exceptional difficulties. Inside the United States many persons feel themselves to be most bruised and restricted by the structures of racism, sexism, etc.; and of course they must and will contest these oppressions. But outside the United States, in Central America and increasingly in West Europe, it appears that it is arrogant American hegemonic nationalism which is most threatening (and which may even descend in the shape of a black F-111 pilot or a female cruise missile launch-officer). American military personnel are now, after all, everywhere: in El Salvador, in the Sudan, in the Lebanon, in England and West Germany, in Turkey and in Greece, in Diego Garcia and South Korea and Honduras and around the Persian Gulf. What are they doing there? And by what right? It is to ask us un-Americans to show superhuman exercises of self-restraint if we are not to pray, at times, for a reversion to good old-fashioned Middle American isolationism. There is nothing wrong with authentic American nationalism, if it is concerned with America's own cultural and historical traditions: but will it please go home and stay indoors?

I have written "hegemonic nationalism" and not "imperialism". Of course United States imperial interests and strategies palpably exist (albeit with some inner contradictions). Yet the naming of power in the United States simply as "imperialism" may make the whole problem seem too tidy. For it suggests that the problem may be easily isolated, as a powerful group of interests somewhere over there – the Pentagon, the multinational corporations – whereas what must be confronted is a whole hegemonic "official" national ideology, which permeates not only the state and its organs but also many "liberal" critics of the state (in the Democratic Party), which saturates the media, and which even confuses opposition groups within the society.

People in vast regions of Middle America know almost nothing of the world across the waters (where their fellow countrymen are being rapidly deployed) except the fragmentary ideological fictions offered in their local media. Yet it is this ideologically-confected national self-image – the herded self-identification with the goals (any goals) of the nation state, the media-induced hysterias during the Iranian hostages crisis and in the Falklands/Malvinas War, the manipulation of the minds of vast publics as to the benign motives of their rulers in invading Afghanistan or the Lebanon or in intruding into Central America in support of "national interests" – which constitutes, quite as much as weaponry, the threat to survival.

The most dangerous and expansionist nationalist ideologies are those which disguise themselves as missions on behalf of human universals. Such ideologies build upon the generous, as well as the self-interested, impulses of the evangelizing nation. (Middle America believes that US militarism is about the export of "freedom".) This was true of the French in Napoleonic times and even of the British at the zenith of empire (the "civilizing mission", the "white man's burden"). These ideologies have now shrunk back (almost) within their own historical and cultural frontiers, and are (almost) content to be simple nationalisms once more, celebrating unique historical experiences and cultural identities.

Today they are the Soviet and the American nationalist ideologies which have become expansionist, and which walk the world in the disguise of universals: the victory of World Socialism or the Triumph of the Free World. And of the two it is the American which is the most confusing. Because the United States population is made up of so many heterogeneous in-migrations, it is possible to fall into the illusion that America is a reservoir of every human universal, rather than a peculiar, local, time-bound civilization, marked by unusual social mobility, competitiveness and individualism, and with its own particular problems, needs and expectations.

Because of the youth and rapid growth of America, nationalist ideology is more artificially-confected there than in any other nation. It did not grow from experiential and cultural roots, fertilized and watered by a watchful imperial ruling-class, as in Britain. It was, like other artifacts of the New World, a conscious ideological construction, in the work of which ruling powers in the state, the media and the educational system all combined. The ideology consists, not in the assertion of the superior virtues of the (German/British/Japanese) race, but in the pretense that America is not a race or nation at all but is the universal Future. It lays arrogant claim to a universalism of virtues – an incantation of freedoms and rights – and asserts in this name a prerogative to blast in at every door and base itself in any part of the globe in the commission of these virtues. As it is doing in Central America now.

Characteristically, in the going rhetoric which still engulfs this nation, "human rights" require no more definition than that these are goods which Americans enjoy (to a superlative degree) and which other guys don't have. This truth is held to be self-evident. It is a cause of immense self-congratulation: and a means of internal bonding, ideological and social control, vote-soliciting, and even attributed identity. If an Other is required, as foil to all this glittering virtue, then this is provided by Communism. But anti-Communism is necessary, less because Communism exists, than because there is an internal need within the ideology to define the approved national self-image against the boundary of an antagonist.

The American peace movement – if it is to have any hope of success – must perforce challenge not only the military-industrial complex (or imperialist interests) but the hegemonic ruling ideology of the nation. It must strive to deconstruct this ideologically-confected national self-image, which now gives a very dangerous popular license to expansion and aggression, and to disclose in its place an authentic self-identity. I will leave it to Americans – to poets and to historians – to say what this self-respecting self-image should be.

When I said that we un-Americans pray for a reversion to isolationism I was of course in jest. What we pray for – and what we can now recognize with delight – is the rebirth of American internationalism: but an authentic internationalism, which conducts its relations with equals, and which conducts them with like-minded popular movements and not with client states or with servile parasitic elites.

It is only the positive of internationalism which will be strong enough to contest and drive out the reigning national (or supranational) ideology. There is a concern and generosity in American radical, labour, religious and intellectual traditions within which this internationalism has long been nourished, although sometimes as a threatened minority tradition. The greatest achievement of the American disarmament movement of the past two years has not been in winning this vote or that, but in raising peace consciousness throughout the continent, in questioning the self-congratulatory official image of America, and in providing a nationwide network of groups within which the level of international discourse and information is continually rising.

Yet this still must rise a little further. The foundation stone of internationalism cannot be guilt: it must be solidarity. We need, in some new form, a . . . vocabulary of mutual aid and of plain duty to each other in the face of power. And we need to hammer out together our international strategies, in which the American movement clearly sees itself and feels itself to be part of a whole "International" of self-liberating impulses from imperialisms and war.

We shall . . . require solidarity in the face of power. It has been apparent that . . . there has been an orchestrated NATO strategy of rolling back the peace movement, on both sides of the Atlantic, by the most careful employment of public relations, media management, the provocation of dissension, and, very probably, the infiltration of agents and provocateurs. The old NATO elites feel more threatened by their domestic peace movements than they do by their purported adversary (the Warsaw bloc); they fear that a whole way of managing the world, and of controlling publics and clients, may be slipping away into some dangerous and unstable unknown. (The Warsaw power leaders are suffering similar anxieties.) In that sense, the most important thing of all to them is not to be forced into defeat by their own domestic opposition. The MX is pointed, not at Russia, but at the Freeze; the cruise missiles at Greenham will be pointed at CND.

This is to say that our peace movements are engaged in one of the sharpest confrontations of our national political lives. They have challenged "the bomb" – and behind it they have found the full power of the State. If they are to reach "the bomb" they must now take on also a whole State manipulated and media-endorsed ideology.

It is ideology, even more than military-industrial pressures, which is the driving-motor of Cold War II. What occasions alarm is the very irrationality – the rising hysteria – of the drive, when it is measured against the "objective" economic or political interests of ruling groups or forces. It is as if – as in the last climax of European imperialisms which led on into World War I, or as in the moment when Nazism triumphed in Germany – ideology has broken free from the existential socio-economic matrix within which it was nurtured and is no longer subject to any controls of rational self-interest. Cold War II is a replay of Cold War I, but this time as a deadly farce: the content of real interest-conflict between the two superpowers is low, but the content of ideological rancor and "face" is dangerously high.

If history eventuated according to the notional rational self-interest of states or of classes, then we could calm our fears of nuclear war. It is scarcely in the interest of any ruling-class or state to burn up its resources, its labour-force, its markets, and then itself. But ideology masks out such interferences from an outer rational world. Just at the moment when the adversary posture of the two blocs is becoming increasingly pointless – and when a cease-fire in the Cold War would be greatly to the advantage of both parties – ideology assumes command and drives towards its own obsessional goals.

We may call this ideology "imperialist" if we wish (and on both sides). But in naming it as "imperialism" we should avoid the reduction of the problem to simplistic preconceptions (for example, Leninist) of what imperialism is. No two imperialisms have ever been the same – each has been a unique formation. And none can be reduced to a mono-causal analysis: for example, the pursuit of markets and profits, guided by some all-knowing committee of a ruling-class. Even in Europe's imperialist zenith (which culminated in the direct occupation and exploitation of subjected territories) a wide configuration of motivations were always at play: markets, missions, military bases, naval and trade routes, competition between imperial powers, ideological zeal, the vacancy left by the dissolution of previous power structures (the Moghul empire), populist electoral hysteria, the definition of frontiers, imperial interest-groups, the expectation of revenue or gold or oil (which sometimes was not fulfilled).

American and Soviet "imperialism" are also unique formations, and of the two it is Soviet which is most threatened and insecure, and American which is most expansionist. The American formation is a whole configuration of interests: financial, commercial (or "corporatist") and extractive (the search for reserves of fossil fuels, uranium and scarce minerals); military (the alarming thrust of the arms trade, with its crazily insecure financial underwriting, which, in its turn, must be underwritten by military guarantees); political (the establishment of hegemony not by direct occupation but through the proxies of parasitic ruling elites in the client nations); and ideological. This formation is plainly beyond the control of any all-wise committee of a ruling-class: the White House is simply rolled around every-which-way as the eddying interests break upon its doors. Now one interest, now another, penetrates into the Oval Office.

In this messy, indecisive formation it is clearly ideology which – in the person of the President and his closest advisors – binds the whole configuration together and gives to it what ever erratic direction it takes. The rhetoric of the Cold War legitimates the whole operation, and therefore the Cold War is necessary to power's own continuance: the Cold War's ideological premises must continually be recycled and its visible instruments of terror must be "modernized". But since this direction increasingly defies the self-interest of any of the participants, ideology itself becomes more hysteric: it combines into one mish-mash the voices of militant Zionism, born-again fundamentalists, traumatized émigrés from Communist repression, careerist academics and bureaucrats, Western "intelligence officers" and the soothsayers of the *New York Times*.

There is now (in this sense) a "war movement" in Washington and in London. It is made up of (a) particular military-industrial interest groups – searching for bases, fossil fuels, new weapons, markets for arms; (b) New Right ideologues and publicists, and (c) the confrontational rhetoric and policies of populist politicians of the right. The climate thus engendered is sheltering nakedly militarist adventures and interventions, legitimated within the cant of Cold War apologetics.

But the military confrontation between the blocs has less and less rational strategic function: nuclear missiles are now becoming symbolic counters of political "posture" or "black mail", negotiations are about political "face". Both SS-20s and Euromissiles are superfluous to any sane armoury. And the cruise and Pershing missiles have got to come because they are symbols of US hegemony over its own clients, and their acceptance is demanded as proof of NATO's "unity". They must be put down in noxious nests in England, Germany and Sicily, in order to hold the old decaying structures of life-threatening power together. The rising military appropriations are all for glue to paste over the places where the post-war political settlement ("Yalta") is beginning to come apart.

Nuclear weapons are not designed for the continuation of politics by other means: they are already the suppression of politics, the arrest of all political process within the frigid stasis of "deterrence", and the substitution of the threat of annihilation for the negotiated resolution of differences. And within this degenerative process, the simulated threat of the Other becomes functional to the tenure of power of the rulers of the rival blocs: it legitimates their appropriation of taxes and resources, it serves to discipline unruly client states, it affords an apologia for acts of intervention, and it is a convenient resource for internal social and intellectual control. Increasingly the symbolism of State terror is employed to menace domestic opposition within each bloc. Like a curving ram's horn, the Cold War is now growing inwards into the warriors' own brains.

The war movement in the West encourages – according to the Cold War's "law of reciprocity" (whether missiles or ideology) – an answering ideological response in the East. In Moscow as in Washington and London the tattered scripts of the early 1950s are dug out of the drawers and the lines are rehearsed. (Some of the old actors are still around.) The *New York Times* and *Pravda* recite the old crap: "KGB agents", "agents of Western imperialism", "un-American activities" and "peace loving forces" ... indeed, some of the malodorous agents (of both sides) are actually sent in. But the true adversaries which power fears, and seeks to hem in within the old ideological controls, are now not without but within their own blocs and spheres of influence: the real enemies of United States and NATO politicians are Central American insurgency, European "neutralism", and domestic peace and radical movements; the real enemies of the Soviet power elite are Solidarnosc, Afghan insurgents, and the growing desire of East European peoples for greater autonomy.

There could be two ways back from the precipice to which a threatened and dying ideology is conducting us. One would simply be a reassertion within the power-elites of the United States and the Soviet Union of the claims of rational self-interest. What is the point of burning up the world when, with a little loss of rhetoric, it could be managed and exploited to their mutual advantage? If this way is taken, then the elites will back away from war: they will come to some agreement (above the peace movements' heads); go back into arms control and SALT; and draw up a new "Yalta" for the entire globe, dividing it up between them according to agreed rules: Afghanistan and Poland for you, El Salvador and Nicaragua for me.

This way is so much preferable to nuclear war that it seems churlish to call it in question. Indeed, it has some support in the United States, among elite groups which are now getting into discourse with their Soviet analogues, and who give some backing to the Freeze. Yet the trouble with this superpower settlement, this Orwellian "Yalta 1984", is that it could never be more than a brief interim arrangement: it might last for five or ten years. For the superpowers can no longer command that the rest of the world, that Poland and Nicaragua, Europe and Latin America – stand still. It is indeed a question as to how much longer the elites can command their own domestic publics.

So that, even if the elites of both superpowers snuggle up together and attempt to take this first way back from nuclear suicide, it can only afford a brief interval before the second way to human survival is resumed. This is the way of the "International" of peace movements, of non-aligned nations, and movements for civil rights and for liberation, working out – through many complexities – common strategies of mutual support and solidarity. Their common aims will be the enlargement of spaces for national autonomy, the peaceful break-up or melding of the blocs, and the refusal of every syllable of the vocabulary of nuclear arms. No one can draw an accurate map of this way and show where it leads. We must find out together as we go along.

Cartoon 16 El Salvador military (by Auth)
Resistance to local powers backed by geopolitical hegemons is always dangerous. In November 1989, death squads linked to El Salvador's right-wing government murdered six Jesuit priests who preached "liberation theology" in the tiny Central American state. By claiming that the El Salvadorian government will bring the murderers to justice, President Bush managed to dissuade the US Congress from cutting the $85 million that government received from Washington in military and economic aid.
Source: AUTH, Universal Press Syndicate

36 GEORGE KONRAD

"Antipolitics: A Moral Force"

from *Antipolitics* (1984)

ANTIPOLITICS: A MORAL FORCE

The political leadership elites of our world don't all subscribe equally to the philosophy of a nuclear *ultima ratio*, but they have no conceptual alternative to it. They have none because they are professionals of power. Why should they choose values that are in direct opposition to physical force? Is there, can there be, a political philosophy – a set of proposals for winning and holding power – that renounces a priori any physical guarantees of power? Only antipolitics offers a radical alternative to the philosophy of a nuclear *ultima ratio*.

Antipolitics strives to put politics in its place and make sure it stays there, never overstepping its proper office of defending and refining the rules of the game of civil society. Antipolitics is the ethos of civil society, and civil society is the antithesis of military society. There are more or less militarized societies – societies under the sway of nation states whose officials consider total war one of the possible moves in the game. Thus military society is the reality, civil society is a utopia.

Antipolitics means refusing to consider nuclear war a satisfactory answer in any way. Antipolitics regards it as impossible in principle that any historical misfortune could be worse than the death of one to two billion people. Antipolitics bases politics on the conscious fear of death. It recognizes that we are a homicidal and suicidal species, capable of thinking up innumerable moral explanations to justify our homicidal and suicidal tendencies.

We shouldn't shy away from the suspicion that the generals think of war with something other than pure horror. It's inconceivable that the American President or the Russian President is not pleased at the thought of being the most powerful man in the world. His pleasure is only disturbed, perhaps, by the fact that he can't be quite sure about it, since his opposite number may very well think the same thing. . . .

[. . .]

The career of Adolf Hitler was an extreme paradigm of the politician's trade. He rose from the ranks of the feckless lumpen-intelligentsia to become Gotterdammerung incarnate over the bodies of fifty million people, like a wayside angel of death. When he addressed his followers, a veritable frenzy of verbal aggression gripped the speaker himself and suffused the glowing faces of his listeners as they breathed "Sieg Heil!" in response.

I am afraid of a third world war because, to my mind, there lives in every politician more or less of the delirium that was Hitler's demon. More exalted than the others, he could find exhilaration in pure unbridled power abstracted from all other considerations, from economic rationality or cultural values. We have to be wary of them because in all politicians worth their salt there is present, albeit in more sober form, some of the dynamite that came out in Hitler with such savage brutality; if there were not, they would not have chosen the politician's trade.

No matter what ideology a politician may appeal to, what he says is only a means of gaining and keeping power. A politician for whom the exercise of power is not an end in itself is a contradiction in terms. In culturally stable societies this kind of cynicism runs up against strong social and ethical inhibitions, and any observer who discerns this cynicism behind the inhibitions is himself called cynical. In less well-balanced societies the relentless instrumentalism of political power may come to the fore in hysterical crises of identity, exacerbating hidden suicidal tendencies by hazarding the greatest risk of all, the doomsday gamble.

Politicians have to be guarded against because the peculiarity of their function and mentality lies in the fact that they are at times capable of pushing the button for atomic war. There is in them some of that mysterious hubris that would like to elevate the frail and mortal "I" into a simulacrum of the Almighty. If this psychic dynamite should go off, it could draw all mankind into a global Auschwitz.

Why should I as a writer stick my nose into political matters? Because they frighten me. I feel mortally threatened by them, because there is more and more talk in political circles of rearmament and the likelihood of war. If the other side doesn't back down, they say, there will be war, and the responsibility will be wholly theirs.

All right, I look at the other side: they don't back down, and they say exactly the same thing. All right, I say, neither one will back down, now what? Are they going to have it out, or are they bluffing? Are they just trying to scare us, or are they serious after all?

I am speaking out because I feel confined by the Iron Curtain and the web of censorship restrictions that has grown up along with it. I know I may be locked up if tensions mount and the regime becomes more stringent. Most of the world is poor and military waste infuriates me. I loathe a culture that represents preparations to kill millions of people as a patriotic obligation. Thanks to the whims of politicians, I have more than once been in a fair way to depart violently from this scene, where otherwise it is still possible to live a good life. On the basis of their public statements, I suspect that politicians still think of war, even in the nuclear age, as a possible political action – "politics with bloodshed," to quote Mao Tse-tung's more graphic version of Clausewitz's aphorism.

I don't like it when they want to kill me. I don't like it when the agents of the politicians hold a gun on me. To me it doesn't matter much whether a bomb kills me or a death squad. To die by war is no better than to die by terror. War is terror too; the possibility of war is terror, and those who prepare for war are terrorists. The prospect of war and the absence of democracy are two sides of the same reality: politicians threatening defenseless people. If reality means people working at their own deaths for fatuous reasons, then I am bound to think reality even more absurd than deadly.

Escalation is the rule when weapons are put to use, yet no manner of social conflict can be solved by atom bombs. Our entire mythology of revolution and counterrevolution is an anachronistic shadow from the days of simple firearms. I am convinced that the redeeming doctrine of war as a continuation of power politics – the doctrine of the balance of terror – doesn't work any longer.

The abject stupidity of the flower of our intellectuals has contributed to the killing of millions in the big and little wars of this century. The ideologue is responsible because it is possible to kill with ideologies. In order to make war, drop bombs, build concentration camps, and dispose efficiently of the bodies, the skills of intellectuals are required.

I am repelled by men of ideas who chatter in tune with the military's propaganda machinery, who never lifted a finger against the butchery, who are left with only the sad excuse of declaring afterward that they were not in agreement with the terror to which they paid homage. A disturbingly large proportion of our thinkers have become experts in the service of our leaders. They are at pains to depict in rational colors something that is deathly irrational. The intellectual specialists in the logic of atomic and ideological war get their money for deceiving others, for leading them like lambs to the slaughter.

I was in a slaughterhouse once – I saw the lambs. A sly faced black ram led them. Just before reaching the block he slipped to one side, escaping from that corridor of death through a trapdoor. The others, following in his tracks, kept on going – right up to the block. They called the black ram Miska. After each of these performances he would go up to the canteen, where he was given a roll with salami and some cake, and he would eat. For me, the scholars of ideological war are so many Miskas, except that they themselves have no way of slipping through any trapdoor to safety...

THE POWER OF THE STATE AND THE POWER OF THE SPIRIT: POLITICS AND ANTIPOLITICS

Antipolitics is the political activity of those who don't want to be politicians and who refuse to share in power. Antipolitics is the emergence of independent forums that can be appealed to against political power; it is a counter power that cannot take power and does not wish to. Power it has already, here and now, by reason of its moral and cultural weight. If a notable scholar or writer takes a ministerial post in a government, he thereby puts his previous work aside. Henceforth he must stand his ground as a representative of his government, and in upholding his actions against the criticisms of democratic antipolitics he may not use his scholarly or literary distinction as either a defense or an excuse.

Antipolitics and government work in two different dimensions, two separate spheres. Antipolitics neither supports nor opposes governments; it is something different. Its people are fine right where they are; they form a network that keeps watch on political power, exerting pressure on the basis of their cultural and moral stature alone, not through any electoral legitimacy. That is their right and their obligation, but above all it is their self-defense. A rich historical tradition helps them exercise their right.

Antipolitics is the rejection of the power monopoly of the political class. The relationship between politics and antipolitics is like the relationship between two mountains: neither one tries to usurp the other's place; neither one can eliminate or replace the other. If the political opposition comes to power, antipolitics keeps at the same distance from, and shows the same independence of, the new government. It will do so even if the new government is made up of sympathetic individuals, friends perhaps; indeed, in such cases it will have the greatest need for independence and distance.

In his thinking, the antipolitician is not politic. He doesn't ask himself whether it is a practical, useful, politic thing to express his opinion openly. In contrast with the secrecy of the leadership, antipolitics means publicity; it is a power exercised directly over society, through civil courage, and one that differs by definition from any present or future power of the state.

Antipolitics means perspicacity; it means ineradicable suspicion toward the mass of political judgments that surround us. Often these judgments are simply aggression in another form. We shouldn't forget that older men whose physical and nervous energies are failing are especially prone to intellectual aggression of the most savage and relentless kind, though always in the name of noble ideals. Spiritual authority is the practice of this kind of antipolitical understanding.

But what does spiritual authority have to offer that is positive? How is it anything more than sheer negativity? It asserts the worth of human life as a value in itself, not requiring further justification. It respects human beings' fear of death. It views the lives of people of other countries and cultures as equal in value to those of our countrymen. It refuses to license killing on any political grounds whatever. I regard the commandment "Thou shalt not kill" as an absolute command. I have never killed, I want to avoid killing, yet it's not impossible that situations may arise in which I will kill. If I do, I will be a murderer and will consider myself one. Murderers must expiate their crimes.

Antipolitics looks kindly on the ecumenical variety of religions and styles and doesn't believe that the condition for the existence of one cultural reality is the extinction of another.

Antipolitics prefers qualitative competition to silly quantitative questions about who is stronger. Who is stronger is really of no interest. For the antipolitician, it is more interesting to know whether a community produces an intelligent and honest portrait of itself, not how much technical power it commands.

Antipolitics asserts the right of every community to defend itself, with adequate defensive weapons, against occupiers. It is a great misfortune to have to fire on occupiers. We would become murderers ourselves in so doing, but it may happen that we will decide we have to be murderers.

37 ABOUALI FARMANFARMAIAN

"Did You Measure Up? The Role of Race and Sexuality in the Gulf War"

from *Genders* (1992)

INTRODUCTION

In its preparation as well as in its conduct, the flash war in the Gulf functioned as a machine realizing a national fantasy, with all its fears, anxieties, desires, and excitements. This national fantasy, insofar as it was related to the identity formation of America, circulated primarily in the sphere of sexuality and consequently of race. For the history of imperial and colonial ventures – and the Gulf War was one – has dictated a racial connection wherever sexuality appears.

I examine the role of sexuality in the Gulf War not as a metaphor, but as a determinant.

Concerns about Iraq and the desirability of war were mediated through notions of family and sexuality – always with a racial link that implicitly emphasised western values – and only thereby managed to generate a unanimity in outrage against an outside evil, Iraq.

THE APPEAL

By positioning the United States as the righteous protector of the world and Iraq as an evil destructive force, George Bush managed to rally and unite public opinion in favour of a military strike. . . . Bush justified the likelihood of war by speaking directly on "the immorality of the invasion" of Kuwait by Iraq. The campaign to paint immorality on Iraq and moral righteousness on the United States started . . . with a sudden and massive infusion of reports on Iraqi violations, focusing particularly on *sexual* atrocities.

The images and concomitant fears of rape were present from the outset. As Bush desired American outrage to escalate, the "violation of Kuwait's sovereignty" became increasingly tied to *sexual* atrocities committed by the Iraqis, and infanticide, rape, and torture became the main focus of attention. Thus, repeatedly the concepts of sovereignty and violation in the international arena were linked to sexual counterparts of integrity and rape.

From mid-December to early January, the media were inundated with such reports. They were extremely successful . . . in shifting national consensus in favour of war. Hiroshima-on-the-Tigris was endorsed as "just" and indeed "necessary" because the Other side could be perceived as rapist and barbaric.

[M] any of the emotions, fears, and images justifying war . . . moved the issue of war outside the realm of real causal threats and appealed to notions and anxieties that are an intricate and integral part of American consciousness: the image of an Other and the boundaries set against that Other in favour of the Self. The war took place against that image, against the representation of Iraq, not Iraq itself.

I do not want to imply that the terrors unleashed mainly on migrant female workers in Kuwait by the Iraqi army were imaginary events. However, to imagine that rape is not one of the facets of military life everywhere in the world is an illusion.

Locating danger and deviancy in distinct groups of Others allows for an exonerating and coherent, though false and illusory, conception of self.[1]

Such is the *international division of attributes*: The label of rape only sticks to the racial Other. Europe and America must displace themselves as, and hence believe themselves to be, paragons of respectability and order. Chaos, irrationality, and violence are the constructed realm of the racial Other – the "Orient," Africa, but, significantly, not the ex-arch enemy the "Russians," whose constructed identity as European significantly erased the Muslim and Asian populations of the Soviet republics and did not carry the labels of irrational or rapist but merely represented an abstract and quite recent ideological evil called communism. The international division of attributes rose out of the dynamics of colonialism and with the development and domination of mass media and western information networks it has taken on a central role in the present international dynamics. The divisions remain unquestioningly accepted. That is why it proved so easy to build a case around Iraqi immorality and atrocity, equally easy to spoon-feed it to the nation, and then go on to annihilate Basra without the blink of an eye.

WHITENESS, COLONIALISM, AND IDENTITY

To understand how and why the above consensus took shape so easily in the United States we need to examine the historic construct of race, with particular attention to the Middle East as a uniformalised Other.

In ... European colonialism ... [a] well-defined outline of a Self was drawn against projected notions of the Other – and actions, motivations, and decisions emerged in relation to safeguarding that definition and maintaining its boundaries.... [T]he overriding concern for the government of empire, whether Dutch, British, or French, eventually became the drawing of absolute racial boundaries in an attempt at self-definition. From then on, the violation – perceived or real – of any of these mental social boundaries would be cause enough for violent retribution.

In the period of administration, rigid physical and mental boundaries of Whiteness also developed within notions of family ... all served to strictly separate the ruling race from the ruled, and in particular White women from native men.

While boundaries were drawn along racial lines, the terrain on which they were drawn was sexuality. If empire as a system of government were to survive it had to reinforce its boundaries in such a way as to ensure the continued separation between Self and Other, ruler and ruled. The colonial administrator was obsessed with self-reproduction: he had to reproduce the White colonial family.[2] He, therefore, attempted to control White womanhood by sanctifying it and thereby justifying its protection from impurity, evil intent, lasciviousness, rape – all these being qualities and motivations attributed to non-Whites. But it was precisely the fabrication of a threat that served to define the boundaries of Whiteness. White womanhood could only be, and indeed was, sanctified in contrast to a transgressive Other, namely, the native rapist and the promiscuous native woman. If the myth of the Other Rapist fades, White womanhood as an ideal collapses, and vice-versa. The construct of the White Self and the construct of Other cannot be evoked without each other. Therefore, racial engagements with the Other have rarely been more than internal battles in the definition of self. America's evocation of the Other Rapist in the Gulf War by definition involved the ideal of White womanhood, and functioned to reinforce the construct of the American Self at a time when the American economy, quality of life, and values were being globally and internally questioned.

American Whites developed boundaries with respect to their race at a quicker pace than Europeans in the colonies.

The direct contact with Black people brought about by an increase in slavery in America necessitated stronger physical and psychological boundaries that could maintain and carry hierarchy through the high-contact life on the plantation as well as in later northern urban interaction between mistress and domestic. As in European projects of empire, America erected the sanctity of White womanhood to define its own boundaries. As an ideal that ensured continuity and secured the definition of home, family, and nation, White womanhood was inviolable. Any perceived threat to its sanctity could lead, in America as well as in the colonies, to an unleashing of violence. By the time of the urban race riots of the 1920s, the protection of White womanhood from "rapist Black men" occupied the principal platform in the justification of attacks on and lynching of Black men.

Historically, then, "rape" as such has been constituted as a threat only in relation to White women. The word itself does not, in common consciousness and usage, signify the rape of non-White women, since such rapes have never been made the centre of attention; on the contrary, the rape of Black women has consistently been condoned, denied, or disregarded. Thus, in the context of the Gulf War ... the word rape, without any specific referents, triggered fears that could only be connected to the constructs of Black rapist/White woman. Calculated or not, it was, as was shown above, the concentration on this particular fear that mobilised the American public behind the war effort. If in the 1920s the evocation of "rape" would lead to mass vigilantism and participation in violent lynching rituals, in 1991 the majority of the American people participated by voting for, and then watching, live on the networks, the exercise of military violence.

Although for White Americans the occasion for the expression of violence has come principally in relation to African Americans, constructs of superiority and White womanhood have been upheld against other peoples too. Otherwise, the Gulf War would not have been so popular and as easily condoned. In world fairs, which gained massive popularity at the end of the nineteenth century, the American imagination fed on images of European colonies, and developed its own notions of expansionism.

Beyond the overriding intention of justifying empire through the visual display of colonies, the fairs served to uniformalise racist notions of the Other and spread these constructs between Europe and America.

The "Orient's" main particularity appeared in the image of the "Oriental despot" that not only permeated the travelogues, booklets, and films, but had found a firm place in western sociology through the likes of Marx and Weber. A continuous line gets drawn between those conceptions of a blood thirsty despot cutting off the heads of courtiers as well as citizens and today's media depictions of Saddam Hussein. "Barbaric" is another construct attached largely to that region and used extensively by President Reagan to describe Arabs at the height of hostage-takings in the mid-1980s. Over the past decade and a half, starting with Palestinians and moving onto other Arabs and finally covering all Islamic peoples, the options of political struggle in the region of the Middle East have been depicted as inherently "barbaric" and "uncivilised" in accord with the conceptions of that region passed down from the nineteenth century. In all other capacities, peoples of that region are absent from the machinery of public knowledge, so that when a threat appears, the steadfast notions of two centuries past reappear in White consciousness. During periods in which the focus is not on the Middle East these labels may be dormant, but they remain ready for use at will.

In terms of sexuality and race, White consciousness of the Middle East took shape in the nineteenth century primarily around the idea of licentiousness and "endless sexual gratification."[3] In the White mind, sexuality in the "Orient" inevitably appeared laced with violence: the despot cutting off the head of his lover, the exchange of women slaves for arms, and so on. These .... positioned the Middle Eastern man not only as licentious (like the Middle Eastern female) but also violent in his sexuality. Thus ... the basic mythology of rapist survived through other projected ideas precisely because that mythology had already been erected in relation to other Others in empire.

LEADING INTO THE GULF

As most studies of colonial constructs correctly point out,[4] the myths and portrayals of the racial Other had much more to do with those who created them than with those who were its objects. This would help explain, for example, the use of genital mutilation after the ritual of lynching in American culture. If the penis is cut off – and, as happened occasionally, stuffed in the victim's mouth – after the lynching, then clearly the ritual had little to do with the actual use of the penis or the person attached to it; it had much more to do with the concerns of White men who were in charge of the rituals.[5] By castrating the racial Other, they proved their virility and momentarily escaped the castration anxiety produced, ironically, by the force of White constructs of non-White men as sexually aggressive. It is interesting to note that although most lynchings did not even involve the charge of rape,[6] the most popular and valid explanation of lynchings became and continues to be the sexual one: Regardless of the real circumstances surrounding the event, the sexualised construct has gripped the American imagination.

On occasions like the Gulf War, when the threat of rape is evoked, it is inevitably accompanied by a similar sexual anxiety which necessitates a display of virility. These anxieties are indisputably products of the White consciousness and the constructs that have fed its imagination. The conflicts are not with a real foe, but with erected opponents who fit within the boundaries of White consciousness.... [T]he anxieties and the national reaction rise out of the same history: a threat to the constructed boundaries of White manhood. Aside from the castration anxiety produced inside US borders mostly in relation to constructs and boundaries erected around African Americans, American virility had for some time been ridiculed in the international arena.

The adamancy with which the Gulf War was pursued over a five-month period marked it as a *desire*, not merely a necessary *option*. The desire was to restore a lost potency to a nation that, despite its massive and well-advertised prowess, was saturated in public humiliation. In the international arena, America's military machine had been frustrated continuously. Despite extensive military prowess it had not been able to display its power since World War II. This was the fear of impotence that permeated the whole nation, the fear brought on by possessing the largest military machine of the world in theory, but remaining unsure of its ability to rise to the occasion.

America's largest overt colonial endeavour – Vietnam – ended not merely in a military defeat but in national humiliation. The Vietnam War ... was the largest symbol of impotence for a relatively new White colonial power.

The "recovery" from Vietnam has been slow. Castration anxiety lived on in "the ghost of Vietnam," haunting America until the Gulf War. America's forte in the post-Vietnam War era has been covert operations. But these cannot serve the purpose of restoring national virility. National virility ... requires a public display. Furthermore, the military embodies the virility of a nation. Covert wars can not fulfil this purpose, for the military's achievements are by definition out of public view; their images, their discourse, and their knowledge are also covert.

From Vietnam, America moved to the hostage crisis of the 1980s and again witnessed "a travesty of its manhood" paraded across the globe. The 1979 hostage drama left America impotent, unable to wield its might. The small, confused, rather desperate attempt at freeing the hostages led to a humiliating catastrophe in the desert near Tabas, Iran. While the US army was looking pitiful in the sand, White American masculinity – since all African-American and White-women hostages were released by the Iranian captors – was gagged, tied, and put on display for the world to see. America ... stood obsessed but paralysed watching the humiliation of their male citizens. That these most recent evocations of American impotence arose out of the Middle East in part explains the national enthusiasm for the military lynching ceremony we witnessed in the Gulf War. Indeed, the Gulf War started precisely with such a panic over captured, humiliated citizens. It reached a hysterical climax with the airing of images of bruised prisoners of war, and caused a traumatic self-doubt when news of women prisoners and casualties hit the air.

During the years prior to and in the months leading up to the Gulf War, the fear of impotence was "a consciously held fear,"[7] in part because of the historically unique role castration has played in the American consciousness. As Winthrop Jordan has pointed out, America was the only country that used castration as a legal punishment, and only in relation to Black men. It was mostly ordered for alleged or attempted attacks on White women, but sometimes appeared as punishment for basic threats to White manhood such as striking a White person.[8] So castration, and hence castration fear, began as a conscious and actualised theme in American minds. Impotence was even an issue in the presidential elections, first in all the talk about Ronald Reagan's age and white hair, then over Bush's "wimp" image. But the flood of macho international-relations movies in the 1980s made up the largest and most open manifestation of the desire to overcome this fear, particularly in an international arena.

As the debate over the Gulf War went public, one of the main articulated fears was a fear of humiliation *à la* Vietnam. From the outset, the words that rang loudest were the emphatic negation of a Vietnam-style castration: "This will not be another Vietnam" became a favourite phrase of American leaders and their public.

The military movement toward war and the behaviour during the war bear out the above points. The effort to restore potency to the United States probably began with Grenada. The invasion of Grenada was the first small but successful military operation engaged in by the US military. Although there was no direct media coverage, its victory chant permeated the media and fed American consciousness. Then came Panama, a larger, more ambitious operation, and again a successfully executed one. Finally, it seemed, the American military machine was coming through. Grenada and Panama were calculated, sequential, and escalatory reclaimings of virility; post-Vietnam power was cautiously pulled out of the closet and put on display.

This movement climaxed in the Gulf War. The first week's outburst of virility, the explosion of military images onto the world's television screens, constituted the most impressive collection ever of "surgical strikes" flawlessly delivered. In a sense the first week of war was the most important week for the American psyche: "Intimate details" of bombs going down chimneys, breaking open doors and exploding into bunkers was the display.... A new and improved manhood. The missiles that missed, we never saw. American casualties, we never saw. This was not to be another image of failure; it had to be an undefeatable machine, not another Vietnam.

Meanwhile, a White brotherhood formed at the helm, pressing onto the American consciousness the notion of a male community, impenetrable, inviolable, and virile. In press briefings and meetings with ambassadors, Bush never appeared without his top aides ... Sununu, Cheney, Baker, and Bush formed a privileged and inviolable chain. The weak link of Danforth Quayle was noticeably absent, and General Powell rarely figured within the chain, at most appearing on the sides.

Press briefings from the front conveyed much the same images. From maverick director of operations, father-figure for all soldiers, and father-hero for the nation, Norman Schwarzkopf, to Brigadier General Richard Neal, there was an exchange, an exclusive flirtatious dance around information ... and a constant switching of partners, spokespersons, and experts, which included Saudi officials. An impenetrable unity took shape ... which strengthened the brotherhood.

Occasionally, too, we caught glimpses of soldiers' collective virility in their group rituals, with one in particular outlining a dear link to territorial penetration. Cameramen filmed a group of Black and White GIs demonstrating their war chant. Like all war chants, it was to give the soldiers a sense of camaraderie and invincibility. The ritual involved the usual chanting, but this time with each of their hands tightly gripped on their crotches which were then thrust forward from their waists in a motion of penetration (towards Iraq?). A World War II testimony of a soldier's feelings during war offers something of an explanation: "The tank.... It protrudes shafts of cold metal with which to fuck a landscape and, by fucking, raze it...."[9] The important difference in the Gulf soldiers' fucking of a landscape lies in its collectivity – the same manifestation of brotherhood as in the press briefings; the use of a collective phallus with which to penetrate.

A look at the discourse of war only confirms the overwhelming castration anxiety that had beset the nation. Under various quite transparent guises "virility" and "impotence" were openly expressed as central issues. In an interview with Barbara Walters on March 22, 1991, Stormin' Norman Schwarzkopf admitted his main concern as the top military planner:

> NS: . . . you want desperately to measure up!
> BW (seductively): Did you measure up?[10]

He did measure up, but this war was much more about "staying power" than about size. The two key terms leading up to war were "staying power" and "withdrawal." Withdrawal was the first and only condition set forth by the coalition. Use of this term may seem either natural or coincidental, but the common consciousness of its double entendre was demonstrated by the oft-heard joke told about both Saddam and Bush: "Withdraw, like your father should've." The term becomes more striking when contrasted with its opposing one: If Bush asked Iraq to withdraw, he asked America to have "staying power." Only staying power could win the war. "Staying power" as a term appears much less natural or coincidental than "withdrawal" for it is a term rarely used outside the context of virility.

FIGHTING FOR THE IDEAL

To paint the Gulf War as principally a war concerned with a recovery of lost virility . . . means to see it as a war of desire, a desire rising out of a sense of lack. Historically, castration anxiety emerged out of constructs that defined racial and sexual boundaries, embodied by and contained in the same boundaries that defined the ideal of White womanhood. Consequently, threats to those boundaries that call up sexual fears will also evoke those ideals around which the fears originated. This war was no exception and in it White womanhood played a defining role in America's consciousness.

The endless number of t-shirts and posters picturing over-sexualised, exclusively blonde women over the inscription "Desert Storm" point to a generalised connection in the American imagination between sex, race, and war. The main message . . . seems to be that running parallel to the fighting, to the war itself, is a sexual fantasy based on a racial ideal, as in those idealised poster images.

[W]omen have a role in war as incarnations of all that men must fight to protect . . . the woman referred to must be White, as amply shown on the posters . . . a fantasy, a longing, an ideal that is internal to manhood itself and thus renders the battle an internal one: A fantasy to fight for and not merely fight to protect, it embeds itself as a permanent motivation rather than a contextual one. In specific contexts, such as war, the fantasy only takes on greater and more emphasised proportions.

The American troops, despite all the talk about Saudi censorship, imported their share of idealised White women to fight for. Pin-ups, pictures, and posters, along with entertainment like Brooke Shields, supplied the iconographic representation of the White fantasy. A jeans-clad pin-up of a White woman also became extremely popular in the Gulf, so much so that a spokesman for the Navy said "Most every marine seems to be aware of the poster."[11]

The ideals of America, the ideals to fight for, were summarised in the body and the circulation of that pin-up.

The participation of women in the war does little to overturn arguments centering on virility. In connecting the nation, manhood, and family, the much-lauded participation of women, rather than castrating the army, appeared as the much-needed link that would familiarise the military. The military needs to become a family in order for the nation to remain one. Thus, not only is the military no longer dominated by single men, but more than ever we saw images of wives and husbands as soldiers.

CONCLUSION

Any project linking sexuality and imperial enterprises in a concrete, rather than merely symbolic, fashion, suffers from lack of information. Sexuality is hidden and often censored.

History assumes an eminently important role, for by tracing constructs, events, images, and words to a connected accumulation in which sexuality, race, and imperialism form a coherent system we provide ourselves with an interpretive tool.

The physical participation of African Americans in the army does not automatically overturn the arguments about White boundaries and fantasies of White consciousness. Physical integration does not mean a change of constructs or signifiers. Even with a Black general or Black CEOs, the association of rape to Blackness will not and has not disappeared – that would mean the dissolution of the White Self which relies on such associations to define itself. In fact, participation merely means acceptance of those boundaries. This war, unlike Vietnam which took place at a time of worldwide decolonisation, coincided with an unprecedented era of American hegemony. Into this glory, some African Americans were assimilated. Unconditionally. Although a majority of African Americans still opposed the war,[12] a large percentage endorsed it and some soldiers in the front even used the term "sand niggers" in reference to Iraqis and possibly to all Arabs since sand has a geological presence in the whole region. This is the final stamp of participation in the project of nation, for when African Americans, too, begin to define themselves by the same constructs and against the other "rapists," "deviants," and "barbarians," they can join the nation; they can enter its boundaries as long as they bury their own past.

Bush himself encouraged the act of "forgetting" in his victory speech by claiming that the "the ghost of Vietnam was purged" and "national self-doubt" eliminated. We might see an end to the ceaseless flow of Vietnam movies, but that will be it. Irrespective of the number of victories, irrespective of any real events, as long as the described constructs and their psychological boundaries are kept alive, the associated anxieties and ideals will also continue to live. Since the very boundaries that define Whiteness depend on the construct of the Other, and since the roots of the anxieties and ideals lie in those constructs, White consciousness will continue this internal battle through engagements with the Other until it understands, faces, and radically redefines its own historical identity. "Forgetting" or "purging" will do nothing towards this understanding and redefinition. They are merely pre-conditions to joining an entity, such as the nation, which will continue to reproduce those same patterns of identity, fantasy, and violence.

AUTHOR'S NOTE

Although I am aware of the problems inherent in the use of "America" when speaking of the United States, I am using the term because of its particular connotations. A specific rhetoric is attached to "being American" which cannot be captured by any other term. Since I view this construct of "America" as tightly linked to other notions in the essay – such as "family," "Whiteness," "Nationhood" – I judged its use to be appropriate here.

NOTES

1 Outside the scope of this paper but related to it, one can also argue that this conception is a phallocratic project. Leslie Wahler Rabine, "A Feminist Politics of Non-Identity," *Feminist Studies* (Spring 1988).

2 See Irvin Cemil Schick, "Representing Middle Eastern Women: Feminism and Colonial Discourse," *Feminist Studies* (Summer 1990): 359–60.

3 Rana Kabbani, *Europe's Myths of the Orient* (London: Macmillan 1986); Schick, "Representing."

4 For the construct of the Orient see Kabbani, op. cit; Malek Alloula, *The Colonial Harem* (Minneapolis: University of Minnesota Press 1986).

5 Two things are of note here. First that there appeared to be a widespread fascination with the penis of a Black man and on occasion the possession of one(!): "It is generally said that the penis in the Negro is very large. And this assertion is so far borne out by the remarkable genitory apparatus of an Aethiopian which I have in my anatomical collection," from a German anthropologist quoted in Jordan, op. cit., 158. The second point, shown by the nationality of this anthropologist, is the way these ideas filtered back from the colonies to mainland Europe.

6 According to a study of lynching, between 1889 and 1929 only some 23 per cent of lynch victims were charged with rape or attempted rape, quoted in Angela Davis, *Women, Race and Class* (New York: Vintage 1983): 189. After Frederick Douglass and Ida B. Wells, Angela Davis argues that lynching did not start as an act of sexual retribution, but gained enormous propaganda value once the cry of rape and the

protection of White women were attached to the event.

7 Theweleit, *Male Fantasies, Vol. 1* (Minneapolis: University of Minnesota Press 1987): 89.
8 Winthrop D. Jordan, *White over Black: American Attitudes Towards the Negro, 1550–1812* (New York: W.W. Norton): 154–5.
9 John Costello, *Love, Sex and War, 1939–1945* (New York: Pan Books, 1985): 140. I use testimonies of World War II because of their honesty and availability as collected in Costello's book.
10 Television interview, 20/20, March 22, 1991.
11 Robert Reinhold, "Policewoman in Denim is Betty Grable of Gulf," *The New York Times*, Feb 15, 1991, A16.
12 *CBS/New York Times* poll, December 14, 1990.

38 SUBCOMMANDANTE MARCOS

"Chiapas: The Southeast in Two Winds, a Storm and a Prophecy"

from *Anderson Valley Advertizer* (1994)

THE FIRST WIND

Suppose you decide to get to know the southeast of the country, and suppose that in the southeast you choose the state of Chiapas. Suppose you take the Tansistmica Highway. Suppose that you ignore the federal army barracks on the highlands above Matias Romero and you go on to Ventosa. Suppose you don't notice the Government Ministry's Immigration checkpoint (which makes one think one is leaving one country and entering another). Suppose you take a left, and move decisively toward Chiapas. A few kilometers ahead you will leave Oaxaca and find a large sign which reads, "WELCOME TO CHIAPAS."

Did you find it? Good, let's suppose so. You got here (to this southeast corner of the country) by one of the three existing roads: the road from the north, the one along the Pacific coast, or the one you supposedly have just taken. But the natural wealth that leaves these lands doesn't travel over just these three roads. Chiapas is bled through thousands of veins: through oil ducts and gas ducts, over electric wires, by railroad cars, through bank accounts, by trucks and vans, by ships and planes, over clandestine paths, third rate roads, and mountain passes.

Billions of tons of natural resources go through Mexican ports, railway stations, airports, and road systems to various destinations: the United States, Canada, Holland, Germany, Italy, Japan – but all with the same destiny: to feed the empire. A handful of businesses, among them the Mexican State, take the wealth of Chiapas and in exchange leave their mark of death and disease....

[...]

In Chiapas there are 86 fangs of Pemex sunk into the municipalities of Estacion Juarez, Reforma, Ostuacan, Pichucalco, and Ocosingo. Every day they suck out 92 thousand barrels of petroleum and 516.7 billion cubic feet of natural gas. They take the gas and oil, and leave the trademark of capitalism: ecological destruction, agricultural waste, hyper-inflation, alcoholism, prostitution, and poverty. The beast is not satisfied, and extends its tentacles to the Lacandon jungle: eight oil fields are now under exploration. The jungle is opened with machetes, wielded by the very same *campesinos* whose land has been taken away by the insatiable beast. Trees fall and dynamite explodes in lands where only the *campesinos* are prohibited from felling trees to plant crops. Every tree a *campesino* cuts can cost him a fine worth ten day's salary and send him to jail. Poor people can not cut down trees, but the oil company, more and more in the hands of foreigners, can. The *campesino* cuts a tree in order to live, the beast cuts to plunder.

Chiapas also bleeds coffee: 87,000 Chiapans work in the coffee industry; 35 per cent of Mexico's coffee production comes from this region. Forty-seven per cent of that is sold on the national market, the other 53 per cent is exported, primarily to the United States and Europe. More than 100,000 tons of coffee leave Chiapas to fatten the bank accounts of the beast: in 1988 a kilo of pergamino coffee was sold abroad at an average price of 8,000 pesos [$2.50], but the producers in Chiapas were paid 2,500 pesos [about 80 cents a kilo or less].

Fifty-five per cent of the nation's hydroelectric power comes from this state, as well as 20 per cent of all the electric energy of Mexico. Nevertheless, only a third of all Chiapan houses have electricity....

Despite the current popularity of ecology, Chiapan forests continue to be destroyed. From 1981 to 1989, 2,444,700 cubic meters of precious woods, conifers, and tropical trees were taken from Chiapas and sent to Mexico City, Puebla, Veracruz, and Quintana Roo. In 1988 the exploitation of the forest produced 23,900,000,000 pesos [almost $8 million] in profit, 6,000 per cent more than in 1980.

Seventy-nine thousand Chiapan beehives are fully integrated into the European and American honey market: 2,756 tons of honey and wax produced every year in the countryside are converted into dollars that the people of Chiapas will never see.

Half of the corn produced here goes to the national market. Chiapas is one of the largest producers of corn in Mexico. Ninety per cent of the tamarind goes to Mexico City and other states. Two-thirds of the avocados are sold outside Chiapas.... Sixty-nine per cent of the cocoa goes on the national market, and 31 per cent goes to the United States, Holland, Japan, and Italy.

WHAT DOES THE BEAST LEAVE, IN EXCHANGE FOR EVERYTHING IT TAKES?

Chiapas ... is the eighth biggest state in Mexico ... its greatest wealth is the 3.5 million people of Chiapas, of whom two-thirds live and die in the countryside. Half of the people do not have potable water, and two-thirds have no sewage systems. Ninety per cent of the people in rural areas have little or no income.

Education? The worst in the country. Seventy-two out of every hundred children do not finish the first grade. Half of the schools go no higher than the third grade, and half of them have only one teacher to teach all the courses. The true drop-out figures are even higher, as the children of indigenous peoples are forced to enter the system of exploitation in order to help their families survive. In every indigenous community it is common to see children carrying corn or wood, cooking or washing clothes during school hours. Of the 16,058 school rooms in Chiapas in 1989, only 96 were in indigenous areas.

Health? Capitalism leaves its mark: a million and a half Chiapans have no medical services whatsoever. There are 0.2 clinics for every thousand people, five times less than the national average; there are 0.3 hospital beds for every thousand Chiapans, three times less than in the rest of Mexico; there is one operating room for every 100,000 people, two times less than in the rest of the country; there are 0.5 doctors and 0.4 nurses for every thousand persons, two times less than the national average.

Health and nutrition go hand in hand with poverty. Fifty-four per cent of the Chiapan population is malnourished, and in the mountains and jungles, 80 per cent of the people are hungry.

WELCOME! YOU HAVE ARRIVED IN THE POOREST STATE OF THE COUNTRY: CHIAPAS

Suppose that you continue driving and from Ocosocoatla you go down to Tuxtla Gutierrez, the state capital. Don't plan to stay long, Tuxtla Gutierrez is just a big warehouse for the state's products.... You pass Chiapas de Corzo, ignoring the Nestle's factory, and begin climbing into the mountains. What do you see? You must have entered another world: an indigenous one....

This indigenous world is made up of 300,000 Tzeltales, 300,000 Tzotziles, 120,000 Choles, 90,000 Zoques, and 70,000 Tojolabales. Even the federal government acknowledges that ("only") half of these people are illiterate.

Continue on the interior mountain road and you arrive at what is called the Chiapan Highlands.... Go on, and you reach San Cristobal de las Casas.... Welcome to the great marketplace.... Here you can buy or sell anything, except the dignity of the indigenous people. Here everything is expensive, except death. But don't stay long, keep going up the road, appreciating what has been built for the tourists: in

1988 Chiapas had 6,270 hotel rooms, 139 restaurants, and 42 travel agencies . . .

Did you add it up? Yes, that's right. While there are seven hotel rooms for every thousand tourists, there are 0.3 hospital beds for every thousand Chiapans.

Fine, forget the figures and move on, taking care to avoid the three lines of cops in camouflage berets trotting along the side of the road. . . . Leaving the "bowl" of San Cristóbal, right along the same road you will see the famous grottos surrounded by lush forests. Did you see that sign? No, you are not mistaken, this natural park is administered by . . . the army! Without letting go of your confusion, continue on. . . . What do you see? Modern buildings, nice homes, paved roads. . . . Is it a university? Workers' housing? No, look carefully at the sign next to one of the cannons. "General Barracks, Military Zone 31." With that painful olive-green image still in your eye, you arrive at the crossroads and decide not to go to Comitan. Thus you avoid the pain of seeing, some meters ahead on the hill they call "The Foreigner's," North American military personnel operating, and teaching their Mexican counterparts to operate, a radar station.

You decide that it is better to go to Ocosingo, as ecology and other nonsense is all the fashion these days. Look at the trees, take a deep breath . . . now do you feel better? Yes? Then keep on looking to your left because if you don't, at the 7 kilometer mark you will see a magnificent edifice with the noble SOLIDARITY logo on the front. Don't look, I tell you, turn your head away, you dont want to know that this new building is a . . . jail.

Don't get discouraged. The worst will always be hidden: too much poverty would scare the tourists. Continue on down to Huixtan . . . drive on to Ocosingo: "The Door to the Lacandon Jungle."

OK, wait here for a while. Take a quick trip around the city. . . . The main points of interest? Well, the two big buildings at the entrance to town are whorehouses, the next one is a jail, and across the street is the church. Next is the office of the Ranchers Association, followed by the Federal Army barracks, the state police office, the City Hall, and finally Pemex headquarters. The rest of the buildings are little houses all on top of each other that rattle and shake as the giant trucks of Pemex and the richest ranchers pass by.

We better move on . . . at the next intersection take a left. . . . Look, we are arriving at Palenque. A quick visit through the city? Those are hotels, over there restaurants, over here City Hall, the troopers, army barracks, and over there. . . . What? No, never mind, I know what you are going to say. Don't say it. . . . Tired? OK, let's stop for a while. Don't you even want to see the pyramids?

No, OK. How about Xi'Nich? It is something different. A march of indigenous people. They are going all the way to Mexico City. Uh-huh, walking. How far is it? 1,106 kilometers. Results? Their petitions were received. Yes, just that.

Are you still tired? OK, let's wait here some more. How about Bonampak? The road is very bad. Let's go anyway, we will take the scenic route. . . . Over there is the military reserve, and here is the navy, and now the state police, and at last the Government Ministry.

Is it always like this? No, sometimes you run into *campesino* protest marches.

So long and good luck. If you need a tourist guide don't forget to call me, I am at your service. Oh, and one other thing. It won't always be like this. Another Mexico? No, the same one. I am talking about something else – how other breezes begin to blow, how another wind is rising.

There is nothing to struggle for. Socialism is dead. Long live resignation, reformism, modernity, capitalism, and a whole list of cruel etceteras. . . . Radio, television, and the newspapers proclaim it, and some ex-socialists, now sensibly repentant, repeat it.

But not everybody listens to the voices of hopelessness and resignation. Not everyone has jumped onto the bandwagon of despair. Most people continue on; they can not hear the voice of the powerful and the faint hearted as they are deafened by the cry and the blood that death and misery shout in their ears. But in moments of rest, they hear another voice, not the one that comes from above, but rather the one that comes with the wind from below, and is born in the heart of the indigenous people of the mountains, a voice that speaks of justice and liberty, a voice that speaks of socialism, a voice that speaks of hope ... the only hope in this earthly world. And the very oldest among the old people in the villages tell of a man named Zapata who rose up for his own people and in a voice more like a song than a shout, said **!Land and Liberty!**

And these old folks say that Zapata is not dead, that he is going to return. And the oldest of the old also say that the wind and the rain and the sun tell the *campesinos* when they should prepare the soil, when they should plant, and when they should harvest. They say that hope also must be planted and harvested. And the old people say that now the wind, the rain, and the sun are talking to the earth in a new way, and that the poor should not continue to harvest death. Now it is time to harvest rebellion.

THE SECOND WIND – THE ONE FROM BELOW

Collective work, democratic thought, and majority rule are more than just a tradition among indigenous people, they have been the only way to survive, to resist, to be proud, and to rebel.

It has been said, quite wrongly, that the rebellion of the people of Chiapas has its own tempo, which does not correspond to the rhythms of the nation. It is a lie.... If the voices of those who write history are not accurate, it is because the voice of the oppressed does not speak ... not yet. There is no historical calendar, national or regional, which records all the rebellions and protests against this bloody system, imposed and maintained by force throughout every region of the country.

In Chiapas, the voice of rebellion is heard only when it shakes up the little world of the powerful.... If the rebellions of the southeast lose, as they lose in the north, the center, and east, it is not because they lack numbers and support, it is because wind is the fruit of the earth, and it has its own season, and matures not in books filled with regrets, but rather in the breasts of those who have nothing more than their dignity and their will to rebel. And this wind from below, the wind of rebellion and dignity, is not just a response to the wind imposed from above, it is not just a brave answer, but rather it carries within itself something new. This wind promises not only the destruction of an unjust and arbitrary system; it is, above all, a hope that dignity and rebellion can be converted into dignity and liberty.

This wind, born below the trees, will come down from the mountains; it whispers of a new world, so new that it is but an intuition in the collective heart.

The indigenous Xi'Nich ("march of the ants") made by the *campesinos* of Palenque, Ocosingo, and Salto de Agua, demonstrates the absurdity of the system. These indigenous people had to walk 1,106 kilometers in order to be heard; they went to the capital of the Republic so that federal authorities would get them an interview with the viceroy back in Chiapas.... They walked back the same 1,106 kilometers with their pockets full of promises. Nothing happened ...

In the town of Betania, on the outskirts of San Cristobal de las Casas, indigenous people are regularly detained and fined by the state police for cutting wood to use in their homes. The police are only complying with their duty to protect the environment, they say. Some indigenous people decide to end their silence and kidnap three state troopers. Not stopping there, they take over the Pan-American highway and cut off communication to the east of San Cristobal.... Business is bogged down; tourism collapses.... Negotiating committees come and go. The conflict seems to resolve itself, the matter subsides, and an apparent calm returns.

In the municipal seat of Ocosingo; 4,000 indigenous *campesinos* march from different points in the city to the ANCIEZ. Three of the marches converge on the municipal palace. The president of the municipality does not know what is happening and flees; a calendar left on the floor of his office shows the date: April 10, 1992. Outside the indigenous *campesinos* of Ocosingo, Oxchuc, Huistan, Chilon, Yajalon, Sabanilla, Salto de Agua, Palenque, Altamirano, Margaritas, San Cristobal, San Andres, and

Cancuc dance in front of a giant image of Zapata, painted by one of them. They recite poems, sing, and speak. They are the only ones there to listen. The large land owners, the big businessmen and the police are all closed up in their houses and businesses, and the garrison seems to be deserted. The *campesinos* shout that Zapata lives and that their struggle continues. One of them reads a letter to Carlos Salinas de Gortari accusing him of destroying the agrarian reform won by Zapata, of selling out the country through the Free Trade Agreement. . . . They forcefully declare that they do not recognize Salinas' changes of Article 27 of the Constitution. At two in the afternoon, the demonstration dissolves, the matter subsides, and an apparent calm returns.

Absalo is an *ejido* in the municipality of Ocosingo. For a long time, *campesinos* there have taken land that legally and naturally belongs to them. Three leaders of their community have been taken prisoner and tortured by the government. The indigenous people decide to end their silence and seize the road between San Cristobal and Ocosingo. Negotiating committees come and go. The leaders are released. The conflict seems to resolve itself, the matter subsides, and an apparent calm returns.

Antonio dreams that the land that he works belongs to him. He dreams that his sweat earns him justice and truth; he dreams of schools that cure ignorance and medicines that frighten death. He dreams that his house has light and that his table is full; he dreams that the land is free, and that his people reasonably govern themselves. He dreams that he is at peace with himself and with the world. He dreams that he has to struggle to have this dream, he dreams that there has to be death so that there might be life. Antonio dreams and wakes up . . . now he knows what he has to do. He sees his wife squatting to poke the fire, he hears his son crying, he looks at the sun greeting the east, and he smiles as he sharpens his machete.

A wind comes up and everything stirs. Antonio rises, and walks to meet the others. He has heard that his desire is the desire of many, and he goes to look for them.

The viceroy dreams that his land is agitated by a terrible wind, and that everything rises up; he dreams that all he has stolen has been taken away from him, he dreams that his house is destroyed and his government overthrown. He dreams and he doesn't sleep. The viceroy goes to the feudal gentlemen and they tell him that they are dreaming the same thing. The viceroy can't rest, he goes to his doctors, and among them they decide that he is suffering from Indian witchcraft, and only blood will free him of its spell; so the viceroy orders murder and imprisonment and the building of more jails and barracks, but his dreams continue to keep him awake.

In this country everyone dreams. Now it is time to wake up.

THE STORM . . .

It will be born out of the clash between the two winds, it will arrive in its own time, the coals on the hearth of history are stoked up and ready to burn. Now the wind from above rules, but the one from below is coming, the storm rises . . . so it will be . . .

THE PROPHECY . . .

When the storm subsides, when the rain and the fire leave the earth in peace again, the world will no longer be the world, but something better.

– *The Lacandon Jungle, August, 1992*

39 JEREMY BRECHER AND TIM COSTELLO

"Reversing the Race to the Bottom"

from *Global Village or Global Pillage* (1994)

Globalization and the Corporate Agenda have engendered a new perspective or paradigm that we have called "globalization-from-below." They are now generating an alternative global agenda – a Human Agenda to counter the Corporate Agenda.

This Human Agenda is emerging from common interests, shared pain, and evolving global norms of human rights, economic justice, and environmental sustainability.

These common interests are not well represented in existing institutions – nation states, corporations, the UN, the IMF, World Bank, and GATT/WTO. So a Human Agenda corresponding to common human interests is more likely to emerge from a dialogue among social movements. The programs they produce inevitably and properly express the common interests of specific coalitions.

We have drawn heavily on proposals that have emerged from dialogue among social movements – particularly dialogues that cross national and issue boundaries. We have tried to put them together as a coherent alternative agenda.

DIVERSE ECONOMIC PHILOSOPHIES

A few years ago, the greatest barrier to a common Human Agenda might well have been the "ideological" conflict of capitalism vs. communism. With the end of the Cold War, this highly oversimplified dichotomy has dissolved into a variety of alternatives that no longer necessarily take the form of choices between total systems.

There is ... no reason that different groups and areas should not follow different "economic models" as long as they do so within a global framework that protects the environment, shares resources justly, and forestalls a race to the bottom. There may be nothing incompatible, for example, between some groups and/or regions following economic practices based on indigenous traditions and others pursuing ecologically corrected versions of Western industrial development.

RECONSTRUCTING THE GLOBAL ECONOMY FROM THE BOTTOM UP

Downward leveling results from the extraction of wealth, power, and productive capacity from communities and the environment and their transfer to global corporations. A program for economic reconstruction needs to replace such downward leveling with upward leveling.

Upward leveling requires, first of all, empowering collective action. This means democratizing government at every level from the global to the local. Such democratization entails far more than simply periodic elections. It means, for example, eliminating the hold of wealthy contributors over election finance and the power of the IMF and World Bank over poor countries' economic policies. It means creating vehicles through which people can act on their common interests, such as local economic development programs. And it means holding corporations, banks, and other private economic actors accountable to the public, for example by means of enforceable corporate codes of conduct.

Second, upward leveling requires the transfer

of resources – power, wealth, knowledge, organization – from haves to have-nots. This may be done in a great variety of ways, from protecting workers' right to organize to international commodity agreements stabilizing markets for Third World products.

Third, upward leveling requires ways to ensure that resources are used to meet the most important needs, not allowed to languish or be devoted to luxury and waste. That requires supporting global demand, cutting Third World debt, and increasing the purchasing power of those at the bottom. And it also requires redirecting resources from financial speculation and luxury cars to such pressing needs as the conversion to environmentally sustainable forms of production.

National institutions are not adequate for realizing this agenda, but neither would be a centralized global or a fragmented local system. Such a program has to be implemented at multiple levels. The decaying nation state-based economic system needs to evolve toward a multi-level, one-world economy in which public institutions regulate economic forces and allocate resources at multiple levels from local to global. These levels will no doubt include local, state/provincial, and national units in their historically evolved forms. They may also, however, include newly emerging formations, such as bio-regions and regional entities like the European Union. They may even involve non-territorial groups, such as ethnic or religious communities scattered across many lands. But however decentralized the system that emerges, it will not be able to prevent downward leveling if it does not have a global dimension.

Globalization has affected every economic structure from the World Bank to local governments and workplaces. Correcting its devastating impact will take changes in each of these interlocking structures.

AN AGENDA FOR UPWARD LEVELING

Democratize

As long as democracy remains exclusively national it will remain largely powerless to address the economic problems of ordinary people. It will take democratization at each level from the local to the global to implement an effective alternative economic program. And it will take continuing grassroots mobilization to see that such a program actually works.

The demands of the Zapatistas in Mexico illustrate what it means for social movements to project democratization at multiple levels. They simultaneously demanded autonomous self-government for indigenous people in southern Mexico; free elections not dominated by wealth for Mexico as a whole; and an end to what they called the "neo-liberal project" in Latin America.

Democratization requires the redistribution of power. It currently has four principal fronts:

Democratize global institutions

The past decade has concentrated enormous power in such global institutions as the IMF, World Bank, and GATT. Yet these institutions are virtually unaccountable to those who are affected by their decisions. Today, these organizations are dominated by the United States and a few other rich countries; their governance needs to be opened up to include the world's poor, represented by their governments and citizen organizations. Their operations are conducted with enormous secrecy; they need to be made open to public scrutiny. They are formally accountable only to national governments; they should be made more accountable to the United Nations and to non-governmental organizations representing citizen interests. They make decisions without the consent of local communities affected by them; their plans should be made in consultation with and require the approval of local communities they affect.

End "preemption" of democratic decision-making

A principal function of global institutions and agreements has become to prevent governments from doing things their people want them to. The effect of these restrictions is almost always to "preempt" governments from doing things that would raise labor, social, and environmental conditions.

Such negative "conditionalities" should be ended. Rather than punishing countries for spending on education, health, and welfare, the conditions governments and international institutions require for loans, investment, aid, and trade advantages should encourage them.

Recapture governments from global corporations

All over the world, national, provincial, and local governments have become the pawns of global corporations and the Corporate Agenda. Coalitions of popular movements and organizations, utilizing tactics adapted to the political context at hand, need to challenge this domination. People need to reassert the right to use governments to regulate corporations and markets in the public interest.

Establish the right to self-organization

Such basic human rights as freedom of speech, assembly, publication, political participation, unionization, cultural expression, and concerted action are crucial supports for resistance to downward leveling. Yet they are widely denied, not only in authoritarian governments, but also in workplaces, schools, and other institutions of supposedly democratic countries. Democratic organization in and control of such institutions can be a crucial vehicle for resisting downward leveling. The self-organization and empowerment of discriminated-against groups, such as racial and ethnic minorities, women, immigrants, and migrants is particularly crucial for countering the race to the bottom.

Coordinate global demand

Ironically, as the economy has become more globalized, international cooperation to encourage adequate global economic demand has been virtually abandoned.

In the past, minimum labor standards, welfare state programs, collective bargaining, and other means to raise the purchasing power of have-nots did much to counter recessions and depressions within national economies. So did the tools of monetary and fiscal policy. Similar instruments increasing the buying power of those at the bottom and providing economic stimulus are now required in the global economy.

Expanded demand will primarily increase the consumption of the wealthiest unless it is combined with global redistribution. The International Confederation of Free Trade Unions has recently proposed a "trade union strategy for world development" that links coordinated recovery in the industrialized countries with jobs and poverty reduction in the developing world. It proposes expanded currency reserves for developing countries and Central and Eastern Europe; debt relief; and redesign of structural adjustment programs to emphasize reducing poverty and creating jobs.

Establish global rights and standards

To prevent competition among workforces and communities from resulting in a "race to the bottom," we need minimum global standards for human, labor, and environmental rights. The European Community's "Social Dimension" provides one possible model for minimum standards in such matters as job security, occupational safety, unemployment compensation, union representation, and social security benefits. Such rights and standards need to be incorporated in a wide range of international economic agreements and institutions.

Enforce codes of conduct for global corporations

Global corporations should be made accountable by means of codes of conduct. Such codes might require corporations to report investment intentions; disclose hazardous materials imported; ban employment of children; forbid environmental discharge of pollutants; require advance notification and severance pay when operations are terminated; and require companies not to oppose union organization. While such codes should ultimately be enforced by the United Nations and by agreement among governments, global public pressure and cross-border organizing can begin to enforce them directly.

Reverse the squeeze on the global poor

Globalization has been marked by the extraction of wealth from poor countries and communities. The first step to reversing this process is to end the structural adjustment and shock therapy programs that the IMF and World Bank have been forcing on poor countries and countries emerging from state-run economies.

Second, new arrangements should be made so that these countries do not have to run their economies to pay the interest on their debt. Debts for the poorest countries should be written off. Debts for other developing countries should be reduced, with the remaining parts paid in local currencies into a fund for local development.

Third, large-scale resource transfers should be provided so that "developing" countries can in fact develop. The Third World Network proposes commodity agreements to improve and stabilize poor countries' terms of trade; opening rich country markets to poor countries; and preferential treatment for underdeveloped countries.

Encourage grassroots development

Deregulation and austerity policies have meant the drain of resources out of local communities. The forced opening of markets to global corporations has created conditions in which small local enterprises are unable to compete. We need instead to foster local, small-scale businesses and farms and a growing "third sector" of grassroots, community- and employee-owned cooperative enterprises designed to mobilize poorly utilized resources to address unmet needs.

Grassroots-controlled enterprises

The last few years have seen an enormous range of experiments in new forms of employee- and community-controlled enterprises. Initiatives in poor communities in Brooklyn, NY and Waterbury, Connecticut, for example, have established employee-owned home health aide companies which provide a needed service to local communities and jobs to a workforce made up primarily of women of color. Such efforts provide a way ordinary people in local communities can control and benefit from productive activity.

Public development authorities

Local, regional, and national development authorities can serve as a vehicle for a proactive economic strategy. An example is the recently created Connecticut Community Economic Development Program. Created by the state government and jointly controlled by the government, representatives of poor communities, and private investors, it provides funding and technical assistance for private, public, and cooperative enterprises in poor communities. Its goals include creation of jobs and development of skills, particularly for people who are unemployed, underemployed, or receiving public assistance; community participation in decision-making; establishment of self-sustaining enterprises; improving the environment; promoting affirmative action, equal employment opportunities and minority-owned businesses; and coordination with environmental and economic planning.

Development banks and credit unions

Various forms of community-based and cooperative banking have developed in the Third World and in poor communities in the United States. For example, over the past few decades, as most banks collected deposits in poor and middle class communities and channeled them into unproductive speculative investment, Chicago's South Shore Bank reversed this process, dedicating its resources to rebuilding a poor, majority African-American neighborhood which had been cut off from credit by other area banks. By providing residential mortgages and small business loans and organizing initiatives in commercial development and housing rehabilitation, South Shore financed and redeveloped the neighborhood's infrastructure and services, funding the renovation of nearly 30 per cent of the neighborhood's apartments.

Sweat equity and labor exchange

Sweat equity converts labor into a right to a share in the product. It lets people build houses and thereby acquire a share of their ownership or work in enterprises and thereby acquire a proportion of their stock. Labor exchange allows people with different needs and abilities to help each other. A ... "service credit" program lets people work as volunteers in meeting community needs and receive for each hour of service a "service credit" which entitles them to one hour of service for themselves, their family, or organization from others in the program. Such programs allow people to make use of resources which the mainstream economy leaves to languish.

Community-based development organizations

Solving economic problems requires mobilization of diverse segments of the community. In many parts of the world, citizen-based organizations and coalitions are playing a crucial role in representing the needs and mobilizing the capabilities of grassroots people and organizations. Perhaps the most famous is the Mondragon network of banks, social service organizations, technical education institutions, and producer cooperatives in the Basque region of Spain.

Rebuild the public sector

Structural adjustment programs and the desire to reduce business taxes have led to sharp cutbacks in public sector activities all over the world. The constant attack on government and the privatization of formerly public functions have led to worldwide decay of education, healthcare, infrastructures, environmental protection and enhancement, and services for the young, the old, and the disabled. It has also led to unemployment and aggravation of the downward spiral. An expansion of education, health, infrastructure, environmental, and similar public sector activities is an essential element of economic reconstruction.

Convert to sustainable production and consumption

The current industrial system is already destroying the earth's air, water, land, and biosphere. Global warming, desertification, pollution, and resource exhaustion will make the earth uninhabitable long before every Chinese has a private car and every American a private boat or plane.

The solution to this dilemma lies in converting the system of production and consumption to an ecologically sound basis. The technology to do this exists or can be developed, from solar energy to public transportation and from reusable products to resource-minimizing production processes. However, a system in which the search for ever-expanding profits has no regulation or limits will continue to use environmentally destructive processes to produce luxuries, pollutants, and waste.

This malappropriation of resources is exacerbated by the huge share of human wealth squandered on the military. Despite the end of the Cold War, global military spending is more than $1,000 trillion per year – nearly half of it by the United States. This is justified in large part by the need to control economic rivals and the revolts of poor and desperate peoples.

The energies now directed to the race to the bottom need to be redirected to rebuilding the global economy on a humanely and environmentally sound basis. Such an approach requires limits to growth – in some spheres, sharp reductions – in the material demands that human society places on the environment. It requires reduced energy and resource use; less toxic production and products; shorter individual worktime; and less production for war. But it requires vast growth in education, health care, human caring, recycling, rebuilding an ecologically sound production and consumption system, and time available for self-development, community life, and democratic participation.

The vehicle for realizing the Human Agenda

is not something that pre-exists; it is a social movement under construction. Those who seek to realize their own interests by working with others to advance the common human interest are part of it. To correct David Rockefeller's refrain, "Broad human interests" are "being served best" when human cooperation is "able to transcend national boundaries."

CONCLUSION

GEOPOLITICS, KNOWLEDGE AND POWER AT THE END OF THE CENTURY

Simon Dalby

Many of the geopolitical writers in this volume have had substantial influence on the politics of the twentieth century. As the introductory sections in this book have noted, some have also had their lives dramatically altered by the processes of which they were a part. Adolf Hitler's nightmare geopolitical vision produced a terrible war, systematic genocide and eventually his own suicide. Karl Haushofer, whose pan-regional ideas were ignored by the Nazis in 1941 when they invaded the Soviet Union, committed suicide in 1946 a year after the SS shot his son. George Kennan's ideas codified US foreign policy for decades. Martin Luther King was assassinated, possibly because of the stand he took against the Vietnam War. Many members of The Committee on the Present Danger were influential in the Reagan administration. Václav Havel went from being a playwright, political prisoner and dissident intellectual to being president of a state that, in an extraordinary geopolitical event, subsequently quite peacefully separated into two. Robert Kaplan's bleak vision has influenced the Clinton administration and the debate in Washington about how to respond to rapidly changing events.

Geopolitical texts, like their authors, both reflect on and influence changing understandings of the political world at the biggest scales. The writings of the political world in this volume are diverse and, in many cases, have been, or still are, influential forms of power/knowledge. Reading them shows that geographical ideas are important component parts of state strategic doctrines, politicians' policy-making processes and rhetoric, dissident critiques and academic discussions of world affairs. They show clearly that the twentieth century has had a persistent discourse of politics understood in terms of global processes which can be specified in the supposedly objective geographical language of geopolitical reasoning. As Ó Tuathail and Agnew's argument (Reading 11) made clear, many of these geopolitical concepts are anti-geographical in the sense that they obscure complex realities with simple concepts.

Reading these texts also shows how rapidly particular places' political significance can change and how wary one has to be about geopolitical predictions, in part, because what often seems solid and permanent turns out to be temporary and fluid. This is often the case because "natural attributes" that are used to structure geopolitical languages are social constructions. Mackinder's assumptions (Reading 1) of the natural seats of power have been overtaken by technological and economic changes which suggest that a more complex understanding of global political geography has to account for the changing spatial arrangements of economic activity and the importance of changing resource uses, transportation links and technological innovation.

Mackinder's suggestion that there are geographical causes to history, a powerful argument that survived through the Cold War reinterpretations of geopolitics, may not be completely wrong if very loosely interpreted in some ways. But dramatic changes in global patterns of power in the twentieth century, driven by economic and technological changes, and the outcomes of huge wars, suggest that the keys to understanding global politics require a much more sophisticated theory than one that concentrates on the question of who controls the "heartland." Clearly focusing on the processes of rapid change rather than on eternal geographical patterns may provide more appropriate tools of analysis. These, however, are not of much use to conservative intellectuals who hanker after eternal verities as the basis for their politics.

In the last decade the pace of change has been especially dramatic. In the mid-1980s the re-evaluation of the Soviet Union's geopolitical situation, and the resulting change of foreign and security policy in Moscow by the Gorbachev administration, led to the end of the Cold War and numerous diplomatic innovations that were considered impossible by Washington's geopoliticians a few years earlier (MccGwire, 1991). The subsequent collapse of the Soviet Union has changed matters further although it has often served to obscure the events of the 1980s when interpreted in the simplistic Western rewritings of history that suggest that the West simply won the Cold War. Viewed in terms of geopolitics, George Bush's claims to a New World Order suggested both an American victory in the Cold War, and at the same time a very unsettling reminder of Nazi aspirations to another "new order" earlier in the century.

The end of the Cold War geopolitical order brought about dramatic reductions in the levels of troops and armaments stationed in Europe. Treaties have outlawed whole categories of nuclear missiles. The negotiation of a comprehensive nuclear test-ban treaty and other innovations in international diplomacy have extended moves to disarmament. These developments, which are consistent with the demands of European Nuclear Disarmament in the 1980s, were thought to be wholly unrealistic a decade ago by most "security experts." But while nuclear disarmament arrangements by the major nuclear weapons states are slowly moving ahead, geopolitical worries about the proliferation of nuclear weapons, and other threats, in the hands of non-state organizations and terrorists are now the subject of considerable concern (Sopko, 1996).

Ironically, disarmament in some parts of Europe has flooded the world market, and most obviously Croatia and Bosnia, with cheap weapons rendering struggles elsewhere in the world all the more bloody. But most of these struggles are now within state boundaries rather than in line with the traditional concern of geopolitics with struggles between states. The end of the apartheid regime in South Africa has also suggested the possibility of dramatic political change being possible without massive bloodshed. However, the large numbers of victims of political violence there since the accession of the African National Congress to power also suggests the need for caution in any prognosis of easy success in economic and political reform.

Since the end of the Cold War a number of international conferences on environment, development, human rights, women and population have suggested that there is now a new possibility for large scale cooperation on many other matters. High profile commissions on Global Governance (1995) and Population and the Quality of Life (1996) have outlined international programs of cooperation to deal with many global problems. But critics of the United Nations conferences in particular have cautioned that the apparent promise of such events belies the way elites have formulated the "global" problems in such ways as to ensure that power and wealth are not substantially redistributed (Chatterjee and Finger, 1994). Meetings of the Group of Seven and the emergence of new economic institutions like the World Trade Organization and agreements like the North American Free Trade Agreement, have worked to shore up the primacy of neoliberal economic arrangements and the importance of transnational corporations in international politics.

Amid all the geo-economic discussion of free trade and globalization, one important theme that is often excluded from explicit discussion is the rise of transnational corporations and the political consequences of their actions (Barnet and Cavanagh, 1994). In terms of sheer economic power, corporations dwarf many states. Comparing corporate sales to gross domestic product, Anderson and Cavanagh (1996) note that of the 100 largest "economies" in the world, fifty-one are corporations but only forty-nine are countries. Mitsubishi, the largest corporation, has sales greater than the gross domestic product of Indonesia, Denmark and

Thailand. Ford Motor Company is larger than Turkey, South Africa and Norway. Wal-Mart – the twelfth largest corporation in terms of sales – is bigger than 161 countries, including Poland, the Ukraine and Portugal. Overall, the combined sales of the top 200 corporations accounted for 28.3 per cent of the world's GDP. In the geo-economic discourses of transnational liberalism such developments are taken for granted as an appropriate organization of power and wealth and hence not usually remarked upon. But as has been especially clear for many years in the case of oil companies, large corporations have had very substantial impacts on the foreign policies and the understandings of the geopolitical interests of even the most powerful states (Yergin, 1991).

As the readings in Part 3 of this volume make clear, the dominant themes in global politics in the last decade have been a combination of celebration of the supposed victory of the West in the Cold War coupled to redoubled support for transnational "liberal" trading practices in the global economy. But these celebrations have been contrasted by anxieties that the whole system will come apart as a result of economic rivalry, spreading crime, destabilizing migrations, environmental degradation or civilizational clashes. Fears of geopolitical "vertigo" or "chaos" have appeared in numerous forms, many of which suggest a "need" for geopolitical certainty in which the basic components of global politics can be clearly demarcated, known and hence controlled. Lack of clear geopolitical concepts as well as precise boundaries often seems to render many intellectuals of statecraft disorientated and fearful.

The demand for conceptual certainty links to the long standing imperial and Cold War geopolitical assumptions that security is about controlling an external environment. Such thinking is part of the process whereby the North Atlantic Treaty Organization (NATO) is attempting to expand into Eastern Europe despite the clear indications that this will alarm Russian politicians and provide arguments for militarists in Moscow to exploit what can be portrayed as a threatening military alliance. Given the perilous state of the Russian political and economic situation, such actions on the part of NATO suggest a continued need by military planners to interpret danger in their geopolitical terms and to act accordingly to prepare to provide a form of "security" that might be extremely counterproductive.

Which vision, whether optimistic or pessimistic, celebratory or fearful, a writer thinks is most important is interconnected with the institutional context in which they write and the related geographical choice of where they choose to look for evidence to support their geopolitical theory. While attention is often focused on disasters in Africa and central Asia, parts of China's economy have been growing at historically extraordinary rates. Globalization by transnational corporations continues apace and the international financial markets are beginning to flex their muscles in ways that suggest that economic sovereignty for large powers other than the Germans and Americans is a historical matter. Whether it is the peso crisis in Mexico or discussions of a new Euro currency, in many cases assumptions about state institutions like central banks being central are no longer tenable. Meanwhile, at the other end of the geographical scale peasants continue to be dispossessed from lands in many villages around the world, feeding massive urban slums in the growing megacities of the "South." Similarly, while much speculation and political activity continues on the process of European integration, on the other hand very local appeals to ethnic nationalism have been linked directly to bloody territorial wars in the Balkans and in the Caucasus.

Benjamin Barber's (1996) analysis of the current situation suggests that both visions, of global economic integration, and of local tribal conflict, are essential to understanding the future. Both local and global processes are in play he suggests. They are obviously interconnected in many ways, but there is much more to these processes than can be summarized in simple spatial models. Perhaps more important for our consideration of the current state of geopolitics as knowledge is, as Timothy Luke points out (Reading 19), the crucial point that, the conventional geopolitical vocabularies that are the legacy of the Cold War, and the related spatial assumptions of the nation-state, may not be useful tools of analysis at all. The problems of terminology are very considerable and once again warnings about the use of simplistic spatial concepts are germane. Globalization, a term much used to summarize contemporary changes, is itself heavily contested with numerous interpretations being brought to bear on the term (Kofman and Youngs, 1996).

In re-thinking the contemporary political economy, Manuel Castells (1996) has suggested that the complex interconnections of contemporary global life in the information age might better be understood in the terms of the rise of a "network society." Networks suggest linkages and connections rather than spatial barriers. The "internet" is the buzzword for the innovative entrepreneur as business moves onto the information highway. The Cable News Network (CNN) coverage of events around the world sets the political agenda in many circumstances. To some writers this integration of the global economy and networking of information flows is supposed to ensure that the threat of war between large industrial states becomes a thing of the past. Other people have speculated that all these technological possibilities have changed the importance of geopolitics to such an extent that contemporary politics is now better understood in terms of "chronopolitics" – a politics of speed and time rather than territory and distance. In a world of intercontinental missiles and satellite communications, Paul Virilio (1986) has suggested that territory has lost its significance and that speed is more important in politics than place; space is a matter of electronics rather than territory.

But to others geopolitical considerations remain of prime importance and military power is now sometimes understood in terms of information power. The ability to have real time intelligence from round the globe suggests that at least the United States has the ability to construct a global panopticon of total surveillance (Nye and Owens, 1996). In discussing the possibilities of information power, it has to be remembered that military, political and economic intelligence is only as useful to its possessors as the analysts who compile the information, and they cannot help but use their geopolitical understandings of the significance of what they see on their Visual Display Units and television screens to organize the information and suggest policy actions. Through the use of satellite imaging, American mediators in the Dayton Ohio peace talks about Bosnia in 1995 had better real time knowledge of the dispositions of Serbian, Croatian and Bosnian military forces than the representatives of these forces at the negotiating table. But while this information was important in arranging a cease fire agreement with clearly defined zones of control for each side, all this technological capability could not produce a lasting peace settlement and a political arrangement that is likely to be stable in the long term once foreign troops are withdrawn from the region. Geo-informational power alone does not guarantee either geopolitical power or peace.

Whatever the fascination with, or the extraordinary capabilities of, technology, and while it may open up new opportunities and change the dynamics of politics by dramatically speeding up events, information flows and reaction times, technology alone is not responsible for political decisions and actions. What is often in danger of being forgotten in all this discussion of information technology is that the information society still depends on fuel supplies to run the vehicles and provide electrical power for the computers and telephone exchanges. Oil supplies from Saudi Arabia or elsewhere in the Middle East are still essential to the economies of the post-industrial world. In addition, pessimistic security analysts remind us that the very openness and interconnectedness of electronic sources of information make a society dependent on them potentially vulnerable to either the depredations of terrorist "hackers" or the possibility of a more concentrated "cyberwar" assault by a hostile power. Propaganda and disinformation take on new possibilities when considered in the terms of electronic sabotage or manipulation of government and corporate data banks.

The question of which of the current geopolitical understandings will be dominant in the immediate future is an important part of how the politics of the early twenty-first century will be constituted. Just as at the beginning of this century Halford Mackinder made the world visible as a global stage with a pivot and marginal crescents, landpower and seapower rivalries, and crucially, territorial states covering all the earth's surface, so current writers are constructing the geopolitical categories through which contemporary events are understood and through which politicians, executives and mutual fund managers interpret how to make their decisions. But while there is currently widespread general support among policy makers for the geo-economic discourses of transnational liberalism, it is not at all clear which, if any, of the current popular geopolitical theories will become the hegemonic understanding of the post-Cold War period. This is partly because the efficacy of these forms of knowledge are not judged by which is the best theory on some independent scholarly standards.

The debate about geopolitical visions is a political one, and if a framework is adopted by a broad segment of the policy makers in Washington and other industrial states, it will be a political decision. Like the doctrine of containment in the Cold War, the arguments will be ones that fit into the larger political discourse of the time in ways that are obviously useful in supporting many powerful economic and political institutions and interests. This is especially clearly stated in an article in *Foreign Affairs* in 1996 where three Yale University scholars make their case for an American strategic policy of giving aid to a number of countries they term (following Mackinder) "pivot states" (Chase, Hill and Kennedy, 1996). Concerned that the US needs to maintain the political stability of large Third World states that have the potential to cause substantial disruptions to the international system if they collapse, they justify such a policy on the grounds that it both works in America's interests, and that it is a politically acceptable idea in the current circumstances.

Much of the contemporary writing, whether about geo-economics, the clash of civilizations, the differences between wild and tame zones, the politics of climate change, demographic upheavals or the dynamics of economic globalization, argues that the territorial state is declining in importance and that military matters are now of much less significance in the affairs of the largest powers (Ó Tuathail, 1997a). But to declare that the end of the Cold War, and the rise of the current global economic system, means the end of history or the end of twentieth-century geopolitics is to miss some of the key points about geopolitics that the readings in this volume highlight. This is both because many of the current theories have elements of traditional geopolitics in their writings and because there is much more to geopolitics than only a focus on states and the geographical aspects of great power rivalry.

Perhaps most obviously, Samuel Huntingdon's "clash of civilizations" argument (Reading 21) can be seen as being very similar to the focus on national state rivalries, but merely a scale change up from supposedly permanent states with defined borders separating them to supposedly permanent civilizations with defined borders separating them. This follows the imperialist geopolitics of Mackinder by positing essential geographical entities in rivalry with one another as the basis of global affairs. Robert Kaplan's additional assertions about the environmental causes of conflict also suggest, in a different manner, that many political factors are determined by so-called "natural" phenomena (Reading 23). Geographical determinism may not be popular amongst academic geographers, but clearly such ideas still have considerable influence in political and journalistic circles. Arguments about rogue states suggest a clear imperial obligation to keep politically dissident states in line wherever they may challenge the international order of the powerful industrialized states (Reading 20). Some of the geo-economic arguments suggest that economic rivalries have replaced military competition between states and that the old rivalries between identifiable geographic entities continue by other means of competition (Reading 16). Ethnic nationalism and territorial conflict look very like traditional boundary warfare.

But geopolitics persists beyond these continuities because it is more broadly about ways of reading and writing global political space. It is about the assumptions and geographical codes that politicians and policy makers use to specify the significance of places in the construction of, and arguments legitimating, policies

(Agnew and Corbridge, 1995). As a number of the readings in this volume have made clear, geopolitics is also part of the processes of constructing identity and specifying geographical threats to that identity by specific strangers in particular places. But now geopolitics is also increasingly related to the ways that mass media frame global space, and how places and locations are caught, or not caught, in the lenses of television cameras and shown daily in our living rooms (Wark, 1994). When lenses probe the disasters and conflicts of our world, they suggest dangers and conflicts are endemic. Such things are often defined as news; while daily economic processes and the routine lives of people around the world are often not part of these understandings of global life.

But the routine operations of economic life are crucially important in understanding contemporary change. As we have tried to make clear in the readings in this volume, especially in Parts 4 and 5, the consequences of geopolitics work themselves out in different ways in the practical lived experience of people in various places around the world. And in economic terms where one lives is, if anything, becoming more important. According to the 1996 United Nations Development Report, the total global Gross Domestic Product was $23 trillion in 1993. Of this, $18 trillion was in the industrial countries and only $5 trillion in the developing world, even though they have nearly 80 per cent of the world's people. The discrepancies have grown in recent decades. The poorest 20 per cent of the world's people saw their share of global income decline from 2.3 per cent to 1.4 per cent in the past 30 years. Meanwhile, the share of the richest 20 per cent rose from 70 per cent to 85 per cent, doubling the ratio of richest to poorest shares from 30:1 to 61:1. The assets of the world's 358 billionaires exceed the combined annual incomes of countries with 45 per cent of the world's peoples. Such amazing wealth in the face of such poverty is also tied to numerous endemic problems of corruption in which politicians and policy makers in many states can be easily influenced by both national and international corporations.

In all the discussion of contemporary accelerations and changing geopolitical arrangements, what is also often left out is the fate of the poor and those marginalized by the economic processes of globalization and acceleration. The poor may not have modems, know the precise details of economic statistics, nor understand the military significance of the reconnaissance satellites that pass over their heads, but they often pay the price of military actions in many parts of the world and know very directly of the dramatic economic changes that are concentrating fabulous wealth in the hands of the global economic and political elites who are literally plugged into the global circuits of capital. The poor and marginal don't usually read geopolitical treatises either, which is why the politics of how they are represented, or as is often the case, not represented, is an important part of any critical analysis of geopolitics as a mode of power/knowledge. But more than this, thinking through the consequences of specific geopolitical understandings of the world for the poor and marginalized in both the "South" and the "North" sheds important light on the politics of geographical knowledge.

Discussions of the "war on drugs" in the US make this especially clear. Understood as an external threat to the US, policy actions in response to drugs that are appropriate include helping governments in Latin America and elsewhere undertake military actions against drug producers in remote areas. Using military style actions against smugglers and their aircraft and boats is also thought appropriate. So too is a policy of incarceration for people charged with drugs-related offenses. Such reasoning was also used to legitimate the invasion of Panama in 1989 (Weber, 1995). The source of the problem in this geopolitical specification is primarily of externally-sourced threats.

But it is possible to argue that the drugs problem in the US, and in other Western states, is not an external problem, but one better understood as a series of social and health problems among the poor in the city ghettos and elsewhere within these regions, which indirectly lead to an economic demand for drugs. Understood as a matter of "internal" poverty and lack of economic and social opportunity for young men in particular, the policy options that are then up for discussion and implementation are obviously very different. Reducing the demand for drugs would also reduce drug-related crime. Community economic initiatives, improved health services for the poor, education and employment programs are very different responses to those suggested as appropriate by geopolitical specifications of external threats.

In a similar manner discussions of population migrations and environmental degradation as potential security threats in the near future also specify the causes of the problems as outside the West (Dalby, 1997). Seen in terms of threats to the Western order, the causes are related to environmental change or birth rates by non-Western populations. Obviously these are factors in what is going on, but other analyses suggest that these situations are complicated by economic factors having to do with the development models of recent decades. Looking at the effects of International Monetary Fund induced structural adjustment programs, which have emphasized reduction of government spending and the promotion of export agriculture, suggests that international economics is an important part of these processes that conventional geopolitical writings usually either ignore, or incorporate in geoeconomic discourses of emerging markets and global financial volatility (Ó Tuathail, 1997b).

Peasants displaced by the commercial export agriculture schemes of modernization are clearly an important part of the rebellion in Chiapas, Guerrero and Oaxaca in southern Mexico (Oppenheimer, 1996). Across Central America land distribution problems, where rich land owners use plantations to grow crops for export markets mainly in North America, while peasants cannot get access to plots to grow food, are an important cause of migration and political instability. When these people reach the borders of the US, they are then often treated as an external threat. Again, which geopolitical framework is invoked to understand these "threats," and consequently how they are to be dealt with, is crucial to policy formulation.

Intellectuals write within institutional and ideological frameworks; understanding geopolitical writing in these terms makes the arguments and the claims of the texts all the more important in understanding contemporary events. Thinking carefully about the geopolitical claims made by politicians, business leaders, journalists, academics, policy makers, as well as dissident activists, will continue to shed light on politics and on the modes of reasoning used in the contemporary political process. As citizens and humans on a planet facing rapid change, and where the old Cold War geopolitical categories obviously do not fit current situations, critical evaluation of the taken for granted assumptions of how we think about geography, politics, knowledge and their interconnections is clearly very much needed.

This conclusion has not offered any suggestions as to which of the many contemporary geopolitical theories is in some sense correct or best. Most readers will have their own theories dependent on the political questions they ask about their places of living and working, personal roles and social responsibilities, whether they think of these questions as political or not. The point of this volume has not been to suggest that there is a simple "right answer" to any of the questions about the future of global politics. Rather we have tried to make clear that geopolitical knowledges are modes of power and influence with political consequences.

Geopolitical texts are not "neutral" writings from some detached position outside politics, history or geography, attempting to answer a single commonly agreed upon "question." Indeed, we have shown that precisely these claims to a "god's eye" view of the world is a rhetorical tactic repeatedly used to convince readers that the author of the text has just this kind of "answer." Not surprisingly, given what has been discussed in this book, "answers" of this kind usually are the precursor to political decision-making and the exercise of power by some government, business or agency.

What we have also tried to do in the readings, the section introductions and in this conclusion is make it clear that geopolitical reasoning usually operates by simplifying complex realities and focusing on particular ways of describing the world. Many of the prognostications about the future, discussions of globalization, rogue states, civilizational clashes, environmental degradation, migration as threats, deal with parts of geographical reality. By narrowing the focus and assuming that some identities, like the US or the West, can be taken for granted, they specify the world in particular ways.

Through this volume we have asked hard questions about how these taken for granted geopolitical entities are constructed in the first place, and how it might be possible to imagine different geopolitical arrangements. This is the stuff of critical geopolitics (Ó Tuathail, 1996). So too are the arguments by the dissidents and by those ignored in the scripts of conventional geopolitical reasoning. Listening to dissidents allows us to understand that politics is also about claims to truth, descriptions of geographical realities, and about silencing voices that contest official definitions.

Geopolitics is about complicated contested claims to knowledge. As we have also tried to make clear in these readings, it is also about asking the devastatingly simple question: "how is it possible to think that?" (Foucault, 1973).

REFERENCES AND FURTHER READING

Agnew, J. and Corbridge, S. (1995) *Mastering Space: Hegemony, Territory and International Political Economy*, London: Routledge.

Anderson, S. and Cavanagh, J. (1996) *The Top 200: The Rise of Global Corporate Power*, Washington: Institute for Policy Studies.

Barber, B. (1996) *Jihad vs. McWorld: How Globalism and Tribalism are Reshaping the World*, New York: Ballantine.

Barnet, R.J. and Cavanagh, J. (1994) *Global Dreams: Imperial Corporations and the New World Order*, New York: Simon and Schuster.

Castells, M. (1996) *The Rise of the Network Society*, London: Blackwell.

Chase, R.S., Hill, E.B. and Kennedy, P. (1996) "Pivotal States and U.S. Strategy," *Foreign Affairs*, 75(1): 33–51.

Chatterjee, P. and Finger, M. (1994) *The Earth Brokers: Power, Politics and World Development*, London: Routledge.

Commission on Global Governance (1995) *Our Global Neighbourhood*, Oxford: Oxford University Press.

Dalby, S. (1997) "The Threat from the South: Global Justice and Environmental Security," in Deudney, D. and Matthew, R. (eds) *Contested Grounds: Security and Conflict in the New Environmental Politics*, Albany: State University of New York Press.

Foucault, M. (1973) *The Order of Things: An Archaeology of the Human Sciences*, New York: Vintage.

Independent Commission on Population and Quality of Life (1996) *Caring for the Future: A Radical Agenda for Positive Change*, Oxford: Oxford University Press.

Kofman, E. and Youngs, G. (1996) *Globalization: Theory and Practice*, London: Pinter.

McCGwire, M. (1991) *Perestroika and Soviet National Security*, Washington: Brookings.

Nye, J.S. and Owens, W.A. (1996) "America's Information Edge," *Foreign Affairs*, 75(2): 20–36.

Oppenheimer, A. (1996) *Bordering on Chaos*, Boston: Little, Brown and Company.

Ó Tuathail, G. (1996) *Critical Geopolitics: The Politics of Writing Global Space*, Minneapolis: University of Minnesota Press.

——(1997a) "Emerging Markets and Other Simulations: Mexico and the Chiapas Revolt in the Geo-Financial Panopticon," *Ecumene*.

——(1997b) "At the End of Geopolitics? Reflections on a Plural Problematic at the Century's End," *Alternatives*, 22: 35–55.

Sopko, J. (1996) "The Changing Proliferation Threat," *Foreign Policy*, 105: 3–20.

Virilio, P. (1986) *Speed and Politics*, New York: Semiotext(e).

Wark, M. (1994) *Virtual Geography: Living with Global Media Events*, Bloomington: Indiana University Press.

Weber, C. (1995) *Simulating Sovereignty: Intervention, the State and Symbolic Exchange*, Cambridge: Cambridge University Press.

Yergin, D. (1991) *The Prize: The Epic Quest for Oil, Money and Power*, New York: Simon and Schuster.

COPYRIGHT INFORMATION

1 IMPERIALIST GEOPOLITICS

MACKINDER Mackinder, Halford J. "The Geographical Pivot of History," *Geographical Journal*, 23 (1904). Copyright © 1904. Reprinted by permission of *Geographical Journal*.

ROOSEVELT Roosevelt, Theodore "The Roosevelt Corollary," annual address to Congress, 6 December 1904, from J. Richardson *A Compilation of Messages and Papers of the Presidents*, Volume 9 (Washington, DC: Bureau of National Literature and Art, 1905). Copyright © 1904/1905. Reprinted by permission of the Bureau of National Literature and Art.

HAUSHOFER Haushofer, Karl "Why Geopolitik?" from Andreas Dorpalen *The World of General Haushofer* (New York: Farrah and Rinehard, 1942). Original German from Karl Haushofer "Politische Erdkunde und Geopolitik," in *Freie Wege vergleichender Erdkunde. Festschrift füt Erich von Drygalski* (Munchen: Oldenbourg, 1925). Copyright © 1925/1942. Reprinted by permission of Farrah and Rinehard.

HITLER Hitler, Adolf "Eastern Orientation or Eastern Policy?" from *Mein Kampf* (Boston: Houghton Mifflin, 1942), based on Vol. 2 of the original German *Mein Kampf* (1927). Copyright © 1927/1942 by the *Estate of Ralph Manhelm*. Reprinted by permission of Houghton Mifflin and Random House UK Ltd.

HAUSHOFER Haushofer, Karl "Defense of German Geopolitics," from Edmund Walsh *Total Power: A Footnote to History* (New York: Doubleday, 1948). Translation by the US Army of the testimony of Dr Karl Haushofer to the Nuremberg War Crimes Investigation, Nuremberg, Germany, 5 October 1945. Copyright © 1948. Reprinted by permission of Doubleday.

2 COLD WAR GEOPOLITICS

TRUMAN Truman, Harry "The Truman Doctrine," an address to Congress, 12 March 1947, from *Public Papers of the Presidents of the United States. Harry S. Truman. 1947* (Washington, DC: United States Government Printing Office, 1947). Copyright © 1947. Reprinted by permission of United States Government Printing Office.

KENNAN Kennan, George F. "The Sources of Soviet Conduct," *Foreign Affairs*, 25 (1947). Copyright © 1947 by the Council on Foreign Relations, Inc. Reprinted by permission of *Foreign Affairs*.

ZHDANOV Zhdanov, Andrei "Soviet Policy and World Politics," from *The International Situation* (Moscow: Foreign Languages Publishing House, 1947). Copyright © 1947. Reprinted by permission of Foreign Languges Publishing House.

O'SULLIVAN O'Sullivan, Patrick "Antidomino," *Political Geography Quarterly*, 1 (1982). Copyright © 1982. Reprinted by permission of Elsevier Science Ltd, Kidlington, UK.

BREZHNEV Brezhnev, Leonid "The Brezhnev Doctrine," from *The Current Digest of the Soviet Press*, 20(39) (1968, trans. S. Kovalev). Original Russian published as "Sovereignty and Socialist Self-Determination" by S. Kovalev in

Pravda (26 September 1968): 4. Copyright © 1968.

Ó TUATHAIL and AGNEW Ó Tuathail, Gearóid and Agnew, John "Geopolitics and Discourse: Practical Geopolitical Reasoning in American Foreign Policy," *Political Geography Quarterly*, 11 (1992). Copyright © 1992. Reprinted by permission of Elsevier Science Ltd, Kidlington, UK.

COMMITTEE ON THE PRESENT DANGER "Common Sense and the Common Danger," from *Alerting America: The Papers of the Committee on the Present Danger* (Washington, DC: Pergamon Brasseys, 1984). Released to the press on 11 November 1976. Copyright © 1976/1984. Reprinted by permission of Pergamon Brasseys.

END COMMITTEE "Appeal for European Nuclear Disarmament (END)," from E.P. Thompson and D. Smith (eds) *Protest and Survive* (London: Penguin, 1980). Copyright © 1980. Reprinted by permission of Penguin.

GORBACHEV Gorbachev, Mikhail "New Political Thinking," from *Perestroika: New Thinking for Our Country and the World* (New York: Harper and Row, 1988). Copyright © 1988. Reprinted by permission of HarperCollins Publishers, Inc.

3 NEW WORLD ORDER GEOPOLITICS

FUKUYAMA Fukuyama, Francis "The End of History?" *The National Interest*, 16 (1989). Copyright © 1989 by Francis Fukuyama. Reprinted by permission of *The National Interest*.

LUTTWAK Luttwak, Edward N. "From Geopolitics to Geo-Economics: Logic of Conflict, Grammar of Commerce," *The National Interest*, 20 (summer 1990). Copyright © 1990 *The National Interest*. Reprinted by permission of *The National Interest*.

BUSH Bush, George "Toward a New World Order," from an address before Congress, 11 September 1990, from *Public Papers of the Presidents of the United States. George Bush. 1990* (Washington, DC: United States Government Printing Office, 1991). Copyright © 1990/1991. Reprinted by permission of the United States Government Printing Office.

BUSH Bush, George "The Hard Work of Freedom," from the "State of the Union" address before Congress, 29 January 1991, from *Public Papers of the Presidents of the United States. George Bush. 1991* (Washington, DC: United States Government Printing Office, 1992). Copyright © 1991/1992. Reprinted by permission of the United States Government Printing Office.

LUKE Luke, Timothy W. "The Discipline of Security Studies and the Codes of Containment," *Alternatives*, 16(2) (1991). Copyright © 1991 by *Alternatives*. Reprinted by permission of Lynne Rienner Publishers, Inc.

KLARE Klare, Michael T. "The New 'Rogue-State' Doctrine," *The Nation* (8 May 1995). Copyright © 1995 by *The Nation*. Reprinted by permission of *The Nation*.

HUNTINGTON Huntington, Samuel P. "The Clash of Civilizations?" *Foreign Affairs*, 72 (1993). Copyright © 1993 by the Council on Foreign Relations, Inc. Reprinted by permission of *Foreign Affairs*.

Ó TUATHAIL Ó Tuathail, Gearóid "Samuel Huntington and the 'Civilizing' of Global Space," from *Critical Geopolitics* (Minneapolis: University of Minnesota Press, 1996). Copyright © 1996. Reprinted by permission of the University of Minnesota Press.

4 ENVIRONMENTAL GEOPOLITICS

KAPLAN Kaplan, Robert D. "The Coming Anarchy," *The Atlantic Monthly* (February 1994). Copyright © 1994. Reprinted by permission of Robert Kaplan.

DALBY Dalby, Simon "Reading Robert Kaplan's 'Coming Anarchy'," *Ecumene*, 3(4) (1996). Copyright © 1996. Reprinted by permission of Arnold.

HOMER-DIXON Homer-Dixon, Thomas F. "Environmental Scarcity and Mass Violence,"

Current History (November 1996). Copyright © 1996 by Current History, Inc. Reprinted by permission of Current History, Inc.

SMIL Smil, Vaclav "Some Contrarian Notes on Environmental Threats to National Security," *Canadian Foreign Policy*, 2(2) (autumn 1994). Copyright © 1994. Reprinted by permission of *Canadian Foreign Policy.*

PORTER Porter, Gareth "Environmental Security as a National Security Issue," *Current History* (May 1995). Copyright © 1995 by *Current History*, Inc. Reprinted by permission of *Current History*, Inc.

FINGER Finger, Matthias "The Military, the Nation State and the Environment," *The Ecologist*, 21(5) (Sept/Oct. 1991). Copyright © 1991. Reprinted by permision of *The Ecologist*.

SHIVA Shiva, Vandana "The Greening of Global Reach," from Wolfgang Sachs (ed.) *Global Ecology: A New Arena of Political Conflict* (London: Zed, 1993). Copyright © 1993 by Zed. Reprinted by permission of Zed and Vandana Shiva.

VISVANATHAN Visvanathan, Shiv "Mrs Brundtland's Disenchanted Cosmos," *Alternatives*, 16(2) (1991). Copyright © 1991 by *Alternatives*. Reprinted by permission of Lynne Rienner Publishers, Inc.

5 ANTI-GEOPOLITICS

SAID Said, Edward "Orientalism Reconsidered," from F. Barker, P. Hulme, M. Iversen and D. Loxley (eds) *Europe and Its Others, Volume One* (Colchester: University of Essex, 1984). Copyright © 1984. Reprinted by permission of Edward Said.

FANON Fanon, Frantz "Concerning Violence," from *The Wretched Earth* (New York: Monthly Review Press, 1963). Copyright © 1963. Reprinted by permission of Monthly Review Press and Grove/Atlantic, Inc..

KING King, Martin Luther "A Time to Break Silence," from J.M. Washington (ed.) *A Testament of Hope: The Essential Writings of Martin Luther King, Jr.* (San Francisco: Harper and Row, 1986). Copyright © 1967 by Martin Luther King, Jr. Reprinted by arrangement with the heirs to the Estate of Martin Luther King, Jr., c/o Writers House Inc.

HAVEL Havel, Václav "The Power of the Powerless," from *The Power of the Powerless* (Armink, NY: M.E. Sharpe, 1985). Copyright © 1985. Reprinted by permission of M.E. Sharpe.

THOMPSON Thompson, E.P. "America and the War Movement," from *The Heavy Dancers* (New York: Pantheon Books, 1985). Copyright © 1985. Reprinted by permission of Pantheon Books.

KONRAD Konrad, George "Antipolitics: A Moral Force," from *Antipolitics* (Harcourt Brace & Company, 1984). Copyright © 1984. Reprinted by permission of Harcourt Brace & Company.

FARMANFARMAIAN Farmanfarmaian, Abouali "Did You Measure Up? The Role of Race and Sexuality in the Gulf War," *Genders* 13 (spring 1992). Copyright © 1992. Reprinted by permission of *Genders*.

MARCOS Marcos, Subcommandante "Chiapas: The Southeast in Two Winds, a Storm and a Prophecy," *Anderson Valley Advertizer*, 42 (31) (1994). Translated by Frank Bardacke, Leslie Lopoz and the Watsonville, California Human Rights Committee. Copyright © 1994. Reprinted by permission of *Anderson Valley Advertizer*.

BRECHER and COSTELLO Brecher, Jeremy and Costello, Tim "Reversing the Race to the Bottom," from *Global Village or Global Pillage* (Boston: South End Press, 1994). Copyright © 1994. Reprinted by permission of South End Press.

INDEX